TEUBNER-TEXTE zur Informatik Band 31

C. H. Cap

A Calculus of Distributed and Parallel Processes

TEUBNER-TEXTE zur Informatik

Als relativ junge Wissenschaft lebt die Informatik ganz wesentlich von aktuellen Beiträgen. Viele Ideen und Konzepte werden in Originalarbeiten, Vorlesungsskripten und Konferenzberichten behandelt und sind damit nur einem eingeschränkten Leserkreis zugänglich. Lehrbücher stehen zwar zur Verfügung, können aber wegen der schnellen Entwicklung der Wissenschaft oft nicht den neuesten Stand wiedergeben.

Die Reihe „TEUBNER-TEXTE zur Informatik“ soll ein Forum für Einzel- und Sammelbeiträge zu aktuellen Themen aus dem gesamten Bereich der Informatik sein. Gedacht ist dabei insbesondere an herausragende Dissertationen und Habilitationsschriften, spezielle Vorlesungsskripten sowie wissenschaftlich aufbereitete Abschlußberichte bedeutender Forschungsprojekte. Auf eine verständliche Darstellung der theoretischen Fundierung und der Perspektiven für Anwendungen wird besonderer Wert gelegt. Das Programm der Reihe reicht von klassischen Themen aus neuen Blickwinkeln bis hin zur Beschreibung neuartiger, noch nicht etablierter Verfahrensansätze. Dabei werden bewußt eine gewisse Vorläufigkeit und Unvollständigkeit der Stoffauswahl und Darstellung in Kauf genommen, weil so die Lebendigkeit und Originalität von Vorlesungen und Forschungsseminaren beibehalten und weitergehende Studien angeregt und erleichtert werden können.

TEUBNER-TEXTE erscheinen in deutscher oder englischer Sprache.

A Calculus of Distributed and Parallel Processes

By Prof. Dr. Clemens H. Cap
Universität Rostock

B.G.Teubner Stuttgart · Leipzig 2000

Prof. Dr. Clemens H. Cap

Born in 1965 in Innsbruck. Studied mathematics, computer science and physics at Innsbruck. He received his diploma degree in mathematics in 1986 and his PhD in mathematics in 1988. After several years as research assistant, postdoc and assistant professor for formal methods in computer science at the university of Zurich, he moved to Germany where he presently is full professor and Heinz-Nixdorf-Chair for information and communication services at the university of Rostock. His research interests are the specification, implementation and application of distributed communication systems.

Gedruckt auf chlorfrei gebleichtem Papier.

Die Deutsche Bibliothek – CIP-Einheitsaufnahme

Ein Titelsatz für diese Publikation ist bei
Der Deutschen Bibliothek erhältlich

ISBN-13: 978-3-322-86765-0 e-ISBN-13: 978-3-322-86763-6
DOI: 10.1007/978-3-322-86763-6

Preface

It is the good reader that makes the good book.

RALPH WALDO EMERSON,
Society & Solitude.

In the course of two projects, the author of this book was involved in the design of the platforms PARFORM [CS93] and LOLA [Cap94], [CS] for the support of parallel computing in distributed systems. The former system was geared towards the highly efficient use of idle resources in networks of workstations, and the latter system was intended as a scalability study: How many workstations in the global Internet can be used simultaneously for solving a massively parallel problem? In one of the experiments conducted with these systems, up to 800 workstations on all five continents were cooperating for the solution of a search problem from molecular biology [Cap94].

The most important lessons which the author was forced to learn during the course of these projects were not to rely on any documentation of network- and low-level system-calls, to use neither common sense nor mathematical logic during the design of a large distributed system, but to be happy with a working program, and not to ask, why it would work.

This practical experience is in sharp contrast to the reliability that is generally expected from engineering sciences which utilize the rigor of mathematical methods. The consequent application of mathematical technology allows the prediction of physical phenomena ranging from elementary particles to cosmology. There is an endless number of examples of successful applications of mathematics in nearly every engineering and natural science. Presently, society is just starting to realize the tip of the iceberg of erroneous computer programs. The reader is kindly invited to consider this problem when sitting in a computer controlled airplane, train, or elevator. Recently, one of the largest chip manufacturers had to admit a design error in one of its bestselling microprocessors. The company continued selling the processor for a considerable time before informing the customers. Again, the reader is kindly invited to consider this attitude when sitting in a computer controlled airplane, train, or elevator.

Computer science still seems to suffer from two problems: First, for many constructions of applied computer science no satisfactory and successfull mathematical theory is available. There is a long tradition of mathematical concepts for physics but the practical and theoretical problems of correct software have been known only for several decades. Thus, adequate mathematical tools are not yet available. Second, formal concepts have an acceptance problem in the community of applied computer science. The trend to quick-and-dirty solutions prevails systematic approaches, and time-to-market is more important than product quality. To summarize, the tremendous complexity of theoretical computer science and its importance for applications as well as for society is consistently underestimated.

In his own research as well as in the present book, the author wants to contribute to narrowing the present gap between theoretical computer science and computer engineering.

My first encounter with process theory was the book on Communicating Sequential Processes by Prof. TONY HOARE from Oxford. During a meeting with Prof. JAN BERGSTRA at the Marktoberdorf Summer School I decided to undertake own research in process theory. First ideas for this work originated in a joint seminar on process theory with Prof. ROMAN LIEDL in Innsbruck.

A major technical break-through was possible when I spent six weeks in the stimulating atmosphere of the Research Institute for Symbolic Computation (RISC) which is located in an ancient castle in Hagenberg, Upper Austria. I want to thank Prof. BRUNO BUCHBERGER, chairman of RISC, for the opportunity to stay as a guest at his institute and I want to thank the research staff of the institute for all the fruitful discussions.

Several results of this work were presented in the process algebra seminar of Prof. JAN BERGSTRA in Amsterdam and at the logic seminar of Prof. ERWIN ENGELER and Prof. GERHARD JÄGER in Berne and Zurich. I want to thank Prof. JAN BERGSTRA for three fascinating days in Amsterdam during which I regained much confidence concerning my work, during which I was able to learn much about process theory and efficient ways of research management, and during which I got the opportunity to enjoy the relaxed and friendly way of living in the Netherlands. Prof. ERWIN ENGELER drew my attention to cumulative logic as another logic approach to process theory.

Prof. KURT BAUKNECHT and Prof. KARL HANTZSCHMANN, directors of the Departments of Computer Science of the Universities of Zurich and Rostock, the two departments where I worked on this book, are permanently spending all their time and effort to offer and secure an optimal research environment for their departments. Prof. LUTZ RICHTER provided all his guidance and friendly advice through my years as postdoc and young assistant professor. Prof. HELMUT SCHAUER continuously took interest in my work. The financial support of the Swiss government in a special program for supporting young talents provided the financial framework for my work by funding the position of an assistant professor. Prof. GERD WECHSUNG accepted this work in his series "Teubner-Texte zur Informatik".

My friend and colleague Dr. EDGAR LEDERER spent uncounted hours in reading and discussing the topics of this book. CHRISTIAN HIDBER carefully read parts of this work, made highly valuable remarks on my mathematical style, and kept repeating these remarks until I finally managed to make significant progress in my style of presenting mathematical topics. Dr. RUTH MAGDOWSKI shared the ups and downs during the year of writing this book.

Many other friends and colleagues at Zurich and Rostock provided invaluable support for my work. I want to thank in alphabetical order CLAUDINE ACKERMANN for ordering and administrating the scientific journals, WOLFRAM BÜTOW for setting up the technical infrastructure, MAJA EBNER for her secretarial work, NORBERT FUCHS for porting a linear logic proof system to the local PROLOG dialect, WOLFGANG GEHRKE at Hagenberg for discussions and for pointing out important references, ANKE HÄNLER for setting up the technical infrastructure, ANGELIKA JUNGE for her secretarial work, NATALIA IOUSTINOVA for many fruitful discussions, SARA KALVALA at Cambridge for suggestions on an implementation of linear logic in the ISABELLE proof system, LOTTI KÜNDIG for her administrative help and for keeping up my moral, MICHAEL KÜNZLI and RAFAEL KÜNZLI for implementing a subset of the specification language and for fruitful discussions, CORINNE MAURER for her secretarial work, LARRY PAULSON at Cambridge for an introduction to the ISABELLE proof system, BEAT RAGETH for maintaining the workstation environment, KERSTIN REIHER for performing bibliographic research and for allowing me to keep a significant and ever increasing number of library books in my office, MATHIAS RICHTER for discussions and proofreading, PETRA SCHMIDLI for her secretarial work, NATARAJAN SHANKAR from SRI for providing his proof system for multiplicative additive linear logic and for adapting it to my needs, RICO SOLCA for fixing my portable computer, which consistently crashed on Friday afternoon, and VOLKER STRUMPEN for discussions and proofreading.

Special thanks to JÜRGEN WEISS, editor at Teubner-Verlag, for his help and patience during the phase of reformatting and final typesetting.

For typesetting this book I used the TEX system by DONALD KNUTH, the macro packages LATEX by LESLIE LAMPORT, and the diagram and proof style packages from PAUL TAYLOR. The bibliography was produced with the help of BIBTEX by OREN PATASHNIK, the index was generated with the MAKEINDEX system from PEHONG CHEN, several figures were produced with the X-Windows application IDRAW. All of the above programs are part of one of the most advanced scientific typesetting systems. It is most pleasant to find such elaborate and most reliable software in the public domain. Accompanied by the music of JOHANN SEBASTIAN BACH, typesetting was even more enjoyable.

Rostock, September 1999 C. H. Cap

Contents

Chapter 1

Introduction

We have a habit in writing articles published in scientific journals to make the work as finished as possible, to cover up all the tracks, to not worry about the blind alleys or describe how you had the wrong idea first, and so on. So there isn't any place to publish, in a dignified manner, what you actually did in order to get to do the work.

R. FEYNMAN,
Nobel Lecture, 1966.

This chapter gives an overview of the present work. It is organized as follows:

Section 1 provides a *motivation* for process theory.

Section 2 explains the *research problems* of process theory and sketches the *scientific method* which is used in process theoretical research.

Section 3 presents some *elements of process theory* in an informal manner and introduces the reader to the way of thinking in process theory.

Section 4 provides a *classification of existing process theories*. A list of the various theories together with a compilation of the most representative references allows the interested reader to study the respective approaches more closely.

Section 5 contains a *reading guide* through the subsequent chapters of this work.

1.1 What is a Process? – An Introduction

In computer science, a *process* is obtained by combining *elementary states* and *state changes* to a new entity.

A *state* is, for example, the contents of a storage cell in a computer or the position of a robot in a numerically controlled manufacturing plant. Similarly, a *transition* is the modification of the contents of this storage cell by a write statement or a change in the position of the robot by a suitable program. A state or state change is called *elementary*, if it cannot be subdivided into smaller components. This may depend on the chosen level of abstraction:

At the level of a programming language, a single statement represents an elementary state change, but at the register-transfer level this statement is further subdivided into many assembler statements and an even larger number of register transfers.

As it will be outlined in Section 1.4, the *emphasis* of a process theory may be placed either on the states or on the state changes which in the latter case are also called *actions*.

Common operations for combining two actions are the *sequential composition*, the *parallel composition*, and the *non-deterministic choice*. Let a and b denote two elementary actions. The *sequential composition* of a and b is a process which first performs action a and then action b. The *parallel composition* of a and b is a process which performs the actions a and b without[1] any restriction on the order of their execution. The *non-deterministic choice* between a and b yields a process which performs either a or b. There is a large number of similar possibilities for constructing larger entities from elementary actions.

Process theory provides a precise mathematical meaning to the notions of states and state changes and thus *defines* what a computer scientist means when he is talking about a "process". Furthermore, process theory is the arena in which processes are *specified* and all sorts of *questions* about processes are posed and answered. In the remainder of this section, it will be shown why *process theory* is important for the *applied computer scientist*.

A "computational program" reads as input several data elements $i_1, i_2, \ldots, i_m$, performs some calculations, and finally prints as output a number of data elements $o_1, o_2, \ldots, o_n$. The "meaning" of such a program is given by the function which maps the input data elements to the output data elements: $f : \text{Input} \rightarrow \text{Output}$, $f(i_1, i_2, \ldots, i_m) = (o_1, o_2, \ldots, o_n)$.

Many programs not only calculate functions but also *perform certain actions*. Consider, for example, the program inside of the autopilot of an airplane: This program must continuously monitor the altimeter, the compass, the airspeed indicator, and a large number of other navigational instruments, and it must control the position of the rudder, the power setting of the engines, and many other devices, so that the airplane safely reaches the destination whose coordinates were entered by the pilot. Additionally, during all the necessary manoeuvring many parameters, for example the speed of the airplane, must remain within certain operational limits.

The task of such programs is the *coordination* of states and state changes, i.e. these programs realize a *process*. Therefore, if we specify, implement, and analyze such programs we are *applying* process theory. Often the programmer or the user of a system is not aware of this fact, since the process-theoretical concepts are hidden behind programming languages and software engineering environments.

1.2 Process Theory as Research Area

Statement of Research Problems

In process theory, the following research problems are studied:

[1] Maybe the reader expected a simultaneous action of a and b or at least some overlapping condition. As will be outlined in Section 2.2, this approach has a number of problems which are associated with the notion of *time*. The above idea of *action independence* will prove to be a much better concept for parallelism.

- Which *mathematical concept* models our intuitive notion of a process most *adequately*?
- How can processes be *defined* by using suitable *specification languages*?
- What are *interesting properties* of processes and how can they be *formalized* by using a suitable logic?
- Which properties of processes can be *decided* by analyzing an arbitrary specification of the process, how do the decision algorithms look like, and what complexity do they have?
- How can process theory be used to model *real systems*?
- How can the *life-cycle* of real systems be supported with the help of concepts, algorithms, and techniques from process theory?

State of the Art and Open Problems

In the present state of research, a considerable number of process theories have been developed and there is a large number of different formal approaches. Several of the above questions have been answered partially and interesting applications have been studied. From the present experience with process calculi several problems and *new questions* arise which are discussed in the following.

Languages, Logic Frameworks, and Applications

Although there is a large number of process theories, most calculi have not been developed with *practical applications* in mind: The calculi are usually presented in an abstract mathematical setting and a small number of illustrating examples is given in this setting. Practical applications of process theory require a sound mathematical basis but also an adequate specification language as well as a logical framework with the help of which properties of the specified processes may be derived.

Process algebra ACP [BW90a] and the associated specification framework PSF [MV93], and CSP [Hoa85] with the programming language OCCAM [Inm84] are two examples of a well-balanced connection between a process calculus and a specification language. However, there are specification languages like LOTOS [Hog89] which are difficult to connect with a process theory [Ber], and there are process theories like the event structures [Dro89] or the π-calculus [MPW92] for which an adequate specification language has not yet been developed.

Conclusion: It is necessary to study process theory in a closer connection with specification languages, logical frameworks, and applications.

Specification Languages for Partial Order Theories

Two approaches are known for dealing with parallelism: In *interleaving theories*, for example process algebra ACP [BW90a] or CSP [Hoa85], parallelism is reduced to non-determinism. In *partial order theories*, for example PETRI-nets [Rei91] or event structures [Dro89], a partial order relation is used to represent parallel behaviour. For interleaving theories there are elegant algebraic and logic descriptions, whereas for partial order theories no entirely satisfactory representation is known. Only very recently, linear logic has been used in descriptions of partial order theories: [EW94], [Pra94a], [Kan94].

Conclusion: Develop more adequate specification languages for partial order process theories.

Interaction of Parallelism and Non-Determinism

In interleaving theories it is not possible to distinguish parallelism from non-determinism, since the former is defined by the latter. In partial order theories, ***parallelism*** can be represented in a very elegant manner, but for ***non-determinism*** often additional constructs are required. For example, in event structures [Dro89] an explicit conflict relation has to be added. Only in PETRI-net theory [Rei91] parallel and non-deterministic phenomena have their natural place in the calculus. PETRI-nets allow ***confusion*** which is a complicated form of interaction between parallelism and non-determinism.

Conclusion: Study the interconnection of parallelism and non-determinism and find frameworks which treat both phenomena separately and adequately.

Higher Order Process Theories

Most process theories are ***first order*** theories: They provide operations for coordinating states and state changes but these operations themselves remain fixed. For example, in the PETRI-net formalism[2] operators are provided for defining a net. However, once a net has been defined and the transitions are firing, there is no possibility to change the structure of the net. To provide this possibility, a ***higher order*** process theory is required.

Conclusion: Study the notion of higher order processes and find adequate mathematical models.

Specification of Behavioural and Data Aspects

Many process theories deal only with the ***behavioural*** aspects of a process, i.e. the way how states and state changes are coordinated, but the ***data*** aspects, i.e. the descriptions of the sets of states and state changes, are neglected. The sets of states and state changes initially are taken to be finite sets, and only during later extensions of the theory the data aspects are added. For example, PETRI-nets were extended to coloured nets [Jen91], and in process algebra parametrized actions and state operators were added [GP90], [GP91], [MV93]. Although these extensions provide additional possibilities for specifying processes there remains the danger that as late add-ons they do not fit well into the formalism of the calculus.

Conclusion: Study the interconnection of the behavioural and data aspects of a specification and the joint possibilities of a calculus treating both aspects.

[2] In PETRI-net theory several extensions to the elementary net, e.g. coloured nets [Jen91], annotated nets [BEM90], predicate / transition nets [Gen91], or algebraic nets [Rei91] are called "higher order PETRI-nets". In this work, the term "higher order" is used in the same way as it is used in logic [BM77] and in category theory [LS86], and it denotes a more powerful concept than the above mentioned extensions.

Multiparadigmatic Programming

Presently, a number of different programming paradigms are employed: Imperative [Gor88], [Seb93], functional [FH80], algebraic [GWMF93], logic [Llo87], object oriented [BL94], and others. Every paradigm has a number of specific advantages during certain stages of the software life-cycle and for certain applications. Therefore, it is useful to study how a single specification technique could combine several of these paradigms and how one approach may be translated into the other.

Conclusion: Study process theories which can incorporate several programming paradigms.

Basic and Derived Concepts

With the advent of a larger number of process theories, the following question gained importance: What are the ***basic concepts*** and what are the ***derived concepts*** in process construction?

For computational programs, the answers to this question are well-known and provided by classical recursion theory, [EL92], [Odi92], [Cap93]: Several basic constructions are known, e.g. the formalism of partial recursion, the TURING-machine, or the languages WHILE, GOTO, and RECUR, which all constitute minimal requirements for specifying every computable function. Furthermore, it is well-known how the constructors of one formalism translate into the other.

In process theory, there are several very different approaches: The ***composition operators*** of process algebras, i.e. the sequential, parallel, and non-deterministic composition constitute one choice, the ***causality mechanisms*** of PETRI-nets another. It is not yet fully known which constructors are the best for practical and theoretical purposes, what their exact power is, and how they may be reduced to each another.

Conclusion: Find basic constructors for process design and analyze their power.

The Scientific Method

What kind of a scientific problem is posed by the question "What is a process?"

In theoretical physics, an important question is "What is an atom?". Answers to this question come in the form of mathematical models. This does not imply that an atom ***is*** this mathematical model in an ***ontologic*** sense. It is unlikely that there will ever be a scientific method for verifying – or rather falsifying – ***any*** ontologic answer to this question. However, this answer may be understood in an ***explanatory*** manner. In this case, setting up ***control experiments*** and comparing the scales and gauges of the measuring equipment with the predictions of the theory constitutes a widely accepted method for the ***falsification*** of models [Pop84].

In computer science we face the ***opposite situation***: There is ***no experiment*** which may be used to verify or falsify an answer to the question "What is a process?". As above, the answers come in the form of mathematical models, but now a process ***is*** this mathematical model in the ***strict ontologic*** sense. Providing an answer to this question constitutes a creative act: ***A process is, what we define a process to be.***

And even more astonishing: Answers to the question what a process is can influence the design of computing systems. An excellent example for this claim is the Transputer [Inm88] which is the process theory CSP [Hoa85] crafted in silicon. Thus, the answers to this question are not falsified by "reality", they *create* reality.

There is, however, a ***pragmatic form of falsification*** for process theories: A mathematical model for processes is wrong – or at least deficient – if we may *think* of systems, for which it may reasonably be argued that they constitute processes but which may not be described within this framework. For example, a theory in which it is not possible to define a concept of sequential composition cannot be considered an appropriate theory for processes. This fact leads to an ***evolution of process theories***: The more complex formal tools become, the more general notions of procesess can be formulated and the more powerful are their implementations in soft- and hardware.

One should keep in mind that the most powerful formalism is not necessarily the most useful for practical purposes. This is similar in mathematics where for example the theorem of STOKES on differential manifolds [CBDM89], a highly important result for theoretical physics, is formulated in the language of analysis and not in the considerably more precise and powerful formalism of set theory or logic. Furthermore, its proof is not presented with the same rigor as it is expected in set theory or logic.

1.3 Elements of Process Theory

In this section we will present an ***arbitrary collection*** of process theoretical concepts in order to introduce the reader to the subject.

Sequential Composition

If p and q are two processes, then their ***sequential composition*** is a process which first behaves like the process p and then like the process q. Often the operator $*$ is used to denote this sequential composition: $p * q$. It is reasonable to expect that this operator is associative:

$$(p * q) * r = p * (q * r)$$

Non-Deterministic Behaviour

The term ***deterministic*** denotes that, given the present state of a system, there is only one possibility for its future state or behaviour: The system's future is determined by its present.

Non-determinism, in contrast, means that if the present of a system and possibly even its entire past is known, several different future developments are possible, the concrete choice being beyond any prediction. In computer science several forms of non-determinism are known:

Underspecification can lead to non-determinism if not the entire behaviour of a system is fully specified. This usually is turned into determinism during later phases of the design or by a choice of the implementor.

Complex systems may be deterministic by their nature, but in practice not all influences are known or shall be captured by a model. In such situations those deterministic aspects which are *beyond deterministic prediction* are treated as non-deterministic features. Well-known examples can be found in network protocols and operating systems. Here, the precise behaviour of a process may depend on data packets which were generated by a machine far away from the observed computer.

In *internal non-determinism*, the decision of a process to pick a specific behaviour out of a larger set of possible behaviours is internal to the process and cannot be influenced from the outside. It resembles a radio-active atom which non-deterministically chooses the moment of emitting an α-particle.

In *external non-determinism*, the process is interacting with an environment which may influence the choice of its behaviour. Often the process offers various possibilities and the outcome of the interaction is a single, deterministic behaviour. A typical example is communication where a reader process offers to read the value "0" or the value "1". This process cooperates with a write process which by writing a value of "0" forces the reader to read the same value.

In process specification, non-determinism either is generated explicitly by a *choice operator* or it occurs as a consequence of other constructions.

If p and q are processes, then the *non-deterministic choice* between p and q is a process which non-deterministically chooses to behave like p or like q. Often the operator $+$ is used to denote this non-deterministic choice: $p+q$. The distinction between internal and external non-determinism depends on the way in which processes can interact with their environment.

In a number of process theories, non-determinism is *not generated* by a composition operator but *results* from another construction of the calculus. PETRI-nets [Rei91] for example, are governed by a causality mechanism: The presence of a "token" enables an "action" which consumes the enabling token and produces another token. If two different actions are enabled by the same token of which only one copy is available, then these two actions are competitors for this copy. Competition is resolved by the non-deterministic decision for one of the two actions.

Interleaving Parallelism

In *interleaving theories*, parallelism is reduced to non-determinism. This may be accomplished by introducing an *interleaving operator* $\|$ for which the following equations hold for elementary actions a, b, and c, and for processes p and q:

$$a \| b = (a * b) + (b * a)$$

$$(a * b) \| c = a * ((b * c) + (c * b)) + (c * a * b)$$

$$(a * p) \| (b * q) = a * (p \| (b * q)) + b * ((a * p) \| q)$$

The idea behind interleaving theories stems from multiprocessing on single processor architectures: There is only one CPU, and if two processes p and q shall run on this computer, a scheduler must assign the CPU to p and q. Hence, the order in which the commands of p

and q are executed is preserved for p and q, but at any time, the scheduler may suspend one process and assign the CPU to the other. Thus, every individual execution sequence of the process $p \parallel q$ is obtained by a certain history of task switches of the scheduler. The set of all execution sequences is obtained by considering all possible histories of task switches. In the case of the process $((a * b) \parallel c)$, the three execution sequences

$$(a, b, c) \qquad (a, c, b) \qquad (c, a, b)$$

are obtained.

The process $a * ((b * c) + (c * b)) + (c * a * b)$ is equal to the process $(a * b) \parallel c$. It is a process which *first* chooses non-deterministically whether the left or the right process in $\parallel$ shall obtain a time slice of the CPU. Then the respective action is executed. Finally, *another* non-deterministic choice is performed for the subsequent time slice. However, given this process, it is not possible to determine whether the user wanted to model a non-deterministic process of the above kind or whether he modeled a parallel process $(a * b) \parallel c$ and obtained the above expression by an application of process algebraic equations. In this sense, parallelism and non-determinism *cannot be distinguished* in interleaving theories.

The process $(a*b*c)+(a*c*b)+(c*a*b)$ has the same set of execution sequences as the process $a*((b*c)+(c*b))+(c*a*b)$ but it is a different process: In the process $(a*b*c)+(a*c*b)+(c*a*b)$ first a non-deterministic decision is made between three deterministic processes. Once this decision has been made, three actions execute in a sequential manner. This is a *different choice-structure* than in the process $a * ((b * c) + (c * b)) + (c * a * b)$. A consequence of this observation is that a process is not fully determined by the set of its execution sequences.

Partial Order Theories

In *partial order theories*, the states and/or state changes in a process are subject to a partial order relation $\geq$: If for two state changes a and b the relation $a \geq b$ holds then *first* the state change a took place and *then*[3] the state change b. This denotes a sequential behaviour. If for two state changes a and b neither $a \geq b$ nor $b \geq a$ holds, i.e. if a and b are *incomparable* with respect to $\geq$, then a and b are said to *execute in parallel*.

Equational Specification

Equations may be used to specify processes. For example, let a and b be elementary actions. If the equations

$$R = a \qquad S = b \qquad T = R * S * S$$

hold for the processes R, S, and T, then T is the process $a * b * b$ which first performs a, then b, and finally performs for a second time the action b.

[3]Here, the symbol $\geq$ illustrates the direction of time. $\geq$ can be interpreted as an arrow from the left hand side to the right hand side. If the symbol $\geq$ were interpreted as an order relation on some kind of time stamps of the state changes a and b, just the opposite result would be obtained.

Equations can also be used to specify ***infinite processes***. For example, the only solution of the process equation $X = a * X$ is the process $X = a * a * a * \cdots$ which performs a for an infinite number of times.

Not every process equation has a ***unique solution***. For example, all processes X, Y, and Z are solutions of the equation

$$X * (Y * Z) = (X * Y) * Z$$

since both expressions denote a process which first behaves like X then like Y and finally like Z. We may specify certain conditions under which a process equation has only one solution. Intuitively, this is the case when we can determine the sequences of elementary actions which a process may perform by a successive rewriting of the equations. In this sense, the above equation $X = a * X$ has the unique solution $a * a * a * \ldots$, since it allows the following rewritings:

$$X = a * X = a * a * X = a * a * a * X = \cdots$$

An infinite process $a * a * a * \cdots$ ***cannot be extended***: To show this, let two processes X and Y be defined by the following two equations:

$$X = a * X \qquad\qquad Y = X * b$$

X is the infinite process $a * a * a * \cdots$ and Y is a process which first behaves like X and then performs the action b. By ***equational reasoning*** it will be shown that $X = Y$:

If $X = a * X$ then also $X * b = a * X * b$. Since $Y = X * b$ we obtain $Y = a * Y$ as equation for process Y. This is the same equation as for process X and the equation is of such a form that it has a unique solution. Therefore, $X = Y$.

Deadlock

A ***deadlock*** is a situation which is known from operating systems: Program A owns resource a and must wait for resource b before it may continue its work. Similarly, program B owns resource b and must wait for resource a before it may continue its work. Therefore, both programs have to wait and none is able to continue its execution. See [Ric85], [MOO87] for more details.

In process theory, there are several possibilities to model a deadlock. For example, in process algebra it is represented as an action after which no further action may take place. If δ denotes the deadlock, this is represented by the equation

$$\delta * a = \delta$$

which holds for every action a. In PETRI-net theory every marking, i.e. every global state in which no transition can fire, is called a deadlock.

For many applications it is a very important question whether a process may eventually reach a deadlock situation, since if a deadlock is reached no further processing is possible.

Implementation and Abstraction

After a process has been defined with the help of a suitable specification language we want to *implement* it on a computer. This means that we must *represent* the abstract states and actions of the process by the states and actions of the computer. It is possible that certain states or actions may not be represented directly by a single state or action of the computer. For example, an action of a process might consist of the calculation of the factorial of a natural number. Many computers do not provide a single command for this operation and so the calculation of the factorial is performed by a long sequence of additions, multiplications, and similar operations which are available on the computer.

Hence the implementation of the above process will not contain a single state change of the kind

$$\texttt{calculate\_fac\_of}(4) \longrightarrow \texttt{fac\_is}(4, 24)$$

but by a sequence of state changes, e.g.

$$\texttt{calculate\_fac\_of}(4) \longrightarrow \texttt{temp}(1, 1, 4) \longrightarrow \texttt{temp}(2, 2, 4) \longrightarrow$$

$$\longrightarrow \texttt{temp}(3, 6, 4) \longrightarrow \texttt{temp}(4, 24, 4) \longrightarrow \texttt{fac\_is}(4, 24)$$

The above two processes are *different* but, from a point of view in which we *abstract* from the employed *temporary or intermediate states*, they may be regarded as *equal*.

1.4 Classification of Process Theories

Since presently there exists a considerable number of process theories it is useful to provide criteria by which they can be classified and to list the most representative references.

Classification by Treatment of Parallelism

As outlined above, process theories may be classified by their approach to parallelism: There are *interleaving theories* and *partial order theories*. Recently, connections between these two theories have been developed, [BMC64].

Classification by Constructor Concept

Process theories may be distinguished by the way how combinations of elementary actions are obtained:

In *compositional theories*, there are composition operators which generate for example the sequential, parallel, or non-deterministic composition of two processes. Based upon these basic operations, a number of more sophisticated combinators can be defined. Examples of compositional theories are process algebras, CSP, and CCS.

In *observational theories*, an underlying mechanism is defined which governs the dynamics of a process. Under certain circumstances the behaviour which is resulting from this mechanism may be interpreted as sequential, parallel, or non-deterministic. More complex phenomena like fair forms of behaviours or confusions may be defined. In observational theories, often no operators are available which combine two processes to a new sequential, parallel, or non-deterministic process. However, by a suitable application of the underlying mechanism, these processes can also be defined. Examples of observational theories are PETRI-nets, temporal logic, or the process specification formalism which will be defined in the course of the present work.

There exist also *hybrid theories* which have elements of both kind. In event structures, for example, parallelism is obtained in an observational manner, and non-determinism is provided by a special incompatibility relation, thus in a compositional way.

Classification by Emphasis on States or Actions

Process theories may be classified whether they emphasize *states* or *state changes* as the more important element for modeling a system.

In *action-oriented* process theories, the main element of the theory is the action, i.e. the state change. For example, let two processes be defined by the following recursion equations

$$X = a * a * R \qquad R = b * b * c * a * c * X$$

The sequence actions of the process X is given by

$$a, a, b, b, c, a, c, a, a, b, b, c, a, c, a, a, b, b, c, a, c, a, a, \ldots$$

Of course one may assume that the process initially has some state and that the execution of an action changes this state. For example one could speak of the state of the process after the actions a, a, b, b, c, a have executed. This state of the process must be different from the state which is reached after the execution of the actions a, a, since in the first case the next action is c, whereas in the second one it is b. Since an action-oriented process theory does not provide a notion of a state it must be reconstructed by considering the history of executed actions.

The following example shows that the history of executed actions is not sufficient to determine the subsequent actions: In the process $(a * b * f) + (a * b * e)$, first a non-deterministic choice is made between process $(a * b * f)$ and process $(a * b * e)$. Then either the actions a, b, f are executed or the actions a, b, e. The actual state of the process cannot be determined if it is only known that the actions a, b have executed.

In *state-oriented* process theories, the main element of the theory is the state. Most notions of the theory are defined in terms of states. State changes play only a minor role since they may be determined from observations of the states: If a state a is observed at time t, a state b is observed at time $t + 1$, and no other state can be observed in between, then it may be concluded that a state change took place between time t and time $t + 1$. PETRI-nets [Rei91] and event structures [Dro89] are examples for state-oriented process theories.

Communicating Sequential Processes	[Hoa85], [Plo83]
Calculus of Communicating Systems	[Mil89]
Calculus of Mobile Processes	[MPW92], [MPW89a], [MPW89b]
Process Algebra, HENNESSY version	[Hen88], [HI93]
Process Algebra, BERGSTRA version	[BW90a], [BK90], [BBK87], [BB91], [BB93]
PETRI-Nets	[MM90], [Bau90], [NPW81]
Colored PETRI-Nets	[Jen91]
Algebraic PETRI-Nets	[Rei91]
Predicate Transition Nets	[Gen91]
GIRARD's Geometry of Interaction	[AJ92], [Gir89b], [Gir89a], [Gir90]
Chemical Abstract Machine	[BB92]
Linear Logic	[Lin92], [Abr90a], [MOM91], [Gir87a], [Gir87b]
Guarded Command Language	[AO91]
HOARE-OWICKI-GRIES Logic	[Gri89], [Apt81]
Process Theory by BROY	[Bro86]
Dynamic Logic	[Smi88]
Temporal Logic	[McD82], [Krö87], [Mer92], [Eme90], [Sti92]
Modal Logic	[vBvES93], [Eme90], [Sti92]
Cumulative Logic	[Eng86], [Eng93]
Event Structures	[Win80], [Win87], [NPW81], [GL91], [BMC64]
Power Domains	[Smy78], [Dro89], [NPW81], [Abr91]

Table 1.1: List of Process Theories.

A Small Bibliography

In Table 1.1, a list of several process specification formalisms is provided together with representative references which may serve as an introduction to the specific methods.

1.5 How to Use This Book? – A Reading Guide

This work contains a *new mathematical approach* to process theory which is presented in *three tracks*:

(1) The *relational track* provides *concrete mathematical models* for processes and requires only a relatively modest technical apparatus. It is very near to the practical and intuitive concept of a process and therefore is most useful for the applied computer scientist. In this approach we concentrate on the rich phenomenology of distributed processes.

(2) The *logical track* provides *formal languages* for the specification of processes. It may be connected to the relational and the categorical approaches via model relations. In our process theory the employed formal language is based on the *multiplicative fragment of linear logic* and on *algebraic specification*. This approach is well suited for formal reasoning about processes and thus it will be of interest for the theoretical computer scientist.

(3) The *categorical track* interprets processes in the language of category theory. It uses the same mathematical concepts as the relational approach, but clothes them into a more abstract language. This approach proved extremely helpful in the construction of the higher order aspects of the calculus: Only after I discovered that the relational models of distributed processes are examples of so-called *symmetric monoidal closed categories* it became obvious that and how a higher order theory should be constructed. Its development was closely guided by the well-known constructions of closed[4] categories. Even more fascinating, this approach shows that the rather abstract concept of symmetric monoidal closed categories can be used to describe something of such a high practical importance as a distributed system. The associated open questions which arise from this connection should interest the mathematicians.

The various chapters of this work are organized as follows:

Chapters 2, 3, and 4 belong to the *relational track*. They define ordinary, distributed, and higher order transition systems and give a large number of examples. Not much formal background is required for understanding these chapters. Some of the employed notations are discussed in Appendix A.

Chapter 5 belongs to the *logical track* and presents a linear logic and algebraic framework for specifying processes. The necessary theory on *linear logic* is presented in Appendix B.

Chapter 6 introduces a *specification language* and provides many *practical examples*. It demonstrates that and how this process theory is useful for *practical applications*. Finally, it discusses a *prototypical implementation* of a subset of the specification language.

Chapter 7 belongs to the *categorical track* and points out the connections between processes, linear logic, and categories.

Finally, Chapter 8 draws *conclusions* on the achieved results and outlines the fascinating challenges and thrilling opportunities for *future work*: It discusses several conjectures, projects, and open questions.

The *mathematical apparatus* which is used in this work contains relational structures, algebraic specification, linear logic, and category theory.

Appendix A provides a survey on some *relational and algebraic structures*.

Linear logic is a very new and active research area and thus is not part of the toolset of many computer scientists and mathematicians. Since no concise, sufficiently precise and vivid introduction to this field could be located in the still rather small number of survey articles, it appeared necessary to provide the required information in Appendix B.

Since there is a large number of excellent surveys, articles, and textbooks on *algebraic specification*, and since there is already a quite standardized and well accepted formalism, no remarks are made on this topic. For further information, the reader is referred to the excellent survey article [Wir90] and to other sources like [EGLer], [EM90], [GTW78], [Wec91], [MT92], [MA86], [BHK89], [HL93], [GM89].

Category theory is a very abstract way of analyzing mathematical objects. Its value for the applied mathematical community is controversial since many of its results are very remote from practical applications. Nevertheless, it may provide important insights into construc-

[4] Which in essence means the same as higher order.

tions and concepts, so that it is an important part of the toolkit of theoretical computer scientists. In his *Categorical Manifesto* [Gog91], GOGUEN demonstrates that many important constructions in applied computer science are known in category theory and have a natural and astonishingly easy categorical formulation.

Unfortunately, the *problem of category theory* is in its details. Most texts on category theory provide an elegant approach to categories, but skip these details, thus leaving boring and in no way easy tasks to the frustrated reader who wants to understand categorical concepts including these details. Therefore, in Appendix C an overview on category theory is provided. It is addressed to the reader who knows category theory and has experienced the same frustrations as the author when reading interesting texts which excessively made use of OCCAM's razor, omitting many details.

Digressions from the main stream presentation into the technical appendices may be necessary for a full understanding of the technical contents of this work. Similarly, the consultation of the *catalogue of used symbols* may prove useful.

Chapter 2

Transition Systems

Basic research is when you don't know what you are doing.

C. G. Wilson,
Nature **264**(1976), 100.

Transition systems form the *basis* of our process theory. In this and the subsequent chapters, we will introduce more and more refined variants of transition systems, until we correctly model our intuitive understanding of processes. Each construction is motivated by examples, and the technical details are furnished by definitions. With each step we will obtain mathematical objects which are more expressive, richer in structure, and nearer to our intuition of distributed, parallel processes.

In Section 1, *transition systems* are defined. We will follow the philosophy that a process consists of *states* and *transitions*, both being essential parts of our intuition of a "process". With the example of a *vending machine* we will demonstrate a simple application of this concept.

An abstract transition system in Section 2 will raise the question whether branching points shall be interpreted as *non-deterministic choice* or as *generation of a new parallel process*. We will show why a non-deterministic interpretation is more adequate.

In Section 3, we will discover some *anomalies* which are associated with our present form of transition systems. We will resolve them by imposing *type constraints* on the transitions, and we will introduce *labels* to be able to provide the state changes with additional information. In this section, we will also provide a formal definition for *sequential processes*.

In Section 4, we will discuss the phenomena of *non-determinism* and *causality*. A *process* is a description of what *could* happen. We will define *behaviours* which tell us what actually *did* happen. Determinism and causality will turn out to be the essential properties which distinguish a behaviour from a process.

In Section 5, we will develop a technique for dealing with a *finite number of sequential transitions* at a time. The resulting concept will be presented in two different formalisms. The first one uses the language of transition systems, and the second one shows that processes are a special kind of categories.

In Section 6, we will discuss the notion of a *fair non-deterministic choice*.

Many ideas presented in this chapter are well-known, compare [NRT92]. Due to the youth of the field, there are many different formal approaches and there is not yet a widely accepted terminology. We give a ***unified presentation*** in a formal framework which is compatible to most related approaches and which allows a generalization to distributed and higher order process theories.

2.1 Ordinary Transition Systems

Let us watch a computer, a printer, a traffic light, the navigation system of an airplane, the control unit of a nuclear power station or just any device which is modeled and controlled by an information technological approach. We will see lights going on and off, scales and gauges will display data, and controllers will position steering rods. More formally spoken: We will observe *states* and *state changes*; the latter are also called *transitions*.

1 Definition TRANSITION SYSTEM
A *transition system* $\mathcal{T} = (S, T, \longrightarrow)$ consists of

(1) a non-empty set S, whose elements are called *states*,

(2) a set T, whose elements are called *transitions*, and

(3) a ternary relation $\longrightarrow \;\subseteq S \times T \times S$, the *transition relation*.

◀

The elements of S represent the states of the considered system. If the relation $X \xrightarrow{t} X'$ holds for $X, X' \in S$ and $t \in T$, then the system changes from state X into state X' by a transition named t.

2 Example VENDING MACHINE
A vending machine which accepts a coin and sells coffee or tea, can be modeled as a transition system:

$$S = \{\texttt{idle, coin, coffee\_delivered, tea\_delivered}\}$$

$$T = \{\texttt{enter\_coin, select\_coffee, select\_tea, remove\_cup}\}$$

$$\texttt{idle} \xrightarrow{\texttt{enter\_coin}} \texttt{coin}$$

$$\texttt{coin} \xrightarrow{\texttt{select\_coffee}} \texttt{coffee\_delivered}$$

$$\texttt{coin} \xrightarrow{\texttt{select\_tea}} \texttt{tea\_delivered}$$

$$\texttt{coffee\_delivered} \xrightarrow{\texttt{remove\_cup}} \texttt{idle}$$

$$\texttt{tea\_delivered} \xrightarrow{\texttt{remove\_cup}} \texttt{idle}$$

This vending machine does not allow the transition `select_coffee` when no coin has been thrown in. Since most vending machines have buttons which a user may press at any time, this transition system does not yet completely describe real vending machines. Therefore, the following lines should be added:

$$\texttt{idle} \xrightarrow{\texttt{select\_coffee}} \texttt{idle}$$

$$\texttt{idle} \xrightarrow{\texttt{select\_tea}} \texttt{idle}$$

$$\texttt{coffee\_delivered} \xrightarrow{\texttt{select\_coffee}} \texttt{coffee\_delivered}$$

$$\texttt{coffee\_delivered} \xrightarrow{\texttt{select\_tea}} \texttt{coffee\_delivered}$$

$$\texttt{tea\_delivered} \xrightarrow{\texttt{select\_coffee}} \texttt{tea\_delivered}$$

$$\texttt{tea\_delivered} \xrightarrow{\texttt{select\_tea}} \texttt{tea\_delivered}$$

Furthermore it is possible that a user throws coins into the machine although the machine is not ready to accept (additional) coins. This may be modeled by the following three lines:

$$\texttt{coin} \xrightarrow{\texttt{enter\_coin}} \texttt{coin}$$

$$\texttt{coffee\_delivered} \xrightarrow{\texttt{enter\_coin}} \texttt{coffee\_delivered}$$

$$\texttt{tea\_delivered} \xrightarrow{\texttt{enter\_coin}} \texttt{tea\_delivered}$$

Modern "real" vending machines can give return money, have a "coin-return" button, and are equipped with additional features which require similar extensions of our transition system. ◀

2.2 Interpreting Transition Systems

The example of the vending machine provided us an intuitive idea how to interprete a transition system. In this section we will give a more abstract example whose behaviour can be interpreted parallel as well as non-deterministic.

3 Example INTERPRETATION OF A TRANSITION SYSTEM
The transition system

$$S = \{X, Y, Z, U, V, W, R\} \qquad T = \{a, b, c, d, e, f, g\}$$

$$X \xrightarrow{a} Y \qquad X \xrightarrow{b} Z \qquad Y \xrightarrow{c} U \qquad Z \xrightarrow{d} V$$

$$U \xrightarrow{e} W \qquad V \xrightarrow{f} W \qquad W \xrightarrow{g} R$$

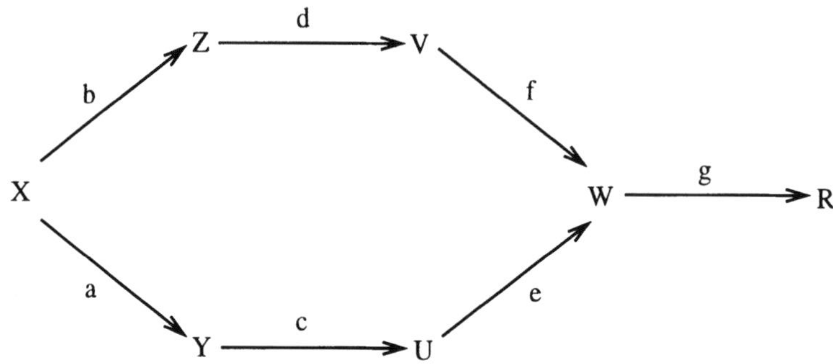

Figure 2.1: Transition System.

is illustrated in Figure 2.1.

This system can be interpreted in two ways:

In the *non-deterministic interpretation* the system is in state X and chooses non-deterministically between two options. If it chooses the "lower" option, it will make the transitions a to reach state Y, then it will make transition c and reach state U and then it will make transition e in order to reach state W. Finally, it will make transition g and reach state R. If the system chooses the "upper" option, it will make the transitions b to reach state Z, then it will make transition d and reach state V and then it will make transition f to reach state W. Finally, it will make transition g and reach state R. In this interpretation *splitting points* like X require a non-deterministic choice, and *joining points* like W may be reached by more than one path of execution. This interpretation yields *sequential behaviour which may be non-deterministic.*

In the *concurrent interpretation* the system starts in state X and immediately generates two processes which may run in *parallel.* There are the following constraints on the order in which transitions may execute: In the "lower" process, a must happen before c and c must happen before e. In the "upper" process, b must happen before d and d must happen before f. There are no further constraints on the transitions. The transitions of the "lower" and the "upper" process may execute in an arbitrary sequence, a transition of the "lower" process and a transition of the "upper" process even may happen at the same time. A sequential order of execution is only required within the "lower" and the "upper" process. In this interpretation, at *splitting points* like X a new process is started or "forked[1]" and *joining points* like W are states where two processes synchronize: One process is terminated, the remaining process takes on state W and executes transition g. This interpretation models *parallel behaviour by underspecifying the order* in which transitions may execute. ◀

Above example suggests that in the present form, transition systems might model parallel situations as well as non-deterministic ones, depending on the mode in which we interprete splitting and joining points. However, it is not possible to model both phenomena *at the same time.* It is therefore necessary to choose between a parallel and a non-deterministic interpretation.

[1]Which is the operating system jargon for starting a new process. In contrast to operating systems, we here are not able to distinguish parent and child process. We therefore cannot answer the question which of the two states represents the *new* process.

The present form of transition systems seems to describe all possible forms of non-deterministic sequential processes which we could think of. However, we will see in the following that there are certain problems with the parallel interpretation. Therefore we will choose the non-deterministic interpretation.

In above example the interpretation of $X \xrightarrow{a} Y$ and $X \xrightarrow{b} Z$ as a fork may be questioned: There is no single "fork" transition which takes one process state (X) and generates two (Y, Z). If in state X transition a took place, $X \xrightarrow{a} Y$, what about the process state X after a took place? If X has vanished, how do we know that we still have to execute b? And if X is still present, how do we prevent another transition a? Or, which mechanism synchronizes a and b? This situation indicates that a "fork" cannot be adequately modeled in this setting.

REISIG [Rei87] provides an example of such a parallel system:

4 Example PARALLEL PROCESS
Suppose a system is described by the PETRI-net in Figure 2.2.

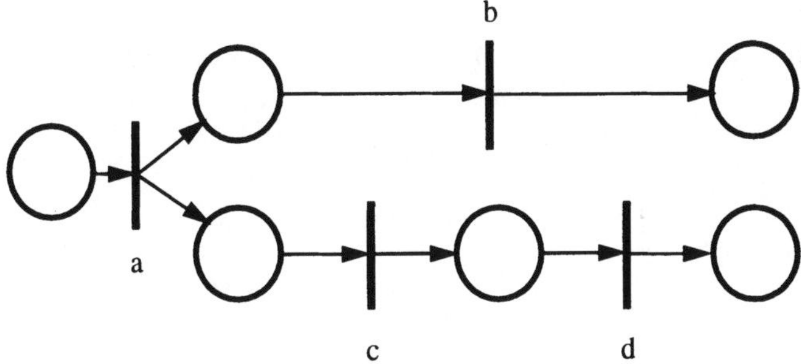

Figure 2.2: PETRI-Net.

In the programming language OCCAM [Inm84], [Wei89] this would correspond to the following program:

```
SEQ
  a
  PAR
    b
    SEQ
      c
      d
```

In process algebra ACP [BW90a] this is denoted with the following algebraic expression:

$$\mathbf{a} * (\mathbf{b} \,||\, (\mathbf{c} * \mathbf{d}))$$

The ultimate goal of all these descriptions is to define an order for the execution of the transitions $\{a, b, c, d\}$. The intended order relation may be derived from above specifications and is given by:

$\mathbf{a} \succ \mathbf{b} \qquad \mathbf{a} \succ \mathbf{c} \qquad \mathbf{c} \succ \mathbf{d}$

The following is a complete list of execution sequences which may be generated by transition systems, and which are consistent with this intended order:

$$1 \xrightarrow{\mathbf{a}} 2 \xrightarrow{\mathbf{b}} 3 \xrightarrow{\mathbf{c}} 4 \xrightarrow{\mathbf{d}} 5$$
$$1 \xrightarrow{\mathbf{a}} 2 \xrightarrow{\mathbf{c}} 3 \xrightarrow{\mathbf{b}} 4 \xrightarrow{\mathbf{d}} 5$$
$$1 \xrightarrow{\mathbf{a}} 2 \xrightarrow{\mathbf{c}} 3 \xrightarrow{\mathbf{d}} 4 \xrightarrow{\mathbf{b}} 5$$

Allowing *multi-transitions*, the following additional sequences may be obtained:

$$1 \xrightarrow{\mathbf{a}} 2 \xrightarrow{\{\mathbf{b},\mathbf{c}\}} 3 \xrightarrow{\mathbf{d}} 4$$
$$1 \xrightarrow{\mathbf{a}} 2 \xrightarrow{\mathbf{c}} 3 \xrightarrow{\{\mathbf{b},\mathbf{d}\}} 4$$

These sequences correspond to the following order relations on the set $\{a, b, c, d\}$ of transitions:

$\mathbf{a} \succ \mathbf{b} \succ \mathbf{c} \succ \mathbf{d}$
$\mathbf{a} \succ \mathbf{c} \succ \mathbf{b} \succ \mathbf{d}$
$\mathbf{a} \succ \mathbf{c} \succ \mathbf{d} \succ \mathbf{b}$
$\mathbf{a} \succ \mathbf{b} \qquad \mathbf{a} \succ \mathbf{c} \qquad \mathbf{b} \succ \mathbf{d} \qquad \mathbf{c} \succ \mathbf{d}$
$\mathbf{a} \succ \mathbf{c} \qquad \mathbf{c} \succ \mathbf{b} \qquad \mathbf{c} \succ \mathbf{d}$

The *intended relation* could not be reproduced by a state transition system approach, since it does not appear in the above list of order relations. The reason behind this is that a transition system approach will always place transition b before c and d, between c and d, or after c and d. If we implement this process on a distributed system in a way that b executes on one processor, and c and d on another, then it is *impossible* to say, whether b was executed before c and d, in between, or after both. A faithful model of this process should therefore lead to an order relation in which b *cannot be compared* to c and d. The intended relation is the intersection of – and at the same time the coarsest relation compatible with – all obtained order relations. ◀

If non-determinism and parallelism shall be modeled at the same time, there are two possibilities. The transition systems can be *significantly extended*. This will be done in the subsequent chapter. Another possibility is to *reduce parallelism to non-determinism*. Then parallel behaviour may be described with the present form of transition systems. In the following, the various aspects of this idea will be discussed.

5 Remark PARALLELISM VERSUS NON-DETERMINISM
The classical *interleaving approach* for reducing parallelism to non-determinism is based on the idea that it is not possible to distinguish a process where two actions a and b may execute in parallel from a process which non-deterministically chooses between executing first a and then b and between executing first b and then a. In basic process algebra BPA [BW90a] where + expresses non-deterministic choice and $*$ expresses sequential composition, this is axiomatized as

$$a \| b = (a * b) + (b * a)$$

This idea is further supported by the observation that parallelism often seems to be associated with non-determinism. This comes from the tradition of "quasi-parallel" processing: Several processes are running on a single time-shared processing unit and give the *impression* that they are executing in parallel, whereas in reality they are executed one after the other: If one process has used up its time slice it is suspended and another process is allowed to execute. Strictly speaking, this switching of a processing unit between several sequential processes *does not constitute a truly parallel behaviour*.

A truly parallel behaviour can only occur in a distributed system[2]. a and b are said to execute in *parallel*, iff[3] there is no mechanism which arranges a specific order of execution between a and b, and iff such an order cannot be determined without any doubt. In this sense, parallelism should be seen in contrast to sequential executions, where the order of execution is fully specified by a linear order relation. The term *concurrent* will be used as a synonym to *parallel*.

This situation can be compared with intuitionistic or multivalued logic. Here the formula $(A \vee \neg A)$ is not always true. Similarly, one may argue that " 'a happened before b' OR 'b happened before a' " is not the same as "there is no restriction on the order of execution of a and b".

From a *modeling point of view*, one would like to separate non-determinism from parallelism. *Parallelism* means that there is no reason to prescribe a certain order of the execution since it is of no importance for the correct behaviour of our system. Furthermore, the actual order in which transitions executed, cannot and should not be determined by observations. *Non-determinism* specifies several possibilities for the behaviour of a system, all of which are considered correct. Therefore, parallelism and non-determinism should be treated as *orthogonal concepts* which do not have anything in common. Furthermore, one phenomenon should be able to occur independently of the other.

Of course one might *implement* a process where two events a and b may take place in parallel by a process where a happens before b – or even by a process which non-deterministically chooses between the two execution sequences $a * b$ and $b * a$. However, these three processes should not be considered equal. ◀

6 Remark DISCUSSION OF TRANSITION SYSTEMS

The present form of transition systems always provides a *linear ordering of the events* of the actual behaviour, i.e. after the non-deterministic choice has been made. For arbitrary events a and b, one of the two events did occur earlier than the other. In distributed systems, this is an unpleasant and not realistic property:

Practical experience with distributed systems tells us that it is extremely difficult to synchronize the clocks of two computers which are located at different sites to a sufficient degree of accuracy, [Mul93], [Lam78]. Therefore, a process theory which shall allow the incorporation of real time and real space aspects should not automatically provide a linear ordering

[2] In our context, a distributed system consists of *several* computing agents instead of one. We do not use this word in the same narrow meaning as in the phrase "Amoeba is a distributed operating system". In our sense also a single computer, even a single chip, may be regarded as a distributed system, since it contains several distinct modules which are capable of an individual and independent state change.

[3] As usual in the mathematical literature, "iff" denotes "if and only if".

of all events. Such a theory would not adequately reflect the day-to-day experiences with distributed systems.

Special relativity theory tells us that it is not always possible to determine which of the two events a and b occured earlier [MTW73a]. A process theory which is based on the present form of transition systems is therefore in contradiction to an established physical theory of time. One might object that such relativistic aspects are only of interest when we consider computers which move relatively to each other at a speed near the speed of light. This is not true. Computers circling the earth in satellites and space shuttles presently are used for navigational purposes and require a precise time base. Although their clocks are not circling the earth at relativistically high speeds, the small time shift which builds up after a longer period can be measured with present technology [SS78].

The present form of transition systems also suggests that there is *only one processor* available for executing the transitions. This is in contradiction to the practical situation of every distributed system.

Proof systems which establish certain properties of processes play an important role for proving the correctness of software [Gri89]. Suppose one wants to show a property of a massively parallel process in which 1.000 events may execute concurrently. If this kind of parallelism is reduced to non-determinism, this leads to the incredibly high number of 1.000! (factorial of 1.000 which is at least 10^{1000}) different execution sequences. The proven property is therefore a property of this extremely large number of traces. It is plausible that a proof system working only with 1.000 parallel events reaches its conclusions faster. However, due to the lack of comparable proof systems for both approaches to process theory, the final evidence for this claim cannot be presented here.

In the present form of transition systems always a *global system state* can be defined. Experiences with distributed systems indicate that such states are not existent for practical purposes. Sophisticated algorithms are required to construct so called global system snapshots [TK87]. ◀

Conclusion:

> Transition systems in their present form must be interpreted as *sequential*, possibly *non-deterministic* processes. A theory of *parallel* or *distributed* behaviour requires an *extension* of this concept.

In the remaining sections of this chapter, the present form of transition systems is studied more closely. In the next chapter, the suggested extensions are provided.

2.3 Typed and Labeled Systems

Transition systems $(S, T, \longrightarrow)$ are used to model state changes which are induced by transitions. These transitions convey the following information:

(1) Information, about *which state changes* a system may exhibit. This is given by the set $\{(X, X') \in S^2 \mid \exists t \in T : \; X \xrightarrow{t} X'\}$ of all state changes.

(2) *Additional information* on individual state changes: If (X, X') is a state change, then there exists at least one $t \in T$ so that $X \xrightarrow{t} X'$. This t may contain additional

information on the state change. Furthermore, if (X, X') is a state change with several transitions $t_1, t_2, \ldots, t_n \in T$ and $X \xrightarrow{t_1} X'$, $X \xrightarrow{t_2} X'$, ..., $X \xrightarrow{t_n} X'$, then the t_i indicate different modes by which the system may change from X to X'.

Unfortunately, our present formalism of transition systems is not well adapted for storing these two kinds of information since it exhibits the following two anomalies:

(1) Transitions *without state change:* It is possible that there is a transition $t \in T$ for which there are no states $X, X' \in S$ such that $X \xrightarrow{t} X'$.

(2) Transitions with *more than one state change:* It is possible that there is a single transition $t \in T$ with several state changes $X_1 \xrightarrow{t} X_1'$, $X_2 \xrightarrow{t} X_2'$, ..., $X_n \xrightarrow{t} X_n$.

Both situations are in contradiction to the reasonable expectation that there would be a one-to-one correspondence of transitions and state changes. Both anomalies are easily resolved if we require that the transitions satisfy the *typing condition*:

7 Definition TYPED TRANSITION SYSTEM
A transition system $(S, T, \longrightarrow)$ is called *typed* or is said to *satisfy the typing condition*, iff for every transition $t \in T$ there is exactly one pair $(X, X') \in S^2$ of states, such that $X \xrightarrow{t} X'$. This pair is called the *type* of the transition t. ◀

It is easy to see that this new definition avoids both anomalies outlined above.

In $X \xrightarrow{t} Y$ the state X is also called the *precondition* and Y is called the *postcondition* of the transition t.

As a consequence of the typing condition, we are no longer free in the choice of transitions. If we still want to supply additional information for transitions in a form that two different state changes may carry identical additional information, we must *add labels to the transitions*:

8 Definition LABELED TRANSITION SYSTEM
A *labeled transition system* $(S, T, \longrightarrow, L, \mu)$ consists of

(1) a *typed transition system* $(S, T, \longrightarrow)$,

(2) a set L, whose elements are called *labels*, and

(3) a function $\mu : T \to L$, the so called *labeling function*.

A labeled transition system $(S, T, \longrightarrow, L, \mu)$ is called *strictly labeled*, iff the labeling function is injective. ◀

Every state transition system may be regarded as a labeled transition system over a singleton set $\{\bullet\}$ of labels. Hence, everything which applies to labeled transition systems also applies to ordinary transition systems.

In a strictly labeled transition system, the name of a transition may be reconstructed from its label.

For the vending machine from Example 2 one has:

9 Example VENDING MACHINE
There are the following sets:

$S = \{$idle, coin, coffee_delivered, tea_delivered$\}$

$T = \{$enter_coin_1, enter_coin_2, enter_coin_3, enter_coin_4,

select_coffee_1, select_coffee_2, select_coffee_3, select_coffee_4,

select_tea_1, select_tea_2, select_tea_3, select_tea_4,

remove_cup_1, remove_cup_2$\}$

The transition relation is as follows:

idle $\xrightarrow{\text{enter\_coin\_1}}$ coin

coin $\xrightarrow{\text{enter\_coin\_2}}$ coin

coffee_delivered $\xrightarrow{\text{enter\_coin\_3}}$ coffee_delivered

tea_delivered $\xrightarrow{\text{enter\_coin\_4}}$ tea_delivered

coin $\xrightarrow{\text{select\_coffee\_1}}$ coffee_delivered

idle $\xrightarrow{\text{select\_coffee\_2}}$ idle

coffee_delivered $\xrightarrow{\text{select\_coffee\_3}}$ coffee_delivered

tea_delivered $\xrightarrow{\text{select\_coffee\_4}}$ tea_delivered

coin $\xrightarrow{\text{select\_tea\_1}}$ tea_delivered

idle $\xrightarrow{\text{select\_tea\_2}}$ idle

coffee_delivered $\xrightarrow{\text{select\_tea\_3}}$ coffee_delivered

tea_delivered $\xrightarrow{\text{select\_tea\_4}}$ tea_delivered

coffee_delivered $\xrightarrow{\text{remove\_cup\_1}}$ idle

tea_delivered $\xrightarrow{\text{remove\_cup\_2}}$ idle

As label set and as labeling function one obtains:

$L = \{$enter_coin, select_coffee, select_tea, remove_cup$\}$

$\mu(\texttt{enter\_coin\_1}) = \texttt{enter\_coin}$ $\qquad$ $\mu(\texttt{enter\_coin\_2}) = \texttt{enter\_coin}$

$\mu(\texttt{enter\_coin\_3}) = \texttt{enter\_coin}$ $\qquad$ $\mu(\texttt{enter\_coin\_4}) = \texttt{enter\_coin}$

$\mu(\texttt{select\_coffee\_1}) = \texttt{select\_coffee}$ $\qquad$ $\mu(\texttt{select\_coffee\_2}) = \texttt{select\_coffee}$

$\mu(\texttt{select\_coffee\_3}) = \texttt{select\_coffee}$ $\qquad$ $\mu(\texttt{select\_coffee\_4}) = \texttt{select\_coffee}$

$\mu(\texttt{select\_tea\_1}) = \texttt{select\_tea}$ $\qquad$ $\mu(\texttt{select\_tea\_2}) = \texttt{select\_tea}$

$\mu(\texttt{select\_tea\_3}) = \texttt{select\_tea}$ $\qquad$ $\mu(\texttt{select\_tea\_4}) = \texttt{select\_tea}$

$\mu(\texttt{remove\_cup\_1}) = \texttt{remove\_cup}$ $\qquad$ $\mu(\texttt{remove\_cup\_2}) = \texttt{remove\_cup}$

In this example we started with the untyped transition system of Example 2 whose transitions contained label-like information but could not be typed. For example, transition `select_tea` was connected with the state changes (`coin`, `tea_delivered`) and (`idle`, `idle`). To obtain a typed transition system we splitted this transition into two, namely `select_tea_1` and `select_tea_2`, one for each state change. The labeling function μ then reversed this splitting. ◀

In this example where we already started with label information encoded in the names of the transitions, the reader could get the impression that typed labeled transition systems were introduced "only for adding underscores and numbers to transitions". This is not the case.

In a labeled transition system the essential information is contained in the *states*, in the *labels*, and in the *structure* of the transitions. The specific names of the transitions are not relevant. This is similar for graphs where the vertices and edges are required only to build the graph structure.

10 Definition ISOMORPHISM OF TRANSITION SYSTEMS
Two labeled transition systems $(S_1, T_1, \longrightarrow_1, L_1, \mu_1)$ and $(S_2, T_2, \longrightarrow_2, L_2, \mu_2)$ are called *isomorphic*, iff the sets of states and labels are equal, $S_1 = S_2$ and $L_1 = L_2$, and iff there exists a bijection $f : T_1 \to T_2$, such that

$$\forall X, X' \in S_1,\ t \in T_1 :\ X \xrightarrow{t}_1 X' \ \Rightarrow\ X \xrightarrow{f(t)}_2 X'$$

$$\forall t \in T_1 :\ \mu_1(t) = \mu_2(f(t))$$

For *unlabeled* transition systems, the second condition is omitted. ◀

The second condition of the above definition may also be expressed by the *commuting diagram* in Figure 2.3.

11 Example ISOMORPHIC TRANSITION SYSTEMS
Assume we want to model a very restricted form of an adder: It has the states 0, 1 and 2 and allows transitions labeled with `add_one` and `add_two`. Such a counter is represented by systems 1 and 2 in Figure 2.4.

System 1 is:

$$S_1 = \{0, 1, 2\} \qquad T_1 = \{x, y, z\} \qquad L_1 = \{\texttt{add\_one},\ \texttt{add\_two}\}$$

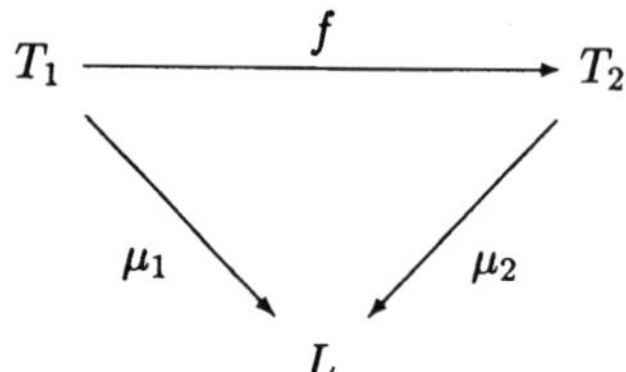

Figure 2.3: Condition for Isomorphic Transition Systems.

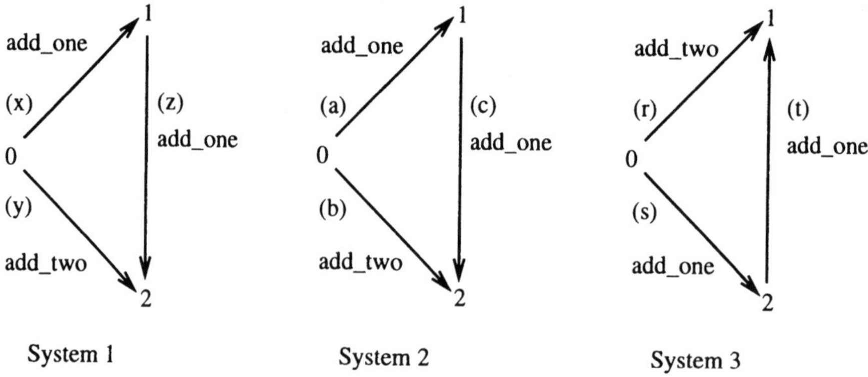

Figure 2.4: Isomorphism of Transition Systems.

$$0 \xrightarrow{x}_1 1 \qquad 0 \xrightarrow{y}_1 2 \qquad 1 \xrightarrow{z}_1 2$$

$$\mu_1(x) = \mu_1(z) = \texttt{add\_one} \qquad \mu_1(y) = \texttt{add\_two}$$

System 2 is:

$$S_2 = \{0,1,2\} \qquad T_2 = \{a,b,c\} \qquad L_2 = \{\texttt{add\_one, add\_two}\}$$

$$0 \xrightarrow{a}_2 1 \qquad 0 \xrightarrow{b}_2 2 \qquad 1 \xrightarrow{c}_2 2$$

$$\mu_2(a) = \mu_2(c) = \texttt{add\_one} \qquad \mu_2(b) = \texttt{add\_two}$$

Systems 1 and 2 only differ in the "names" of their transitions. They are isomorphic via the following isomorphism:

$$f: T_1 \to T_2 \qquad x \mapsto a \qquad y \mapsto b \qquad z \mapsto c$$

System three, see Figure 2.4, is not isomorphic to any of the two other systems. ◀

In the definition of a labeled transition system it is rather complicated to always explicitly mention the transitions. Thus, a ***shorthand notation*** will be introduced. Every ordinary transition system may be "embedded" into a labeled transition system:

12 Proposition EMBEDDING TRANSITION SYSTEMS
Let $(S, T, \longrightarrow)$ be a transition system. Define:

$$T_e := \{(X,t,Y) \mid X \xrightarrow{t} Y\} \qquad L := T$$

$$\mu : T_e \to L \qquad \mu(X,t,Y) := t$$

$$\longrightarrow_e \subseteq S \times T_e \times S \qquad X \xrightarrow{(X,t,Y)}_e Y \;:\Leftrightarrow\; X \xrightarrow{t} Y$$

Then $(S, T_e, \longrightarrow_e, T, \mu)$ is a labeled, typed transition system. ◀

It is possible to define a labeled transition system by specifying an ordinary transition system. The transitions of the ordinary system become the labels of the labeled system.

13 Example SHORTHAND NOTATION
System one from Example 11 was given by:

$$S = \{0,1,2\} \qquad T = \{x,y,z\} \qquad L = \{\texttt{add\_one, add\_two}\}$$

$$0 \xrightarrow{x} 1 \qquad 0 \xrightarrow{y} 2 \qquad 1 \xrightarrow{z} 2$$

$$\mu(x) = \mu(z) = \texttt{add\_one} \qquad \mu(y) = \texttt{add\_two}$$

In shorthand form it[4] may be given by:

[4] i.e. system one or an *isomorphic* system.

$$S = \{0,1,2\} \qquad T = \{\texttt{add\_one}, \ \texttt{add\_two}\}$$

$$0 \stackrel{\texttt{add\_one}}{\longrightarrow} 1 \qquad 0 \stackrel{\texttt{add\_two}}{\longrightarrow} 2 \qquad 1 \stackrel{\texttt{add\_one}}{\longrightarrow} 2$$

◀

In the following, this shorthand notation will be used without always mentioning this explicitly. Furthermore, all transition systems are assumed to be typed.

14 Definition COHERENT LABELED TRANSITION SYSTEMS
A labeled transition system is called *coherent*, iff for two states X, Y there is at most one transition t such that $X \stackrel{t}{\longrightarrow} Y$. ◀

From the modeling point of view, coherency means that there is only one transition for every state change from X to Y.

The transition systems defined so far allow to model non-deterministic, sequential processes:

15 Definition SEQUENTIAL PROCESS
A ***sequential process*** is a typed transition system. A ***rooted*** process consists of a process and an initial state. ◀

Processes may be labeled. *Isomorphic* processes often are considered ***structurally equal***.

There are process theories, for example process algebra [BK90], CSP [Hoa85], and CCS [Mil89], which encode the entire information in the labels. In this case, we may not only abstract from the names of the transitions, but we may extend this process to the labels. This "state-abstract" form of process theory also fits into our framework. For this purpose we need a different notion of isomorphism:

16 Definition S-ISOMORPHISM OF LABELED TRANSITION SYSTEMS
Two labeled transition systems $(S_1, T_1, \longrightarrow_1, L_1, \mu_1)$ and $(S_2, T_2, \longrightarrow_2, L_2, \mu_2)$ are called *S-isomorphic*, iff the sets of labels are equal, $S_1 = S_2$, and iff there exist two bijections $f_S : S_1 \to S_2$ and $f_T : T_1 \to T_2$, such that

$$\forall X, X' \in S_1,\ t \in T_1 :\ X \stackrel{t}{\longrightarrow}_1 X' \ \Rightarrow\ f_S(X) \stackrel{f_T(t)}{\longrightarrow}_2 f(X')$$

$$\forall t \in T_1 :\ \mu_1(t) = \mu_2(f_T(t))$$

◀

17 Example S-ISOMORPHISM
Figure 2.5 shows two S-isomorphic labeled processes.

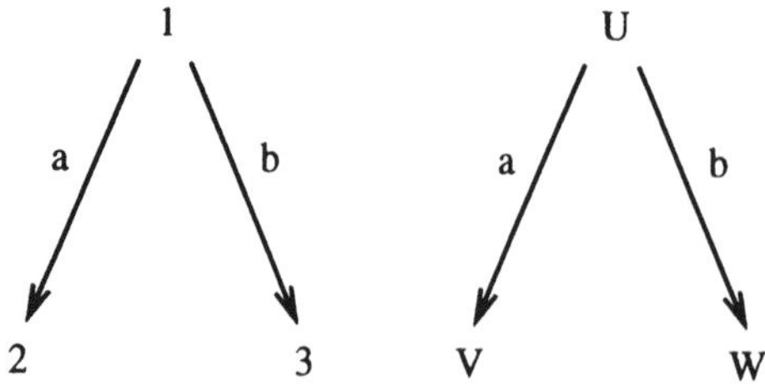

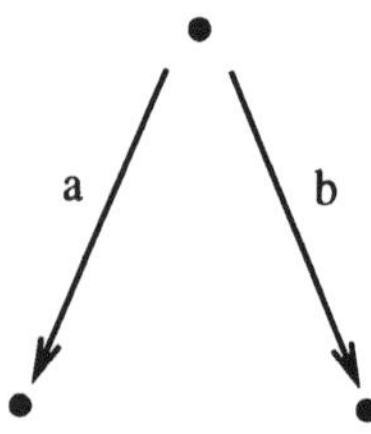

Figure 2.5: S-Isomorphic Processes.

Since the specific states have no meaning for this kind of processes, they may be omitted from the graphical representation, as illustrated in the right hand side of the figure. ◀

In our process theory, we will make a different design decision: Here, the states will be the more important objects.

In the following sections, we will define ***additional transition system structures***. They allow the derivation of important properties of processes which already are encoded in the labeled transition system, but are not readily recognizable.

2.4 Non-Determinism, Causality, and Behaviour

Transition systems describe the ***possibilities*** for the behaviour of a system. For example, a relation with $A \xrightarrow{b} B$ and $A \xrightarrow{c} C$ tells us that if our system is in state A, then it can perform b and reach state B or it can perform c and reach state C.

If a system is observed, then this vagueness ***disappears***: One ends up with observations of transitions and states which ***actually did happen***, and we even may supply our observations with time stamps, for example by writing down the wall clock time of every observation. Such observations of a system may be characterized as transition systems which satisfy the properties of determinism and causality. In this section we will present definitions and applications of these concepts.

First, let us look at an illustrative example:

18 Example TIME-BOMB

The following process, defined by shorthand notation and illustrated in Figure 2.6, models a *time-bomb* which may explode at an arbitrary moment:

$S = \{\texttt{not\_exploded, exploded}\}$ $T = \{\texttt{tick, bang}\}$

$\texttt{not\_exploded} \xrightarrow{\texttt{tick}} \texttt{not\_exploded}$ $\texttt{not\_exploded} \xrightarrow{\texttt{bang}} \texttt{exploded}$

tick
not_exploded
bang
exploded

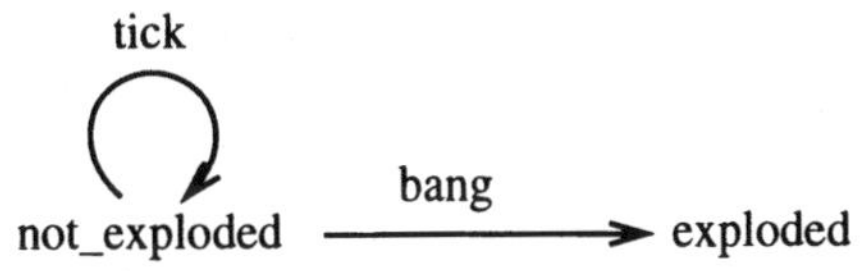

Figure 2.6: Time-Bomb.

If the time-bomb is in the state `not-exploded` we are ***not able to predict the immediate future transitions and states***. The following two "sequences"

$$\texttt{not\_exploded} \xrightarrow{\texttt{tick}} \texttt{not\_exploded} \xrightarrow{\texttt{tick}} \texttt{not\_exploded}$$

and

$$\texttt{not\_exploded} \xrightarrow{\texttt{tick}} \texttt{not\_exploded} \xrightarrow{\texttt{bang}} \texttt{exploded}$$

both represent possible forms of "behaviour" of the time-bomb. ◀

In general, we define non-determinism as follows:

19 Definition NON-DETERMINISM
A transition system $(S, T, \longrightarrow)$ is called *non-deterministic*, iff there exist $X, Y_1, Y_2 \in S$ and $t_1, t_2 \in T$ such that $X \xrightarrow{t_1} Y_1$ and $X \xrightarrow{t_2} Y_2$, and $(t_1 \neq t_2) \vee (Y_1 \neq Y_2)$ is true. The following three forms of non-determinism may be distinguished:

Form	Condition	Example
effective	$(Y_1 \neq Y_2) \wedge (t_1 = t_2)$	$X \xrightarrow{t} Y_1 \quad X \xrightarrow{t} Y_2$
transitional	$(Y_1 = Y_2) \wedge (t_1 \neq t_2)$	$X \xrightarrow{t_1} Y \quad X \xrightarrow{t_2} Y$
general	$(Y_1 \neq Y_2) \wedge (t_1 \neq t_2)$	$X \xrightarrow{t_1} Y_1 \quad X \xrightarrow{t_2} Y_2$

A transition system is called *deterministic*, iff it is not non-deterministic. ◀

A *typed* transition system only can exhibit *transitional* and *general* non-determinism. The former describes a situation, where there are two different transitions t_1 and t_2 of the same type, i.e. $X \xrightarrow{t_1} Y$ and $X \xrightarrow{t_2} Y$. The non-determinism is visible in the choice of the transition but not in the states involved. The latter describes a situation where the non-determinism is visible in the states involved in the change: There is a choice between $X \xrightarrow{t_1} Y_1$ and $X \xrightarrow{t_2} Y_2$. Due to the typing condition this also becomes visible in the transitions.

A *coherent* transition system can only exhibit *general* non-determinism.

The *dual* of a transition system is formed by *reversing the direction of time*:

20 Definition DUAL TRANSITION SYSTEM
If $(S, T, \longrightarrow)$ is a state transition system, then the *dual state transition system* $(S, T, \longrightarrow^*)$ is defined by $X \xrightarrow{t}{}^* X' \Leftrightarrow X' \xrightarrow{t} X$. ◀

21 Definition CODETERMINISM
A transition system is called *codeterministic*, iff its dual is deterministic. It is called *non-codeterministic*, iff its dual is non-deterministic, hence iff it is not codeterministic. ◀

Codeterminism is determinism after time reversal. In a deterministic system every state has only *one possible "future"*. In a codeterministic system every state has only *one possible "past"*.

22 Example CLOCK
A clock which cannot be stopped may be modeled by the following transition system:

$S = \{\texttt{idle}, \texttt{running}\}$ $\qquad$ $T = \{\texttt{tick}, \texttt{start}\}$

$\texttt{idle} \xrightarrow{\texttt{start}} \texttt{running}$ $\qquad$ $\texttt{running} \xrightarrow{\texttt{tick}} \texttt{running}$

Compare Figure 2.7 for a graphical illustration. This system is deterministic but it is not codeterministic, since for the state `running` we are *not able to predict the immediate past transitions and states*.

Abstracting from the differences in states and labels, the systems "clock" and "time-bomb" are dual to each other. ◀

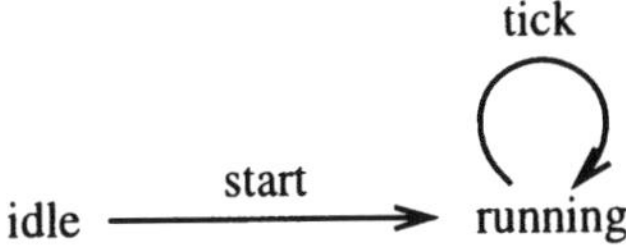

Figure 2.7: Clock.

Determinism and codeterminism guarantee that the time in which the state changes of a transition system are observed is *locally linear*: Due to determinism, time may not split into several different futures, due to codeterminsim several different pasts may not join to form a single present. Our intuitive notion of time also includes a *global ordering*: A moment in the past lies in the past and may never come again in the future. This concept is called *causality*:

23 Example CAUSALITY
Let us analyze the following transition system which is illustrated in Figure 2.8:

$$S = \{A, B, C\} \qquad T = \{b, c, d\} \qquad A \xrightarrow{b} B \qquad B \xrightarrow{c} C \qquad C \xrightarrow{d} B$$

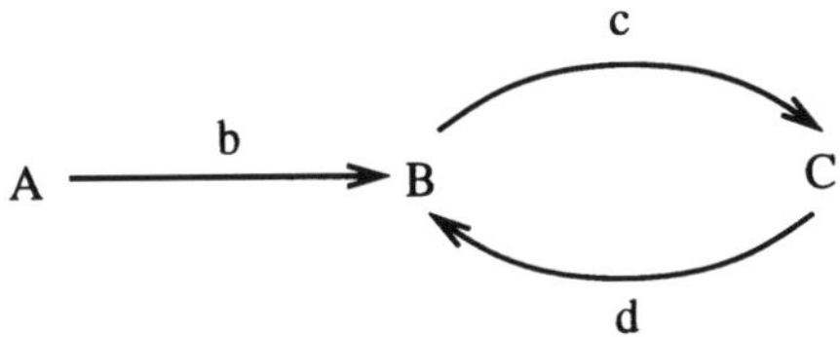

Figure 2.8: Causality.

A is a possible cause for B, since state A always precedes state B. State B, however, cannot be a cause for C, since B itself might depend on C. It is not possible to add time stamps to the states of this system, because it is not clear whether B shall have a higher or a lower time stamp than C. ◀

24 Definition CAUSALITY
Let $(S, T, \longrightarrow)$ be a transition system.

The relation $\gg \ \subseteq S \times S$ of *immediate causality* is defined by $X \gg Y :\Leftrightarrow \exists t \in T : X \xrightarrow{t} Y$. If $X \gg Y$, then X is called an *immediate cause for* Y.

The relation $\succeq \ \subseteq S \times S$ of *causality* is the reflexive, transitive hull of the relation of immediate causality. If $X \succeq Y$, then X is called a *cause for* Y.

A state $X \in S$ is called an *initial cause*, iff there is no state which is an immediate cause for X.

A transition system is called *causal*, iff its immediate causality relation is irreflexive[5] and its causality relation is an order relation on the set of states.

A transition system is called *finitely causal*, iff it is causal, iff every state has a finite number of causes, and iff there is a finite number of initial causes. ◀

[5] so that a state may not be its own immediate cause.

There is no such concept as co-causality, since the dual of a transition system is causal if and only if the transition system itself is causal. Furthermore, the respective relations of the dual transition system are the dual relations of the original transition system.

A process is a description of what *could* happen. Now we will define behaviours which describe what actually *did* happen.

25 Example BEHAVIOUR
Let a process be given as in Figure 2.9 and by

$$S = \{A, B\} \qquad T = \{r, s, t, u\} \qquad L = \{\bullet, *\}$$

$$A \xrightarrow{r} A \qquad A \xrightarrow{s} A \qquad A \xrightarrow{t} B \qquad B \xrightarrow{u} B$$

$$\mu(r) = \mu(t) = \mu(u) = \bullet \qquad \mu(s) = *$$

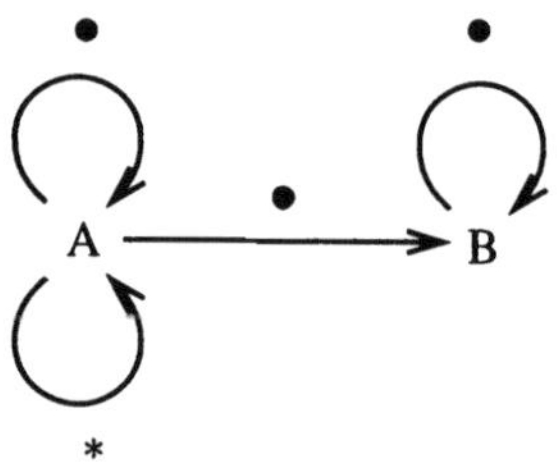

Figure 2.9: Behaviour.

One possible behaviour of this system is

$$A \xrightarrow{\bullet} A \xrightarrow{\bullet} A \xrightarrow{*} A \xrightarrow{\bullet} B$$

This may be reformulated as a (unlabeled, typed) transition system, describing an observer who is watching this system for five state changes:

$$S_b = \{0, 1, 2, 3, 4\} \qquad T_b = \{j, k, l, m\}$$

$$0 \xrightarrow{j}_b 1 \qquad 1 \xrightarrow{k}_b 2 \qquad 2 \xrightarrow{l}_b 3 \qquad 3 \xrightarrow{m}_b 4$$

The link between the observing and the observed transition system is provided by two *observation functions* $f_S : S_b \to S$ and $f_T : T_b \to T$. When the observing system is in state X, then it sees the observed system in state $f_S(X)$. When the observer performs transition t, then the observed system performs transition $f_T(t)$. In this example, these functions are given by:

$$f_S : \{0, 1, 2, 3, 4\} \to \{A, B\} \qquad f_S(0) = f_S(1) = f_S(2) = f_S(3) = A \qquad f_S(4) = B$$
$$f_T : \{j, k, l, m\} \to \{r, s, t, u\} \qquad f_T(j) = f_T(k) = r \qquad f_T(l) = s \qquad f_T(m) = t$$

This connection is illustrated in Figure 2.10.

In a certain sense, behaviours are "unfoldings" of transition systems in time. ◀

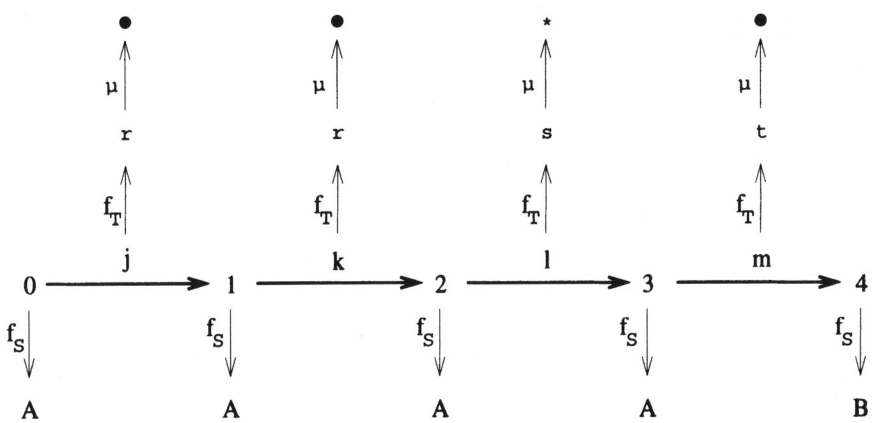

Figure 2.10: Observing and Observed System.

Let us formalize this:

26 Definition BEHAVIOUR
Let $\mathcal{P} = (S, T, \longrightarrow)$ be a process. A behaviour $\mathcal{B} = (S_b, T_b, \longrightarrow_b, f_S, f_T)$ of $\mathcal{P}$ consists of

(1) a *typed transition system* $(S_b, T_b, \longrightarrow_b)$ which is deterministic, codeterministic, finitely causal, and has exactly one initial cause, and

(2) a pair (f_S, f_T) of *functions* $f_S : S_b \to S$ and $f_T : T_b \to T$, such that

$$\forall X, X' \in S_b,\ t \in T_b :\ X \xrightarrow{t}_b X' \ \Rightarrow\ f_S(X) \xrightarrow{f_T(t)} f_S(X')$$

$(S_b, T_b, \longrightarrow_b)$ is called the *observing system* and (f_S, f_T) the *observation functions*.

A behaviour of a *rooted process* is a behaviour of the respective process whose initial cause is mapped to the initial state of the process.

A behaviour is called *initial*[6], iff it contains only one state. ◀

For labeled processes the same definition is used. Labels for the observing system may be obtained from the observed system via the observation function by using the labels of the process.

The restriction to one initial cause is not essential. Allowing more than one initial cause would produce the simultaneous description of several transition systems without any possibility of cooperation. Therefore, nothing new is obtained.

Similar as for graphs, the states and transitions of a behaviour only are carriers for their structure. Behaviours which are identical up to a renaming of states and transitions may thus be identified:

[6]For distributed transition systems, an initial behaviour will play a more important role.

27 Definition ISOMORPHISM OF BEHAVIOURS
Let $\mathcal{P} = (S, T, \longrightarrow)$ be a process. The two behaviours $(S_1, T_1, \longrightarrow_1, f_S^1, f_T^1)$ and $(S_2, T_2, \longrightarrow_2, f_S^2, f_T^2)$ are called *isomorphic*, iff there exist bijections $u_S : S_1 \to S_2$ and $u_T : T_1 \to T_2$, such that

$$\forall X, X' \in S_1,\ t \in T_1 :\ X \stackrel{t}{\longrightarrow}_1 X' \ \Leftrightarrow\ u_S(X) \stackrel{u_T(t)}{\longrightarrow}_2 u_S(X')$$

$$\forall X \in S_1 :\ f_S^2(u_S(X)) = f_S^1(X)$$

$$\forall t \in T_1 :\ f_T^2(u_T(t)) = f_T^1(t)$$

◀

The isomorphism of behaviours is an *equivalence relation*. Hence, we may study its quotient:

28 Definition QUOTIENT
The *quotient of all behaviours* of a process $\mathcal{P}$ is the set of all equivalence classes of its behaviours under the relation of behavioural isomorphism. It will be denoted by $\mathbb{B}(\mathcal{P})$. ◀

The following theorem shows how such behaviours typically look like:

29 Theorem CANONICAL FORM OF BEHAVIOURS
Every behaviour is *isomorphic* to a behaviour, whose set of states either is an initial segment $\{0, 1, 2, \ldots, n\}$ of the natural numbers, or the set $\mathbb{N}_0$ of all natural numbers; its set of transitions is empty, the initial segment $\{0, 1, 2, \ldots, n-1\}$ or the set $\mathbb{N}_0$ of natural numbers; its relation is of the form $k \stackrel{k}{\longrightarrow} k+1$ for all sensible k.

PROOF:
The conditions of determinism, codeterminism and causality, and the existence of exactly one initial cause allows to bijectively assign natural numbers to the states of a behaviour and to construct an isomorphism of behaviours. The relevant conditions are easily checked. ◀

This theorem shows that we could have defined behaviours simply as a sequence of states and transitions. However, this approach cannot be generalized to the distributed situation of the next chapter.

Observing a process $\mathcal{P}$ for some time we obtain a behaviour $\mathcal{B}_1$. If we now continue this observation for some additional time we obtain the behaviour $\mathcal{B}_2$. In a certain sense $\mathcal{B}_2$ "knows about" or "contains" $\mathcal{B}_1$. This is made more precise by the notion of an extension:

30 Definition EXTENSION ORDER ON BEHAVIOURS
Let $\mathcal{P} = (S, T, \longrightarrow, L, \mu)$ be a process with two behaviours $\mathcal{B}_1 = (S_1, T_1, \longrightarrow_1, f_S^1, f_T^1)$ and $\mathcal{B}_2 = (S_2, T_2, \longrightarrow_2, f_S^2, f_T^2)$.

$\mathcal{B}_2$ is called *an extension of* $\mathcal{B}_1$, or $\mathcal{B}_2$ is said to *extend* $\mathcal{B}_1$, iff there exist injective functions $i_S : S_1 \to S_2$ and $i_T : T_1 \to T_2$ which are bijections of the sets of initial causes of the two behaviours, and the following conditions hold:

$$\forall X, X' \in S_1,\ t \in T_1 :\ X \stackrel{t}{\longrightarrow}_1 X' \ \Leftrightarrow\ i_S(X) \stackrel{i_T(t)}{\longrightarrow}_2 i_S(X')$$

$$\forall X \in S_1 :\ f_S^2(i_S(X)) = f_S^1(X)$$

$$\forall t \in T_1 :\ f_T^2(i_T(t)) = f_T^1(t)$$

In this case one also writes $\mathcal{B}_1 \trianglerighteq \mathcal{B}_2$. ◀

The direction of the relation symbol $\trianglerighteq$ illustrates the direction of time. Time stamps increase from the left to the right hand side of $\trianglerighteq$. $\mathcal{B}_1 \trianglerighteq \mathcal{B}_2$ denotes that $\mathcal{B}_2$ is an extension of $\mathcal{B}_1$, and thus $\mathcal{B}_2$ is an observation which could be made, after having observed $\mathcal{B}_1$ and then continuing to observe the system for some additional time.

31 Proposition EXTENSION ORDERING IS ORDER RELATION
The *extension relation* $\trianglerighteq \subseteq \mathbb{B}(\mathcal{P}) \times \mathbb{B}(\mathcal{P})$ is a well defined *order relation* on the set of all behaviours of a transition system, identified by transition isomorphism.

PROOF:
In Definition 30 we defined $\trianglerighteq$ only for behaviours of a process. It is straightforward to check that it may be extended to equivalence classes. The remainder of the proof uses the same techniques as in Section 3.3 for distributed transition systems, and it may be obtained by specialization. ◀

32 Definition BRANCHING STRUCTURE
Let $\mathcal{P}$ be a process. The *branching structure of* $\mathcal{P}$ is the pair $(\mathbb{B}(\mathcal{P}), \trianglerighteq)$, consisting of the quotient of all behaviours and the extension ordering $\trianglerighteq$ on $\mathbb{B}(\mathcal{P})$. ◀

33 Example EXTENSION ORDER AND BRANCHING STRUCTURE
Determine the branching structure of the transition system given in shorthand notation by:

$$S = \{A, B\} \qquad T = \{\bullet\} \qquad A \xrightarrow{\bullet} A \qquad A \xrightarrow{\bullet} B \qquad B \xrightarrow{\bullet} B$$

Let $(0,0)$ denote the initial behaviour. For $m, n \in \mathbb{N}$ let the pair (m,n) denote the behaviour which first performs m times the transition $A \xrightarrow{\bullet} A$ and then n times transitions involving state B (which for the first time is transition $A \xrightarrow{\bullet} B$ and then is transition $B \xrightarrow{\bullet} B$). The pair $(\infty, 0)$ denotes the infinite behaviour which only performs transition $A \xrightarrow{\bullet} B$ and (m, ∞) denotes the infinite behaviour which performs m times transition $A \xrightarrow{\bullet} B$, then once transition $A \xrightarrow{\bullet} B$ and then continues for ever with transition $B \xrightarrow{\bullet} B$. The quotient of all behaviours of our transition system thus may be represented by $\mathbb{B} = \{(m,n) \mid m,n \in \mathbb{N}_0\} \cup \{(\infty,0)\} \cup \{(m,\infty) \mid m \in \mathbb{N}_0\}$. Its branching structure is given by the following extension order:

$$\forall (m,n) \in \mathbb{B} : \ (0,0) \trianglerighteq (m,n)$$

$$\forall m \in \mathbb{N}_0, m' \in \mathbb{N}_0 \cup \{\infty\} : \ m \leq m' \Rightarrow (m,0) \trianglerighteq (m',0)$$

$$\forall m \in \mathbb{N}_0, n, n' \in \mathbb{N}_0 \cup \{\infty\}, n \leq n' : \ (m,n) \trianglerighteq (m,n')$$

The maximal elements of this order structure are represented by the set $\{(n,\infty) \mid n \in \mathbb{N}_0\} \cup \{(\infty,0)\}$. There are no behaviours of the form (∞, n) with $n > 0$. An infinite behaviour may not be extended since this would contradict the condition of finite causality. The order structure itself is partially illustrated by the HASSE-diagram in Figure 2.11: ◀

If a behaviour can be extended, it did not observe the process as long as it would have been possible. If a behaviour is a *maximal element* of the extension order, then it can no longer be extended. Such a maximal behaviour may be *finite*, in which case the process had terminated

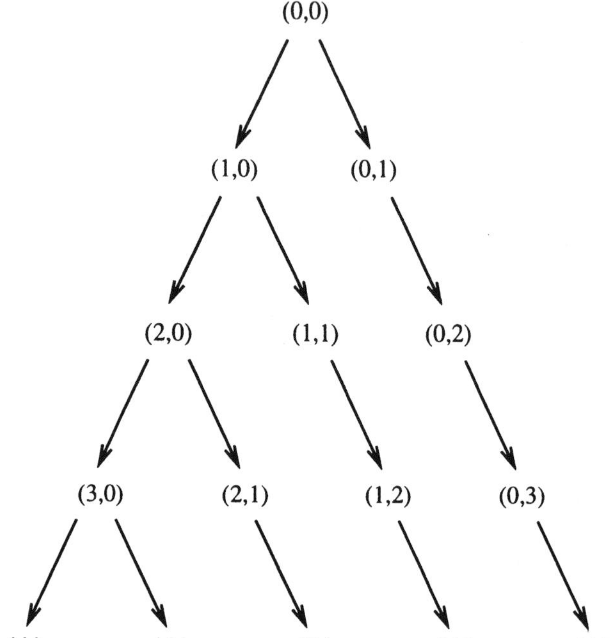

Figure 2.11: HASSE-Diagram of Branching Structure.

after a finite number of steps. Other maximal behaviours are *infinite*. A process may possess finite and infinite maximal behaviours.

34 Theorem EXISTENCE OF MAXIMAL BEHAVIOURS
Every behaviour of a process may be extended to a *maximal behaviour*[7].

PROOF:
Let $\mathcal{B}$ be a behaviour. According the lemma of KURATOWSKY [DP90] this behaviour is an element of a maximal chain in the extension order. Let Γ be such a maximal chain. Then Γ contains equivalence classes of behaviours. In every equivalence class we can find a behaviour in the canonical form of Theorem 29. Let Γ^* be the set of all behaviours in canonical form which may be found in the equivalence classes of Γ. Now let S, T, $\longrightarrow$, f_S and f_T be the union of the state sets, transition sets, relations and respective functions of all the behaviours in Γ. The resulting object $(S, T, \longrightarrow, f_S, f_T)$ is a behaviour of the process in canonical form, it is a maximal behaviour and an extension of $\mathcal{B}$. ◀

35 Theorem NON-DETERMINISM
A process is *non-deterministic*, if and only if there exists an initial state for which there is *more than one*[8] *maximal behaviour*.

PROOF:
The first direction is easy, because for a non-deterministic process we may immediately construct a process with more than one maximal behaviour.

---

[7] i.e. a behaviour whose equivalence class is a maximal element with respect to the extension order.
[8] Isomorphic behaviours only count once.

Now let $\mathcal{B}_1$ and $\mathcal{B}_2$ be two non-isomorphic maximal behaviours. Without loss of generality we may choose them in canonical form. Then, in the observing behaviours there exists a smallest natural number for which the observed states (and/or labels) differ. The state which is observed immediately before this state shows that the process must be non-deterministic. ◀

It may depend on the initial state whether non-determinism actually appears or not.

36 Definition OPERATIONAL SEMANTICS OF A PROCESS
The *operational semantics of a process* $\mathcal{P}$ is given by its set of maximal behaviours. ◀

2.5 Sequential Composition

Transition systems can only model single state changes. If a system contains the transitions $X \xrightarrow{t} Y$ and $Y \xrightarrow{u} Z$, then we want to express by $X \xrightarrow{u \circ t} Z$ that the system may change from state X to state Z by executing the *sequential composition* $u \circ t$, consisting of transition t, followed by transition u. $u \circ t$ is read as "u after t".

The order of the operands was chosen with the aim of being consistent with the convention used for the composition of functions, since in later sections it will be very convenient to denote $X \xrightarrow{t} Y$ by $t : X \to Y$. Unfortunately, this choice makes the order of the operands inverse to the order in which the transitions execute.

The sequential composition only is *partially defined*: Two transitions t and u with $X \xrightarrow{t} Y$ and $R \xrightarrow{u} Z$ may be composed to the transition $u \circ t$ if and only if $Y = R$. Only then transition t generates a state in which transition u may take place.

The sequential composition should be *associative*, since it makes no difference in which order three transitions are composed: $(v \circ u) \circ t = v \circ (u \circ t)$.

In addition to composed transitions we also want a kind of *empty transition*. It shall impose no state change and will serve as neutral element for the composition. Since our transition systems are typed, also the empty transition(s) must possess a type. Hence, there must be an empty transition id_X for every state X.

These observations lead to the definition of a compositional transition system:

37 Definition COMPOSITIONAL TRANSITION SYSTEM
A *compositional transition system* $(S, T, \longrightarrow, \circ)$ consists of a *typed transition system* $(S, T, \longrightarrow)$, and a *partially defined operation* $\circ : T \times T \to T$, the so called *(sequential* [9]*) composition*, such that the following conditions are satisfied:

(def) *Definition of composition*:
For a transition t of type (X, Y) and a transition u of type (R, Z), a composition $u \circ t$ is defined if and only if $Y = R$. In this case u and t are called *composable* (in this order), the type of $u \circ t$ is (X, Z), and $X \xrightarrow{u \circ t} Z$.

(ass) *Associativity of composition*:
Let v, u and t be transitions of types (C, D), (B, C), and (A, B), respectively. Then their composition is *associative*:

[9] In Section 3.8 we will define a *parallel* composition.

$$v \circ (u \circ t) = (v \circ u) \circ t$$

(id) *Existence of identity transitions*:
For every state X there exists an *empty* or *identity transition* id_X, characterized by the property that for all states Y, transitions t of type (X, Y), and u of type (Y, X) the following equations hold:

$$t \circ id_X = t \qquad id_X \circ u = u$$

◀

The transitions of a compositional transition system form a "partially defined monoid".

In the following we give a slightly different definition which catches the same idea by using other technical means. It is nearer to the definition of a category and will facilitate the categorical analysis of transition systems.

38 Definition COMPOSITIONAL TRANSITION SYSTEM, ALTERNATE
A *compositional transition system* $(S, \mathfrak{T}, \circ)$ consists of

(1) a set S of *states*,

(2) for every pair (X, Y) of states a set $\mathfrak{T}(X, Y)$ of *transitions*, and

(3) for all states $X, Y, Z \in S$ a *binary operation* $\circ : \mathfrak{T}(Y, Z) \times \mathfrak{T}(X, Y) \to \mathfrak{T}(X, Z)$ on the transitions,

such that the following conditions are satisfied:

(dis) *Pairwise disjoint*:
The sets $\mathfrak{T}(X, Y)$ of transitions are pairwise disjoint.

(ass) *Associativity of composition*:
For all $v \in \mathfrak{T}(C, D)$, $u \in \mathfrak{T}(B, C)$, and $t \in \mathfrak{T}(A, B)$ the following equation holds:

$$v \circ (u \circ t) = (v \circ u) \circ t$$

(id) *Existence of identity transitions*:
For every state X there exists an *identity transition* $id_X \in \mathfrak{T}(X, X)$, such that for all $t \in \mathfrak{T}(X, Y)$ and $u \in \mathfrak{T}(Y, X)$ the following equations hold:

$$t \circ id_X = t \qquad id_X \circ u = u$$

◀

These two definitions of compositional transition systems are related:

Let $(S, \mathfrak{T}, \circ)$ be a compositional transition system in the sense of the alternative definition. Define $T := \bigcup_{X,Y \in S} \mathfrak{T}(X, Y)$ and $\longrightarrow \ \subseteq S \times T \times S$ by $X \stackrel{t}{\longrightarrow} Y \Leftrightarrow t \in \mathfrak{T}(X, Y)$, then $(S, T, \longrightarrow, \circ)$ is a compositional transition system in the sense of the original definition. Since the $\mathfrak{T}(X, Y)$ are pairwise disjoint, the transition system satisfies the typing condition.

Now let $(S, T, \longrightarrow, \circ)$ be a compositional transition system in the sense of the original definition. Define $\mathfrak{T}(X, Y)$ as the set of all transitions of type (X, Y). Then $(S, \mathfrak{T}, \circ)$ is a compositional transition system in the sense of the alternative definition.

The alternate definition of compositional transition systems may be rewritten once more, using the notion of a *category*:

39 Definition COMPOSITIONAL TRANSITION SYSTEM, CATEGORICAL
A *compositional transition system* is a small[10] category[11] $(\mathfrak{O}, \mathfrak{M}, \circ)$. ◀

40 Definition DECOMPOSABLE AND ATOMIC TRANSITIONS
Let $\mathcal{P} = (S, T, \circ)$ be a compositional transition system.

A transition $t \in T$ is called *decomposable*, iff there exist $t_1, t_2 \in T$ which both are different from identity transitions, such that $t = t_2 \circ t_1$. A transition $t \in T$ which is not decomposable is called *atomic*. ◀

41 Definition FREELY GENERATED COMPOSITIONAL TRANSITION SYSTEM
Let $\mathcal{P} = (S, T, \longrightarrow, L, \mu)$ be a labeled process.

The *compositional transition system generated by* $\mathcal{P}$ is the compositional transition system (and category) $(S, \mathfrak{T}, \circ)$, defined by

$$S := S_c$$

For two states $X, Y \in S$ the class of morphisms is defined as set of sequences:

$$\mathfrak{T}(X,Y) = \{(t_{n-1}, t_{n-2}, \ldots, t_1) \mid t_i \in T,\ \exists X_i \in S:\ X = X_1,\ Y = X_n,$$

$$X_1 \xrightarrow{t_1} X_2,\ X_2 \xrightarrow{t_2} X_3, \ldots, X_{n-1} \xrightarrow{t_{n-1}} X_n\}$$

If $X = Y$, then $\mathfrak{T}(X,Y)$ additionally contains an element id_X representing the identity morphism. $\circ$ is defined as the concatenation of the sequences. On the identity elements it is defined as required by the definition. ◀

The compositional transition system freely generated by a process $\mathcal{P}$ contains all the information of the process $\mathcal{P}$ and additionally allows to study sequences of transitions. The original process $\mathcal{P}$ may be reconstructed by choosing for T the set of all atomic transitions and defining $\longrightarrow$ by

$$X \xrightarrow{t} Y \;\Leftrightarrow\; X, Y \in S,\ t \in T,\ t \in \mathfrak{T}(X,Y)$$

For a *correct interpretation* of the sequential composition, note that the transition $t \circ u$ does *not* mean that transition u *must* execute before transition t. It denotes that transition u *may* execute before transition t, and $t \circ u$ *describes* this transition.

In this sense, the operation of (sequential) composition *does not* constitute a basis of our process theory. Only *after* we defined the notion of a process we discovered that there are certain situations which could be interpreted as a sequential composition. In order to describe these special situations, the operator of sequential composition was introduced *a posteriori*. This is in contrast to many other process theories which use composition operators as the fundament of their approach.

[10]See Definition 319.
[11]See Definition 309.

In the compositional transition system freely generated by a process, only those transition expressions are identified which *must* be identified to satisfy the conditions (ass) and (id). Furthermore, it only contains those states and transitions which it must contain, since they are inherited from the process or required by the definition of the sequential composition. Hence, the compositional transition systems may be considered *freely generated* and *initial*, in the sense of algebraic specification [MG85], [Wir90].

It is possible to study compositional systems which identify more transitions. This provides an interesting approach for ***abstraction from implementation details***:

42 Example NOT FREELY GENERATED COMPOSITIONAL TS
A transition system is given by

$$S = \{\mathtt{add(x,y)},\ \mathtt{is(x)} \mid \mathtt{x,y} \in \mathbb{N}_0\}$$

$$T = \{\mathtt{lz(z)},\ \mathtt{rz(z)},\ \mathtt{li(x,y)},\ \mathtt{ri(x,y)},\ \mathtt{sym(x,y)},\ \mathtt{dc(r,s)},\ \mathtt{null} \mid \mathtt{x,y} \in \mathbb{N}_0, \mathtt{z,r,s} \in \mathbb{N}\}$$

These transitions define methods for calculating the addition by a symmetric, left and right zero (`null`, `lz`, `rz`) rule, by a left and right induction rule (`li`, `ri`), by symmetrization (`sym`), and by direct calculation (`dc`), if both operands are different from zero:

$$\mathtt{add(0,z)} \xrightarrow{\mathtt{lz(z)}} \mathtt{is(z)} \qquad \mathtt{add(x+1,y)} \xrightarrow{\mathtt{li(x,y)}} \mathtt{add(x,y+1)}$$

$$\mathtt{add(z,0)} \xrightarrow{\mathtt{rz(z)}} \mathtt{is(z)} \qquad \mathtt{add(x,y+1)} \xrightarrow{\mathtt{ri(x,y)}} \mathtt{add(x+1,y)}$$

$$\mathtt{add(r,s)} \xrightarrow{\mathtt{dc(r,s)}} \mathtt{is(r+s)} \qquad \mathtt{add(x,y)} \xrightarrow{\mathtt{sym(x,y)}} \mathtt{add(y,x)}$$

$$\mathtt{add(0,0)} \xrightarrow{\mathtt{null}} \mathtt{is(0)}$$

Some of these transitions are illustrated in Figure 2.12.

From this transition system we may freely generate a compositional transition system. Then we identify more transitions than required by Definition 41. In this example we identify all transitions of the same type, thus identifying all methods for calculating the sum of two numbers which may be derived from this transition system.

This means especially that all methods for calculating $2+2$ are identified, and that Figure 2.12 is a commuting diagram. For example, the following equations between transitions must hold:

$$\mathtt{lz(4)} \circ \mathtt{li(0,3)} \circ \mathtt{li(1,2)} = \mathtt{dc(2,2)}$$

$$\mathtt{lz(4)} \circ \mathtt{sym(4,0)} \circ \mathtt{sym(0,4)} \circ \mathtt{li(0,3)} \circ \mathtt{li(1,2)} = \mathtt{dc(2,2)}$$

$$\mathtt{sym(4,0)} \circ \mathtt{sym(0,4)} = id_{\mathtt{add(0,4)}}$$

$$\mathtt{sym(2,2)} = id_{\mathtt{add(2,2)}}$$

◀

Along these lines, a theory of *stepwise refinement* can be developed.

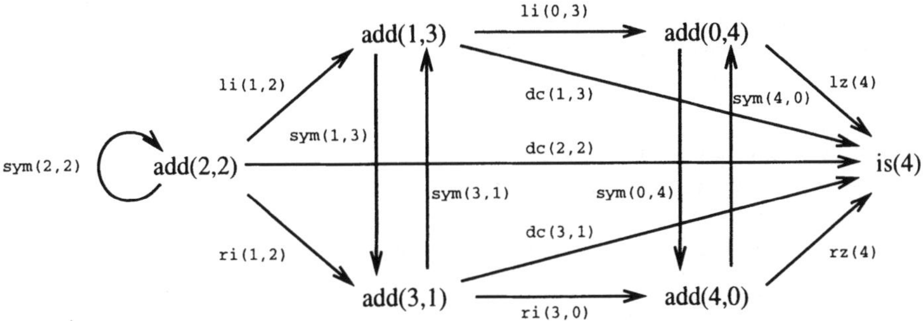

Figure 2.12: Transition System for Additions.

2.6 Fairness

In ordinary transition systems, a repeated non-deterministic choice may always ignore the same transition. A behaviour is called *fair*, iff every transition which repeatedly is enabled will be executed eventually.

In most process calculi there is still much controversy on the apropriate description of fairness. See [Fra86], [Mäh93], [AFL90], and [BBK87] for a review. First, let us consider some examples:

43 Example ERRONEOUS CHANNEL
An unreliable communication channel may be modeled as a non-deterministic system which repeatedly chooses between a correct transmission and the occurrence of an error. As a small and artificial[12] example, we will model a channel which repeatedly shall transmit a clock impulse:

$$S = \{\texttt{send, sent, error}\} \qquad T = \{\texttt{retransmit, ok, err}\}$$

$$\texttt{send} \xrightarrow{\texttt{ok}} \texttt{sent} \qquad \texttt{send} \xrightarrow{\texttt{err}} \texttt{error} \qquad \texttt{error} \xrightarrow{\texttt{retransmit}} \texttt{send}$$

Some maximal behaviours of this process are:

$$\texttt{send} \xrightarrow{\texttt{ok}} \texttt{sent}$$

$$\texttt{send} \xrightarrow{\texttt{err}} \texttt{error} \xrightarrow{\texttt{retransmit}} \texttt{send} \xrightarrow{\texttt{ok}} \texttt{sent}$$

$$\texttt{send} \xrightarrow{\texttt{err}} \texttt{error} \xrightarrow{\texttt{retransmit}} \texttt{send} \xrightarrow{\texttt{err}} \texttt{error} \xrightarrow{\texttt{retransmit}} \texttt{send} \xrightarrow{\texttt{err}} \texttt{error} \cdots$$

In state send this system repeatedly has to make a non-deterministic choice between correctly sending the message or making an error and repeating the transmission. In the first

[12] In Section 6.6 we will apply our approach to a real communication protocol.

two behaviours the transmission eventually is successful, whereas the third behaviour loops, always making errors and performing retransmissions. In this maximal behaviour the non-deterministic choice ***always is made in favour of the erroneous behaviour***, and thus we may consider it to be "***unfair***" with respect to the transition ok. ◀

44 Example SCHEDULING PROCESSES
Multiprocessing in a single processor machine requires a *scheduler*: If a process is ready to run, then the scheduler assigns the CPU to this process until the time slice has elapsed or some other event causes the process to be suspended. If two or more processes are ready to run, the scheduler non-deterministically chooses one. A maximal behaviour of such a scheduler is "*fair*", if a process is ***not always overtaken*** by other processes and hence never is assigned to the CPU. ◀

In the example of the scheduler, a process which ***becomes*** ready to run ***remains*** ready to run until it is assigned to the CPU. In the next example this is different.

45 Example THE CITIZEN AND THE BUREAUCRAT
A citizen needs a signature from a bureaucrat. He enters the office of the bureaucrat and must wait together with many other people. The bureaucrat non-deterministically picks a waiting person and deals with him. Unfortunately our citizen is not picked, so he leaves the room to drink a cup of coffee. In the meantime, the bureaucrat has serviced all people and is waiting. However, just before our citizen returns to his office, three other people enter. When the citizen comes back, he again has bad luck and is not selected. So he leaves for another cup of coffee ...

In contrast to the processes in above example the citizen does not become ready to be served and stays ready afterwards until he is served, but he only repeatedly becomes ready to be served. Above scenario of the citizen and the bureaucrat constitutes an unfair behaviour of a different kind. ◀

We now are able to define fairness of behaviours:

46 Definition ENABLED TRANSITION
Let $\mathcal{P} = (S, T, \longrightarrow, L, \mu)$ be a process and $A \in S$ a state. A transition $t \in T$ is said to be *enabled by* A, iff $A \xrightarrow{t} B$ holds for a suitable $B \in S$. ◀

47 Definition WEAK AND STRONG FAIRNESS
Let $\mathcal{P} = (S, T, \longrightarrow, L, \mu)$ be a process and $\mathcal{B} = (S_b, T_b, \longrightarrow_b, f_S, f_T)$ a behaviour with initial cause X_0 and $X_0 \xrightarrow{t_0}_b X_1 \xrightarrow{t_1}_b X_2 \cdots$

Behaviour $\mathcal{B}$ is called

> *weakly fair with respect to transition* $t \in T$, iff there exists no $n \in \mathbb{N}_0$, such that the transition t is enabled in the states $f_S(X_n), f_S(X_{n+1}), f_S(X_{n+2}), \dots$ of the maximal behaviour, but none of the transitions $f_T(t_n), f_T(t_{n+1}), f_T(t_{n+2}), \dots$ is equal to t.
>
> *strongly fair with respect to transition* $t \in T$, iff there is no infinite set $\{X_{i_1}, X_{i_2}, X_{i_3}, \dots\} \subseteq S_b$ of states, $i_1 < i_2 < i_3 < \dots$, such that t is enabled by all $f_S(X_{i_j})$, but none of the transitions $f_T(t_i)$ is equal to t.

A behaviour is called ***weakly (strongly) unfair***, iff it is not weakly (strongly) fair. ◀

Intuitively, behaviours are ***weakly fair***, if it is not possible that a transition becomes enabled and stays enabled, but does not execute. They are ***strongly fair***, if it is not possible that a transition is infinitely often enabled, but does not execute. For finite behaviours, fairness notions are meaningless.

Strong fairness ***implies weak*** fairness.

In the sense of this definition, the maximal behaviours from Examples 43 and 45 are strongly unfair, and in Example 44 we described a weakly unfair maximal behaviour.

48 Definition FAIR OPERATIONAL SEMANTICS
The ***weakly (strongly) fair operational semantics of a process*** $\mathcal{P}$ is given by the set of all maximal behaviours which are weakly (strongly) fair with respect to all transitions. ◀

The following example may demonstrate the difference between weak and strong fairness:

49 Example GUARDED COMMAND LANGUAGE
A guarded command language shall contain assignments and an iterative command of the following form:

do
$T_1 \rightarrow A_1;$
$[] \; T_2 \rightarrow A_2;$
...
$[] \; T_n \rightarrow A_n;$
od

The T_i denote tests which also are called ***guards***, and the A_i are assignments. This command is executed again and again in a loop-like fashion, as long as there is a test T_i which evaluates to "true". It terminates, iff all tests fail. If exactly one test, for example T_j, is true, then the assignment A_j is executed, and the thread of control returns to the beginning of the command. If more than one test is true, an arbitrary assignment is chosen from all those whose test evaluates to "true".

program 1 **is**
X := +1;
do
$X \geq 0 \rightarrow X := +1;$
$[] \; X \geq 0 \rightarrow X := -1;$
od
endprogram

program 2 **is**
X := +1;
do
$X = +1 \rightarrow X := -1;$
$[] \; X = -1 \rightarrow X := +1;$
$[] \; X = -1 \rightarrow X := +2;$
od
endprogram

Fairness Assumption	program 1	program 2
Strong Fairness	Must terminate	Must terminate
Weak Fairness	Must terminate	May loop
None	May loop	May loop

Table 2.1: Two Guarded Command Programs for Illustrating Fairness.

Consider the two programs in Table 2.1. If there are no assumptions on the fairness of the non-deterministic choice, it is not possible to guarantee that any of these programs will terminate. However, under certain fairness assumptions this is possible. ◀

There are systems where one is not interested in fairness with respect to a single transition but with respect to a *set of transitions*:

50 Example MULTIUSER SCHEDULER
Assume that in a multitasking, multiuser system one wants to share the time slices *equally among the users*. If a user has more than one process, one only cares to assign the CPU to an arbitrary process of this user: The distribution of the CPU to the users must be fair, but in this example one does not care about a fair distribution to the individual processes. In this situation, fairness means that if the property "there exists a process of user A which is ready to run" becomes true, then eventually an arbitrary process of user A must be assigned to the CPU. If this property becomes and remains true, this may be due to a varying set of runnable processes. ◀

This may be formalized as follows:

51 Definition ENABLED SET OF TRANSITIONS
Let $\mathcal{P} = (S, T, \longrightarrow, L, \mu)$ be a process, $A \in S$, is a state and $U \subseteq T$ a set of transitions. U is said to be *enabled by* A, iff there exists a transition $t \in U$ which is enabled by A. ◀

52 Definition FAIRNESS AND SETS OF BEHAVIOURS
Let $\mathcal{P} = (S, T, \longrightarrow, L, \mu)$ be a process, and $\mathcal{B} = (S_b, T_b, \longrightarrow_b, f_S, f_T)$ a behaviour with initial cause X_0 and $X_0 \xrightarrow{t_0}_b X_1 \xrightarrow{t_1}_b X_2 \cdots$.

The behaviour $\mathcal{B}$ is called

> *weakly fair with respect to the set $U \subseteq T$ of transitions*, iff there is no $n \in \mathbb{N}_0$, such that the set U is enabled in the states $f_S(X_n), f_S(X_{n+1}), f_S(X_{n+2}), \ldots$ of the maximal behaviour, but none of the transitions $f_T(t_n), f_T(t_{n+1}), f_T(t_{n+2}), \ldots$ is an element of U.
>
> *strongly fair with respect to the set $U \subseteq T$ of transitions*, iff there is no infinite set $\{X_{i_1}, X_{i_2}, X_{i_3}, \ldots\} \subseteq S_b$ of states, $i_1 < i_2 < i_3 < \ldots$, such that U is enabled by all $f_S(X_{i_j})$, but none of the transitions $f_T(t_{i_j})$ is an element of U.

A behaviour is called *weakly (strongly) unfair*, iff it is not weakly (strongly) fair. ◀

Fairness is a purely *qualitative phenomenon*: If a transition is enabled it *eventually* will execute, but it might be necessary to wait very long. Even worse, the time which one has to wait until a transition executes may increase gradually and could even be unbounded. In many applications also quantitative statements are of interest:

53 Definition FAIRNESS DELAY
Let $\mathcal{P} = (S, T, \longrightarrow, L, \mu)$ be a process, $\mathcal{B} = (S_b, T_b, \longrightarrow_b, f_S, f_T)$ a behaviour and $U \subseteq T$ a set of transitions.

The *fairness delay of $\mathcal{B}$ with respect to U* is defined by

$$d_{\mathcal{B}} = \sup \; \{j - i \mid U \text{ enabled by } f_S(X_i) \text{ but not enabled by } f_S(X_{i-1}),$$

$$f_T(t_j) \in U,\; j \geq i, \text{ and there exists no } k \text{ with } f_T(t_k) \in U \text{ and } i \leq k < j\}$$

◀

The fairness delay of a set of transitions is the *maximal number of transitions we might be required to wait* from the moment where this set is enabled to the next moment where a transition of this set is executed.

54 Definition UNIFORMLY FAIR
Let $\mathcal{P} = (S, T, \longrightarrow, L, \mu)$ be a process, $U \subseteq T$ a set of transitions, and $\mathcal{B}$ a behaviour.

The behaviour $\mathcal{B}$ is called *uniformly fair with respect to* U, iff the fairness delay of $\mathcal{B}$ with respect to U is a finite natural number. ◀

It is easy to show the following:

55 Proposition FAIRNESS DELAY AND UNIFORM FAIRNESS
If a behaviour is *uniformly* fair, then it is *strongly* fair. ◀

The converse of this proposition is not always true for two reasons: If a transition is enabled by only one state of an infinite maximal behaviour and *never executed* during this behaviour, then its fairness delay is $-\infty$, since the set of which the supremum shall be taken is empty. However, this maximal behaviour is strongly and weakly fair. On the other hand, the delay for a transition to execute may *increase beyond every upper bound*. In such a situation a maximal behaviour still may be strongly fair, but it has a fairness delay of $+\infty$:

56 Example TRAFFIC LIGHT AND FAIRNESS
Let a traffic light can be described by the following non-deterministic process:

$$S = \{\texttt{red, green}\} \qquad T = \{\texttt{tick, switch}\}$$

$$\texttt{red} \xrightarrow{\texttt{tick}} \texttt{red} \qquad \texttt{red} \xrightarrow{\texttt{switch}} \texttt{green}$$

$$\texttt{green} \xrightarrow{\texttt{tick}} \texttt{green} \qquad \texttt{green} \xrightarrow{\texttt{switch}} \texttt{red}$$

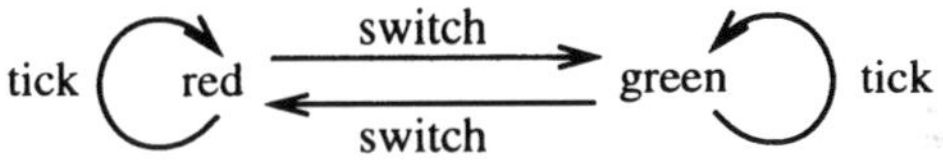

tick red switch switch green tick

Figure 2.13: States of a Traffic Light.

This transition system admits the following *weakly unfair* maximal behaviour:

$$\texttt{red} \xrightarrow{\texttt{tick}} \texttt{red} \xrightarrow{\texttt{tick}} \texttt{red} \xrightarrow{\texttt{tick}} \texttt{red} \xrightarrow{\texttt{tick}} \cdots$$

It also admits the following *weakly fair* but *not uniformly fair* maximal behaviour: It may take longer and longer until the traffic light reacts to the button pressed by the waiting pedestrians:

$$\texttt{red} \xrightarrow{\texttt{switch}} \texttt{green} \xrightarrow{\texttt{switch}}$$

$$\xrightarrow{\texttt{switch}} \texttt{red} \xrightarrow{\texttt{tick}} \texttt{red} \xrightarrow{\texttt{switch}} \texttt{green} \xrightarrow{\texttt{switch}}$$

$$\xrightarrow{\texttt{switch}} \texttt{red} \xrightarrow{\texttt{tick}} \texttt{red} \xrightarrow{\texttt{tick}} \texttt{red} \xrightarrow{\texttt{switch}} \texttt{green} \xrightarrow{\texttt{switch}} \cdots$$

◀

A transition is called *persistent*, iff it stays enabled until its execution:

57 Definition PERSISTENCY
Let $\mathcal{P} = (S, T, \longrightarrow, L, \mu)$ be a process. A transition $t \in T$ is called *persistent*, iff for every

behaviour $\mathcal{B} = (S_b, T_b, \longrightarrow_b, f_S, f_T)$ with initial cause X_0 and $X_0 \xrightarrow{t_0}_b X_1 \xrightarrow{t_1}_b X_2 \xrightarrow{t_2}_b \cdots$ and state $X_k \in S_b$, for which t is enabled by $f_S(X_k)$, the following condition holds: If X_l is a state with $l > k$, so that t is enabled by $f_S(X_l)$ and t is different from all $t_k, t_{k+1}, \ldots, t_{l-1}$, then t is still enabled by $f_S(X_l)$. ◀

This yields the following proposition:

58 Proposition FAIRNESS FOR PERSISTENT TRANSITIONS
For a *persistent* transition t a behaviour is *weakly fair* with respect to t if and only if it is *strongly fair* with respect to t. ◀

Fairness assumptions may be used to *prove properties* of processes: For example, when dealing with communication protocols, we may assume that a channel does not always choose the option of producing an error. This is reasonable, because if the channel always is faulty we might as well say that there is no channel at all. On the other hand, if we design a scheduler for a multiprocessing system, we should not assume that fairness properties hold: It is the very task of a scheduler to guarantee that processes which are ready to run are not always overtaken by other processes. This guarantee must be *established* by a suitable mechanism and must not be introduced by the *assumption*, that the scheduler will automatically make fair choices.

Fair operational semantics assumes that *all* non-deterministic choices are fair with respect to *all* transitions. In larger systems which, for example, might contain scheduler components as well as communication protocols, it is possible that only selected transitions may be assumed to be fair. This situation is called *selective fairness*.

We may analyze non-deterministic scheduling strategies without or even with fairness assumptions. However, if we write a scheduling program, this process will be deterministic – unless you indeed have the computer throw a coin or make a truly and physically random experiment which generates non-determinism. Let us assume that our initial scheduling model was a non-deterministic process which also had unfair maximal behaviours. Our (deterministic) scheduling program which is developed within the framework of this scheduling model will realize a specific maximal behaviour of the model. Of course this program requires additional states and transitions, since as a (deterministic) process it is different from the model which is a non-deterministic process. However, we still may expect some connections between the specific behaviour which is realized by our program, and a maximal behaviour of the model.

The scheduling program is deterministic and therefore – due to the absence of non-deterministic choices – its only maximal behaviour must be fair. On the other hand, this maximal behaviour corresponds to one of the many behaviours of the scheduling model, possibly after abstraction of implementation details. This corresponding behaviour of the model may be fair or unfair. A requirement for the implementor of our scheduling program would therefore read as follows: "Write a scheduling program which *realizes* a *fair* maximal behaviour of the scheduling model!" This requires a precise definition of "realizes", hence some means of abstracting from implementation details.

Chapter 3

Distributed Transition Systems

A distributed system is one that stops you from getting any work done when a machine you've never even heared of crashes.

L. LAMPORT,
in S. MULLENDER, *Distributed Systems*, 1st ed.

Ordinary transition systems form the basis of many process theories, but they only allow a faithful description of *sequential*, possibly *non-deterministic* processes. In this chapter, *distributed transition systems* are introduced which provide a formalism for modeling parallel and distributed systems.

In Section 1, the problems associated with ordinary transition systems are reviewed, and the *design decisions* for distributed transition systems are motivated.

Section 2 formally defines *distributed transition systems* and gives some introductory examples. Types, labels, and isomorphisms of distributed transition systems are introduced. At the end of this section, the notion of a *distributed process* is defined.

In Section 3, *non-determinism, causality, and behaviours* are discussed in a similar way as for ordinary transition systems.

The *branching structure* of a distributed transition system provides a geometric intuition for distributed phenomena. In Section 4, this structure is discussed and exemplified.

In Section 5, a number of *concepts for a more detailed analysis* of the branching structure are provided.

In Section 6, the branching structure is *interpreted geometrically*.

In our approach, non-determinism and parallelism are independent, orthogonal concepts: There are deterministic parallel processes and non-deterministic processes without parallelism. Of course parallelism and non-determinism may both be present in a distributed process. *Confusion* is a special phenomenon which may appear when parallelism and non-determinism are interacting in a special way. Section 7 explains this in detail, illustrates several forms of confusions, and outlines possibilities for their classification.

In Section 8, we deal with the *sequential composition* of transitions which is already known from the previous chapter. Then, *parallel composition* and the relations between these two composition operators are studied.

In Section 9, *fairness* in distributed systems is discussed and the problem of *relative computing speeds* is addressed.

3.1 Motivation

Ordinary transition systems are not able to adequately model parallel and distributed systems:

In ordinary transition systems, the observed events may be *linearly ordered*. As outlined in Section 2.2, this is in contradiction to the practical experiences with distributed systems and to the physics of space-time.

If ordinary transition systems are interpreted as *non-deterministic systems* – as this was done in the previous chapter – then the only possibility to express parallelism is its reduction by an interleaving approach. In Remark 5, we explained the *problems* with this approach.

The global state and time which may be associated with an ordinary transition system may serve as a *starting point* for developing distributed transition systems.

59 Remark GLOBAL STATE AND GLOBAL TIME
Ordinary transition systems allow us to identify a *global state* of the modeled system, and to work with the concept of a *global time*: We may implement a *clock* as an increasing counter to provide time stamps. These time stamps are assigned to every transition and may be used for ordering them: The transition with the lower time stamp has occured earlier. Furthermore, the *global state* of a centralized system at a certain time X may be written to stable storage: At time X all processing is suspended and the state of the agent is written to a disk.

A distributed system consists of geographically dispersed agents which show *local behaviour* and can *communicate* with each other via a network. Every participating process has a local state and performs transitions according to this state. It can partially influence the local state of a remote agent by using the communication network. For physical reasons, it takes some finite amount of time until this influence is visible on the remote site. Often this communication delay is not known precisely and varies randomly, mostly due to the employed network protocols. This random delay and the physical structure of space-time are the reasons why in general there is no possibility of defining a concept of *global time* or of *synchrony* for a distributed system. ◀

60 Remark MOTIVATION FOR DISTRIBUTED TRANSITION SYSTEMS
Local states: There is no concept of a global system state, but collections of one or more elementary states are interpreted as local states, independent of the presence of additional elementary states.

Local causality: Transitions may execute, if certain local states are present. This again is independent of the presence of additional elementary states.

Asynchronous reaction of transitions: A transition which could execute may arbitrarily delay this execution. During such a delay, the execution of another transition may destroy the condition which allowed the execution of the former one. If the reference to a *local* time base is important, a clock can be implemented by incrementing a counter. Then the transitions must be defined so that their execution always increments the counter.

Triggering implies destruction and generation: A transition is triggered by the presence of specific local state information. Since the permanent presence of a local state would continuously trigger the respective transition, the execution of a transition usually will destroy its trigger. Additionally, it also produces new states.

Sharing local states: The local states which influence the transitions are not necessarily in a one-to-one correspondence with the transitions: Two transitions can share local states which they generate, or by which they are triggered.

Conflict: If two transitions are triggered by the same state, and if triggering implies destruction or consumption of this state, then a conflict may arise: If either of the two transitions is triggered before the other, the trigger is destroyed and the remaining transition no longer can be triggered. This may be employed to model a non-deterministic choice between these two transitions.

Multiple local time: Ordinary transition systems provide a single, linear, and global time. For distributed systems, we need a more elaborate concept which uses *several local time bases* between which *no linear ordering* is possible. Only in the case of *communications* there is the requirement that the sending of a message must occur earlier than its receipt; this is similar for certain *synchronization* transitions. This design decision is most important for understanding the operational semantics of distributed systems. ◀

3.2 Distributed Transition Systems

In this section, distributed transition systems are defined. They share design ideas with PETRI-net theory, but they differ in the employed formalism and in the further development of the theory.

61 Definition DISTRIBUTED TRANSITION SYSTEMS
A *distributed transition system* $(S, T, \longrightarrow)$ consists of

(1) a non-empty set S of *elementary states* which generates the free commutative monoid[1] $(S^\circ, \cdot, \varepsilon)$ of *composite states*,

(2) a set T, whose elements are called *transitions*, and

(3) a ternary relation $\longrightarrow \; \subseteq S^\circ \times T \times S^\circ$, the *distributed transition relation*.

A distributed transition system $(S, T, \longrightarrow)$ is called *typed*, iff for every transition $t \in T$ there is exactly one pair $(X, X') \in S^\circ \times S^\circ$ of composite states, such that $X \stackrel{t}{\longrightarrow} X'$. This pair is called the *type* of the transition t.

A *labeled distributed transition system* $(S, T, \longrightarrow, L, \mu)$ consists of

(1) a *typed distributed transition system* $(S, T, \longrightarrow)$,

[1] See Definition 235.

(2) a set L, whose elements are called *labels*, and

(3) a function $\mu : T \to L$, the so-called *labeling function*.

If μ is injective, then the system is called *strictly labeled*. ◀

To get an intuitive understanding of distributed transition systems, we will look at several simple examples:

62 Example DISTRIBUTED TRANSITION SYSTEMS
Distributed transition systems can model sequential, parallel and non-deterministic behaviour:

A *sequential* system:

$$S = \{A, B, C, D\} \qquad T = \{b, c, d\}$$

$$A \xrightarrow{b} B \qquad B \xrightarrow{c} C \qquad C \xrightarrow{d} D$$

If this system is initialized in state A, it performs b to reach state B, *then* it performs c to reach state C and finally it performs d to reach state D.

A *parallel* system:

$$S = \{A, B, C, X, Y, W\} \qquad T = \{\texttt{fork},\ \texttt{ex1},\ \texttt{ex2},\ \texttt{join}\}$$

$$A \xrightarrow{\texttt{fork}} B \cdot C \qquad B \xrightarrow{\texttt{ex1}} X \qquad C \xrightarrow{\texttt{ex2}} Y \qquad X \cdot Y \xrightarrow{\texttt{join}} W$$

If this system is initialized in state A, it performs transition `fork` to reach the composite state $B \cdot C$. The presence of B allows transition `ex1` to take place and the presence of C allows transition `ex2` to execute. There is *no information* which of the two transitions must execute before the other, hence `ex1` and `ex2` execute *in parallel*. When both parallel transitions have executed, the states X and Y are present and transition `join` may execute, resulting in the final state W. The described behaviour is illustrated in Figure 3.1. There is only one behaviour, thus this process is *deterministic*. Furthermore, this behaviour is *parallel*. Therefore, this process is an example for a *deterministic and parallel* process.

A *non-deterministic* system:

$$S = \{A, B, X, Y, C, D\} \qquad T = \{b, e, f, g\}$$

$$A \xrightarrow{b} B \qquad B \xrightarrow{e} X \qquad B \xrightarrow{f} Y \qquad C \xrightarrow{g} D$$

If this system is initialized in state A, it executes transition b and generates state B. Now transition e as well as transition f *could* be triggered. The system has the *choice* of *either* e *or* f, since the execution of one of the transitions consumes the state required for the execution of the other. The distributed transition system and the two possible maximal behaviours are illustrated in Figure 3.2. If this system is initialized in state $B \cdot B$, then there is even more non-determinism, and we may choose between the three different parallel behaviours

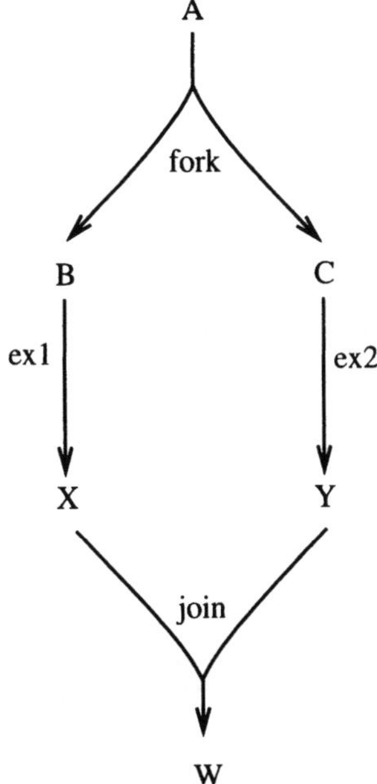

Figure 3.1: Parallel System.

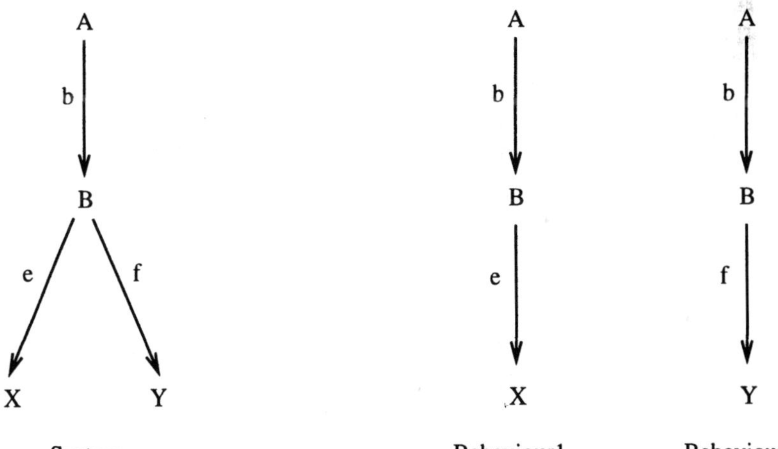

Figure 3.2: Non-Deterministic System with 2 Sequential Behaviours.

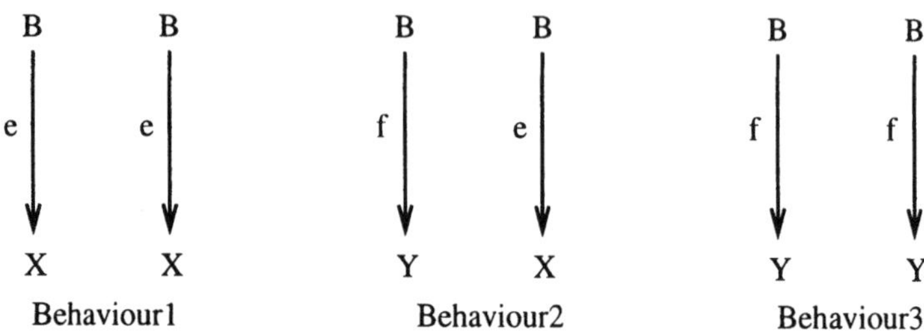

Figure 3.3: Non-Deterministic System with 3 Parallel Behaviours.

shown in Figure 3.3. If this system is initialized in state C, it behaves deterministically and executes transition g to reach state D.

Transitions with *conditions:*

$$S = \{A, B\} \qquad T = \{p\} \qquad A \xrightarrow{p} A \cdot B$$

In this system, transition p is triggered by token A, but token A is regenerated immediately after consumption. Since A is neither deleted nor modified, it may be regarded as a *condition* which must be present in order that the transition may take place. If a second transition q were allowed with

$$A \xrightarrow{q} A \cdot B$$

then this system would exhibit non-determinism but not parallelism: Both p and q consume one token and reproduce it only after the act of consumption. Therefore, they must compete for A. Both transition systems produce an ever increasing supply of tokens B. ◀

Similar as for ordinary transition systems[2], the specific names of the transitions are not important for the structure of a distributed transition system. We therefore define isomorphisms which allow to abstract from them:

63 Definition Isomorphism of Distributed Transition Systems
Two labeled, distributed transition systems $(S_1, T_1, \longrightarrow_1, L_1, \mu_1)$ and $(S_2, T_2, \longrightarrow_2, L_2, \mu_2)$ are called *isomorphic*, iff the sets of states and labels are equal, $S_1 = S_2$ and $L_1 = L_2$, and iff there exists a bijection $f : T_1 \rightarrow T_2$, such that

$$\forall X, X' \in S_1^{\circ},\ t \in T_1 :\ X \xrightarrow{t}_1 X' \ \Rightarrow\ X \xrightarrow{f(t)}_2 X'$$

$$\forall t \in T_1 :\ \mu_1(t) = \mu_2(f(t))$$

For *unlabeled* transition systems, the second condition is omitted. ◀

Above, the second condition can be expressed by the *commuting diagram* in Figure 3.4.

A *shorthand notation* can be introduced for distributed transition systems in precisely the same way as for the ordinary ones[3].

[2] See Definition 10.
[3] Compare Example 13.

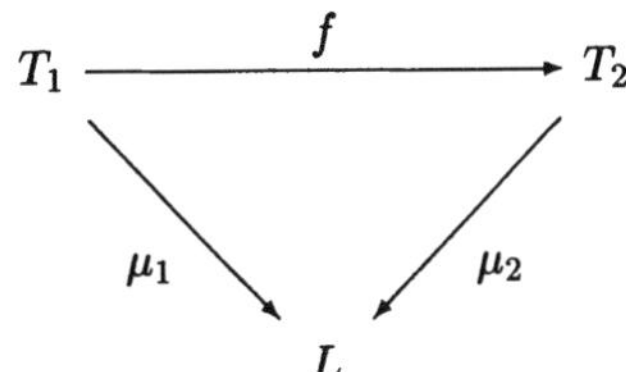

Figure 3.4: Condition for Isomorphic Transition Systems.

The following example shows that the process which could not be described using ordinary transition systems in Section 2.2 can now be modeled with distributed transition systems:

64 Example PARALLEL, DISTRIBUTED PROCESS
We define

$$S = \{X, Y, Z, U, V, W\} \qquad T = \{a, b, c, d\}$$

$$X \xrightarrow{a} Y \cdot Z \qquad Y \xrightarrow{b} U \qquad Z \xrightarrow{c} V \qquad V \xrightarrow{d} W$$

This system is illustrated in Figure 3.5. It is already a maximal behaviour itself. The order relation imposed on the transitions is the one which was required in Example 4. ◀

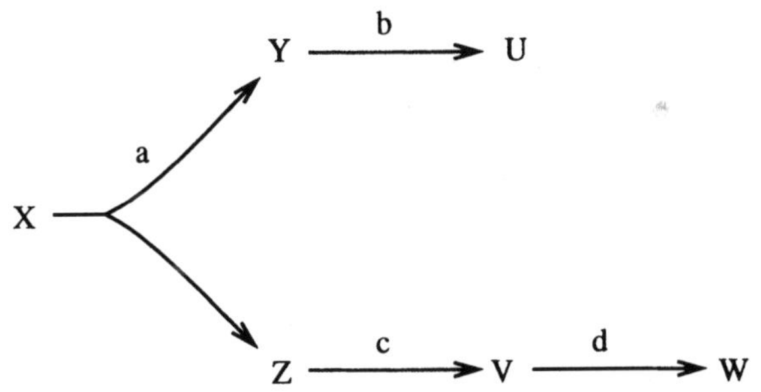

Figure 3.5: Parallel, Distributed Process.

65 Definition DISTRIBUTED PROCESS
A *(distributed) process* is a distributed transition system. A *rooted* distributed process consists of a distributed process and a composite initial state. ◀

Processes may be labeled. Two *isomorphic* labeled processes may be considered *structurally equal*.

A theory of *state-abstract* processes may be developed in a similar[4] way as for ordinary transition systems.

[4]See Definition 16.

3.3 Non-Determinism, Causality, and Behaviour

In this section, a theory of behaviours is developed, similarly[5] as for ordinary transition systems. Numerous examples follow in the subsequent section.

A distributed transition system may exhibit two types of non-determinism. The first is already known from ordinary transition systems[6] and will be called *centralized non-determinism*. It may be subdivided into effective, transitional, and general non-determinism. The second form will be called ***distributed non-determinism*** and occurs only in distributed transition systems. Here an elementary or composite state X meets with a state Y to a transition $X \cdot Y \xrightarrow{u} P$ *or* with a state Z to a transition $X \cdot Z \xrightarrow{v} Q$. This requires the following definition:

66 Definition OVERLAPPING
Let $(M, \cdot, \varepsilon)$ be a commutative monoid.

Two elements $X_1, X_2 \in M$ are called *overlapping*, iff there exists an element $Y \in M$ which is different from the neutral element of the monoid, and elements $E_1, E_2 \in M$, such that $X_1 = Y \cdot E_1$ and $X_2 = Y \cdot E_2$. ◂

Every element of a monoid is overlapping with itself. If M is a commutative monoid, freely generated over a set S, then two elements are overlapping, iff they "contain" a common element of S.

67 Definition NON-DETERMINISM
A distributed process $(S, T, \longrightarrow, L, \mu)$ is called *non-deterministic*, iff there exist overlapping $X_1, X_2 \in S^\circ$, arbitrary $Y_1, Y_2 \in S^\circ$ and $t_1, t_2 \in T$, such that $X_1 \xrightarrow{t_1} Y_1$ and $X_2 \xrightarrow{t_2} Y_2$, and $(t_1 \neq t_2) \vee (Y_1 \neq Y_2)$ is true. Four forms of non-determinism are distinguished:

Form	Condition	Example	
effective	$(X_1 = X_2) \wedge (Y_1 \neq Y_2) \wedge (t_1 = t_2)$	$X \xrightarrow{t} Y_1$	$X \xrightarrow{t} Y_2$
transitional	$(X_1 = X_2) \wedge (Y_1 = Y_2) \wedge (t_1 \neq t_2)$	$X \xrightarrow{t_1} Y$	$X \xrightarrow{t_2} Y$
general	$(X_1 = X_2) \wedge (Y_1 \neq Y_2) \wedge (t_1 \neq t_2)$	$X \xrightarrow{t_1} Y_1$	$X \xrightarrow{t_2} Y_2$
distributed	$(X_1 \neq X_2) \wedge [(Y_1 \neq Y_2) \vee (t_1 \neq t_2)]$	$R \cdot P \xrightarrow{t_1} Y_1$	$R \cdot Q \xrightarrow{t_2} Y_2$

Above, $X_1, X_2, Y_1, Y_2, X, Y, P, Q, R \in S^\circ$, $R \neq \varepsilon$ and $t, t_1, t_2 \in T$. ε denotes the neutral element of the monoid S°.

A distributed process $(S, T, \longrightarrow, L, \mu)$ is called *deterministic*, iff it is not non-deterministic. ◂

The ***dual*** of a distributed transition system and its property of being *codeterministic* is defined in precise analogy[7] to ordinary transition systems.

68 Example DISTRIBUTED NON-DETERMINISM
A distributed transition system is given by

$$S = \{\texttt{cpu\_free},\ \texttt{task\_1},\ \texttt{task\_2},\ \texttt{done\_1},\ \texttt{done\_2}\}$$

[5]See Section 2.4.
[6]See Definition 19.
[7]Compare Definitions 20 and 21.

$$T = \{\texttt{tick}\}$$

$$\texttt{cpu\_free} \cdot \texttt{task\_1} \xrightarrow{\texttt{tick}} \texttt{done\_1} \cdot \texttt{cpu\_free}$$

$$\texttt{cpu\_free} \cdot \texttt{task\_2} \xrightarrow{\texttt{tick}} \texttt{done\_2} \cdot \texttt{cpu\_free}$$

This system models a central processing unit and two ready-to-run tasks which are competing for the CPU.

As in ordinary transition systems, it depends on the initial state whether non-deterministic behaviour actually appears or not: If this system is initialized in the state $\texttt{cpu\_free} \cdot \texttt{task\_1} \cdot \texttt{task\_2}$, then the two tasks indeed compete for the single CPU. If it is initialized with $\texttt{cpu\_free} \cdot \texttt{cpu\_free} \cdot \texttt{task\_1} \cdot \texttt{task\_2}$, then no non-determinism shows up and the two CPUs may execute the two tasks in parallel. ◀

Distributed non-determinism implies that there is a specific state which could trigger either one or another transition, depending on the additional presence of either one or another state information. Which transition finally will be chosen cannot be determined from this specific state alone. *Distributed non-codeterminism* similarly means that there is a specific state which could have been produced by one or another transition – which can only be determined by also considering the context, i.e. other states:

69 Example DISTRIBUTED NON-CODETERMINISM
A distributed system exhibiting non-codeterminism is given by

$$S = \{\texttt{cpu\_1\_free},\ \texttt{cpu\_2\_free},\ \texttt{task},\ \texttt{done}\}$$

$$T = \{\texttt{tick}\}$$

$$\texttt{cpu\_1\_free} \cdot \texttt{task} \xrightarrow{\texttt{tick}} \texttt{done} \cdot \texttt{cpu\_1\_free}$$

$$\texttt{cpu\_2\_free} \cdot \texttt{task} \xrightarrow{\texttt{tick}} \texttt{done} \cdot \texttt{cpu\_2\_free}$$

Whether the task was executed by the first or by the second CPU cannot be decided by just looking at the state done, and it cannot be said by which transition this state was produced. ◀

The notion of *causality* must be expanded a bit, since in a distributed system there may be *multiple* causes. Intuitively, X may cause Y, iff there is a transition $X \cdot R \xrightarrow{t} Y \cdot S$ which generates Y and in which X is participating.

70 Definition CAUSALITY
Let $(S, T, \longrightarrow)$ be a distributed transition system.

The relation $\gg\ \subseteq S \times S$ of *immediate causality* is defined for $X, Y \in S$ by

$$X \gg Y \ :\Leftrightarrow\ \exists A, B \in S^\circ,\ \exists t \in T:\ X \cdot A \xrightarrow{t} Y \cdot B$$

If $X \gg Y$, then X is called an *immediate cause for* Y.

The relation $\succeq \ \subseteq S \times S$ of *causality* is the reflexive, transitive hull of the relation of immediate causality. If $X \succeq Y$, then X is called a *cause for* Y.

An element $X \in S$ is called

an *initial cause*, iff there is no other immediate cause for X.

a *final cause*, iff there is no other element for which X is an immediate cause.

A transition system is called

causal, iff its immediate causality relation is irreflexive and its causality relation is an order relation on the set of states.

finitely causal, iff it is causal, every state has a finite number of causes, and there is a finite number of initial causes.

◀

A deterministic, codeterministic, and finitely causal process is not yet an adequate model for a behaviour. The states in a behaviour shall represent points in a discrete space and time structure which are mapped to a single state of the observed system. The system given by $S = \{A, B\}$, $T = \{x\}$ and $A \xrightarrow{x} B \cdot B$ which is initialized in state A, is deterministic, codeterministic, and finitely causal, yet its final state consists of two instances of B whose coexistence makes above interpretation difficult.

We will call a process *safe*, if such a phenomenon does not occur:

71 Definition SAFETY
A distributed transition system $(S, T, \longrightarrow)$ is called *safe*, iff there are no $X, Y, Z \in S^\circ$, $Y \neq \varepsilon$ and $t \in T$, such that $X \xrightarrow{t} Y \cdot Y \cdot Z$ or $Y \cdot Y \cdot Z \xrightarrow{t} X$, and iff there exist no $X \in S^\circ$ and $t \in T$, such that $\varepsilon \xrightarrow{t} X$ or $X \xrightarrow{t} \varepsilon$. ◀

A behaviour may be defined similarly as for ordinary transition systems[8]. In the distributed case it now makes sense to allow multiple initial causes:

72 Definition BEHAVIOUR
Let $\mathcal{P} = (S, T, \longrightarrow, L, \mu)$ be a process without[9] transitions of the form $\varepsilon \longrightarrow X$ and $X \longrightarrow \varepsilon$. A *behaviour* $\mathcal{B} = (S_b, T_b, \longrightarrow_b, f_S, f_T)$ *of* $\mathcal{P}$ consists of

(1) a *typed distributed transition system* $(S_b, T_b, \longrightarrow_b)$ which is deterministic, codeterministic, finitely causal, and safe, and

(2) a pair (f_S, f_T) of *functions* $f_S : S_b \to S$ and $f_T : T_b \to T$, such that[10]

$$\forall X, X' \in S_b^\circ,\ t \in T_b :\ X \xrightarrow{t}_b X' \ \Rightarrow\ f_S^\circ(X) \xrightarrow{f_T(t)} f_S^\circ(X')$$

[8]Compare Definition 25.

[9]For transitions of this form no entirely satisfactory definition of a behaviour could be found. This restriction is assumed to hold whenever behaviours are discussed in the present work.

[10]f_S° is the continuation of the function $f_S : S_b \to S$ to the function $f_S^\circ : S_b^\circ \to S^\circ$ on the freely generated, commutative monoid – see Definition 244.

$(S_b, T_b, \longrightarrow_b)$ is called the *observing system* and (f_S, f_T) the *observation functions*.

A behaviour of a *rooted process* is a behaviour of the respective process with the following property: If $\{X_1, X_2, \ldots, X_n\}$ is the set of initial causes of the behaviour, then $f_S^\circ(X_1 \cdot X_2 \cdot \ldots \cdot X_n)$ is the initial state of the rooted process.

A behaviour is called *initial*, iff all its states are initial causes, and *finite*, iff its set S_b of states is finite. ◀

Similarly[11] as for ordinary transition systems, we may abstract from the specific names of transitions of the behaviours:

73 Definition ISOMORPHISM OF BEHAVIOURS
Let $\mathcal{P} = (S, T, \longrightarrow, L, \mu)$ be a process with two behaviours $\mathcal{B}_1 = (S_1, T_1, \longrightarrow_1, f_S^1, f_T^1)$ and $\mathcal{B}_2 = (S_2, T_2, \longrightarrow_2, f_S^2, f_T^2)$.

$\mathcal{B}_1$ and $\mathcal{B}_2$ are called *isomorphic*, iff there exist bijections $u_S : S_1 \to S_2$ and $u_T : T_1 \to T_2$, such that

$$\forall X, X' \in S_1^\circ,\ t \in T_1 :\ X \overset{t}{\longrightarrow}_1 X' \ \Leftrightarrow\ u_S^\circ(X) \overset{u_T(t)}{\longrightarrow}_2 u_S^\circ(X')$$

$$\forall X \in S_1^\circ :\ f_S^{2\circ}(u_S^\circ(X)) = f_S^{1\circ}(X)$$

$$\forall t \in T_1 :\ f_T^2(u_T(t)) = f_T^1(t)$$

◀

The last two conditions can be represented by the *commuting diagrams* in Figure 3.6.

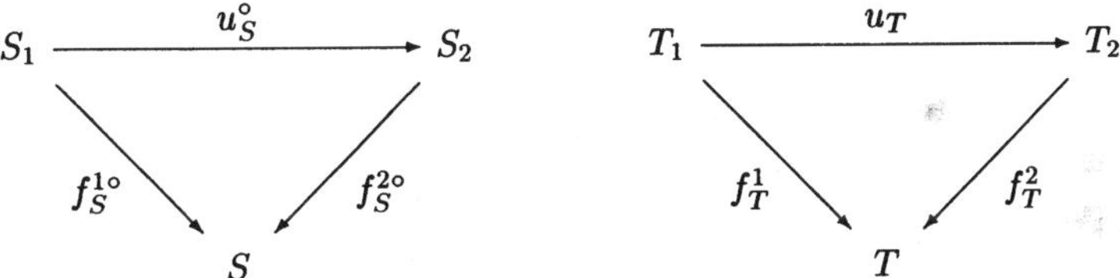

Figure 3.6: Condition for Isomorphic Behaviours

The isomorphism of behaviours is an *equivalence relation*. Hence, we may study its quotient:

74 Definition QUOTIENT
The *quotient of all behaviours* of a process $\mathcal{P}$ is the set of all equivalence classes of its behaviours under the relation of behavioural isomorphism. It will be denoted by $\mathbb{B}(\mathcal{P})$. ◀

Due to the larger number of possibilities in distributed systems, there is no longer a simple canonical form for behaviours and no analogue to Theorem 29.

Similarly[12] as for ordinary transition systems, we may define an extension order:

75 Definition EXTENSION ORDER ON BEHAVIOURS
Let $\mathcal{P} = (S, T, \longrightarrow, L, \mu)$ be a process with two behaviours $\mathcal{B}_1 = (S_1, T_1, \longrightarrow_1, f_S^1, f_T^1)$ and $\mathcal{B}_2 = (S_2, T_2, \longrightarrow_2, f_S^2, f_T^2)$.

[11] Compare Definition 27.
[12] Compare Definition 30.

We say that $\mathcal{B}_2$ is *an extension of* $\mathcal{B}_1$, or that $\mathcal{B}_2$ *extends* $\mathcal{B}_1$, iff there exist injective functions $i_S : S_1 \to S_2$ and $i_T : T_1 \to T_2$ which are bijections of the sets of initial causes of the two behaviours, and if the following two conditions hold:

$$\forall X, X' \in S_1^\circ,\ t \in T_1 :\ X \xrightarrow{t}_1 X' \ \Leftrightarrow\ i_S^\circ(X) \xrightarrow{i_T(t)}_2 i_S^\circ(X')$$

$$\forall X \in S_1^\circ :\ f_S^{2\circ}(i_S^\circ(X)) = f_S^{1\circ}(X)$$

$$\forall t \in T_1 :\ f_T^2(i_T(t)) = f_T^1(t)$$

If these conditions are fulfilled we write $\mathcal{B}_1 \trianglerighteq \mathcal{B}_2$.

$\mathcal{B}_2$ is called an *immediate extension of* $\mathcal{B}_2$, iff there is an extension, $\mathcal{B}_1 \trianglerighteq \mathcal{B}_2$, and if there is no behaviour $\mathcal{B}$, such that $\mathcal{B} \neq \mathcal{B}_1$, $\mathcal{B} \neq \mathcal{B}_2$ so that $\mathcal{B}_1 \trianglerighteq \mathcal{B} \trianglerighteq \mathcal{B}_2$. ◀

The last two conditions can be represented by the *commuting diagrams* in Figure 3.7.

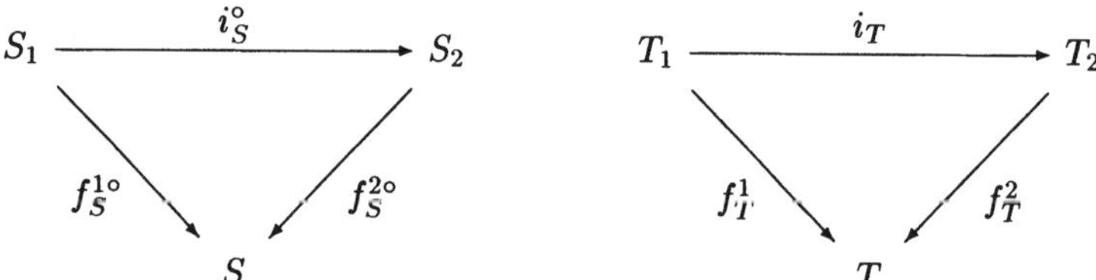

Figure 3.7: Condition for Extension Order.

76 Theorem EXTENSION ORDER IS ORDER RELATION
The *extension relation* $\trianglerighteq \subseteq \mathbb{B}(\mathcal{P}) \times \mathbb{B}(\mathcal{P})$ is a well-defined *order relation* on the quotient of all behaviours of a distributed process, identified by behaviour isomorphism.

PROOF:
In Definition 75, $\trianglerighteq$ was defined only for behaviours of a process. It is straightforward to check that it may be extended in a well-defined form to equivalence classes: $[\mathcal{B}_1] \trianglerighteq [\mathcal{B}_2] :\Leftrightarrow \mathcal{B}_1 \trianglerighteq \mathcal{B}_2$. Furthermore, we easily see that if $\mathcal{B}_1$ and $\mathcal{B}_1'$, and $\mathcal{B}_2$ and $\mathcal{B}_2'$ are isomorphic, then $\mathcal{B}_1 \trianglerighteq \mathcal{B}_2$ iff $\mathcal{B}_1' \trianglerighteq \mathcal{B}_2'$. For the remainder of the proof, we first require several notions and lemmata. The proof will continue after Lemma 80. ◀

77 Definition CAUSALITY CHAINS, INDEX, RANK, AND DEFECT
Let $(S, T, \longrightarrow, L, \mu)$ be a behaviour of a process.

A *causality chain* from X_0 to X_n is a sequence $X_0, X_1, X_2, \ldots, X_n \in S$ of elementary states, such that $X_0 \gg X_1 \gg X_2 \gg \ldots \gg X_n$. n is called the *length* of this causality chain.

The *index* $ind(X)$ of X is the minimal length of all causality chains beginning in an initial state and ending in X.

The *rank* $rk(X)$ of X is the maximal length of all causality chains beginning in an initial state and ending in X.

The *defect* $def(X)$ of X is the difference between rank and index. ◀

Since every behaviour is finitely causal, for every elementary state $X \in S$ there is only a finite number of causality chains terminating in X, and they all are of finite length. Therefore rank, index, and defect are well-defined natural numbers. X is an initial cause, iff its rank, index, and defect are zero.

If we implement our behaviour on a machine, and if every state transition takes one time unit, then the *rank* of state X denotes the minimum[13] number of time steps needed for generating X, the *index* denotes one plus the minimum number of time steps needed for generating the first immediate cause of this state, and the *defect* denotes the minimum number of time steps which a processor on a minimal length causality sequence must wait for synchronisation with a processor on a maximal length causality sequence.

Our processes are not *finitely branching*, since a non-deterministic choice between an infinite number of alternatives is possible. A behaviour may not generate infinitely many states in a finite number of transitions. In this sense our processes are *finitely forking*. This can be made more precise:

78 Lemma FINITE PARALLELISM
Let $(S, T, \longrightarrow, f_S, f_T)$ be a behaviour of a process. Then for every $n \in \mathbb{N}_0$ the set $\{X \in S \mid ind(x) \leq n\}$ of states with index less or equal than n, and the set $\{X \in S \mid rk(x) \leq n\}$ of states with rank less or equal than n are finite.

PROOF:
From geometric intuition, since there is only a finite number of initial causes, and since every transition "consumes" at least one state and "produces" a finite number of states. ◀

79 Lemma INDEX AND RANK ARE EXTENSION INVARIANT
Let $\mathcal{B}_1$ and $\mathcal{B}_2$ be behaviours of a process $\mathcal{P}$, and let $\mathcal{B}_1 \trianglerighteq \mathcal{B}_2$ via the injections $i_S : S_1 \to S_2$ and $i_T : T_1 \to T_2$. Then rank and index of an element $X \in S_1$ equal rank and index of its image $i_S(X)$.

PROOF:
The injective functions i_S and i_T map causality chains of length n to causality chains of length n. ◀

80 Lemma INDEX AND RANK IN POSTCONDITIONS
Let $\mathcal{B}$ be a behaviour and $Y_1 \cdot Y_2 \cdot \ldots \cdot Y_m \stackrel{t}{\longrightarrow} X_1 \cdot X_2 \cdot \ldots \cdot X_n$ a transition. Then the index, rank, and defect of all the X_i are equal.

PROOF:
Immediately from the definition. ◀

Due to above lemma we may also speak of the *rank, index, and defect of a transition* which may be defined as the rank, index and defect of the postcondition states X_i.

We now *continue the proof* of Theorem 76: It is straightforward to show that $\trianglerighteq$ is reflexive and transitive. For antisymmetry assume that $[\mathcal{B}_1] \trianglerighteq [\mathcal{B}_2]$ and $[\mathcal{B}_2] \trianglerighteq [\mathcal{B}_1]$. Hence $\mathcal{B}_1 \trianglerighteq \mathcal{B}_2$ and $\mathcal{B}_2 \trianglerighteq \mathcal{B}_1$. We will show that $\mathcal{B}_1$ and $\mathcal{B}_2$ are isomorphic:

Let $i_S : S_1 \to S_2$ and $i_T : T_1 \to T_2$ be two injections proving $\mathcal{B}_1 \trianglerighteq \mathcal{B}_2$, and $j_S : S_2 \to S_1$ and $j_T : T_2 \to T_1$ be two injections proving $\mathcal{B}_2 \trianglerighteq \mathcal{B}_1$. It suffices to show that i_S is a bijection, then i_T is bijective due to the type condition and the bijection condition on the initial causes.

We define $I_{1,n} := \{X \in S_1 \mid ind_1(X) \leq n\}$ and $I_{2,n} := \{X \in S_2 \mid ind_2(X) \leq n\}$. By Lemma 78, both sets are finite. Since by Lemma 79 i_S and j_S leave the indices invariant, their

[13] For the execution of a transition all precondition states must be present. Therefore the *maximal* length causality chain is responsible for the earliest moment in which a transition may execute.

restrictions are the injective functions $i_S|_{I_{1,n}}: I_{1,n} \to I_{2,n}$ and $j_S|_{I_{2,n}}: I_{2,n} \to I_{1,n}$ of finite sets. Therefore, since domain and codomain are finite, the sets $I_{1,n}$ and $I_{2,n}$ have the same number of elements. Consequently, $i_S|_{I_{1,n}}$ and $j_S|_{I_{2,n}}$ are bijections. Since every element $X \in S_2$ has an index and is in a suitable set $I_{2,n}$, it also is the image of a suitable $Y \in S_1$ under the function $i_S|_{I_{1,n}}$. Therefore, i_S (and j_S) are bijections.

81 Definition BRANCHING STRUCTURE
Let $\mathcal{P} = (S, T, \longrightarrow)$ be a transition system. The *branching structure of* $\mathcal{P}$ is the pair $(\mathbb{B}(\mathcal{P}), \trianglerighteq)$, consisting of the quotient of all behaviours and the extension ordering $\trianglerighteq \subseteq \mathbb{B}(\mathcal{P}) \times \mathbb{B}(\mathcal{P})$. ◀

For *rooted processes* only extensions of the respective initial behaviours are studied.

Maximal behaviours are, similar as for ordinary transition systems, maximal elements of the extension order.

82 Theorem EXISTENCE OF MAXIMAL BEHAVIOURS
Every behaviour of a process may be extended to a *maximal behaviour*.

PROOF:
Since there is no canonical form for behaviours as in the case of ordinary transition systems, the approach used in the proof of Theorem 29 does not work. By using a suitable notation we will mimick the idea of this proof.

First, let us sketch the *intuition* behind the proof: Imagine the behaviours as trees, standing on the ground on their initial causes, and the sequences of executed transitions are growing into the sky. Let behaviour $\mathcal{B}_2$ be an extension of behaviour $\mathcal{B}_1$. Then the trees corresponding to these behaviours have a similar "shape", one seems to be contained in the other. We connect the corresponding states of the behaviour by a horizontal cord. Step by step we connect all corresponding states using such horizontal cords. Finally, the union of all these chords may be used for constructing a maximal behaviour.

Now we will provide a *more formal proof* without being overly technical. Beware of the overloading of symbols which is required to make this proof readable.

Let $\mathcal{B}$ be a behaviour. By the lemma of KURATOWSKY [DP90], $\mathcal{B}$ is an element of a maximal chain $\mathcal{M}$ in the extension order. Let $\mathcal{M}_f$ be the set of all finite behaviours of $\mathcal{M}$. $\mathcal{M}_f$ again is a chain and for every behaviour $\mathcal{B}$ of $\mathcal{M}_f$ there is an immediate extension of $\mathcal{B}$ in $\mathcal{M}_f$, if $\mathcal{B}$ is not already the maximal element of $\mathcal{M}_f$. Beginning with the smallest element of $\mathcal{M}_f$ which is an initial behaviour, we choose for every $\mathcal{B}_1 \vec{\trianglerighteq} \mathcal{B}_2$ two injections $i_S(\mathcal{B}_1, \mathcal{B}_2)$ and $i_T(\mathcal{B}_1, \mathcal{B}_2)$ which prove $\mathcal{B}_1 \vec{\trianglerighteq} \mathcal{B}_2$. Then we define the remaining injections by forming a kind of transitive closure, so that $i(\mathcal{B}_2, \mathcal{B}_3) \circ i(\mathcal{B}_1, \mathcal{B}_2) = i(\mathcal{B}_1, \mathcal{B}_3)$ holds for the injections of the states and of the transitions, and for all $\mathcal{B}_k$.

We define the sets S_M and T_M as the disjoint union of all states and transitions of behaviours in $\mathcal{M}_f$:

$$S_M := \bigcup_{\mathcal{B} \in \mathcal{M}_f,\ X \in S_{\mathcal{B}}} (\mathcal{B}, X)$$

$$T_M := \bigcup_{\mathcal{B} \in \mathcal{M}_f,\ t \in T_{\mathcal{B}}} (\mathcal{B}, t)$$

Now we connect the elements of S_M by cords. Formally, we define an *equivalence relation*: For $\mathcal{B}_1 \trianglerighteq \mathcal{B}_2$ we define $(\mathcal{B}_1, X_1) \sim (\mathcal{B}_2, X_2) :\Leftrightarrow i(\mathcal{B}_1, \mathcal{B}_2)(X_1) = X_2$. We form the symmetric hull and add the diagonal, so that $\sim$ becomes reflexive and symmetric. Due to the construction of i, the relation also is transitive. An analogous construction is made with the transitions. Let Q_S and Q_T be the quotients of S_M and T_M by these equivalency relations.

Now we define a transition relation $\longrightarrow \ \subseteq Q_S^\circ \times Q_T \times Q_S^\circ$. In order to determine whether[14]

$$[(\mathcal{B}_1, X_1)] \cdot [(\mathcal{B}_2, X_2)] \cdot \ldots \cdot [(\mathcal{B}_k, X_k)] \stackrel{[(\mathcal{B},t)]}{\longrightarrow} [(\mathcal{C}_1, Y_1)] \cdot [(\mathcal{C}_2, Y_2)] \cdot \ldots \cdot [(\mathcal{C}_l, Y_l)]$$

holds, we look for the smallest behaviour $\mathcal{D}$ which is an extension of all the behaviours $\mathcal{B}_j$, $\mathcal{C}_j$, and $\mathcal{B}$, embedd all states and transitions into this behaviour using the responsible injections, and check, whether this transition is a transition in this behaviour $\mathcal{D}$. Such a smallest behaviour $\mathcal{D}$ exists in $\mathcal{M}_f$, since we only consider finite behaviours where the order relation is very simple, as outlined above. Furthermore, we can show that $\longrightarrow$ is well-defined over the sets of equivalence classes and does not depend on the choice of the representatives of the equivalence classes. Similarly, functions f_S and f_T may be defined on the sets Q_S and Q_T.

The resulting behaviour $(Q_S, Q_T, \longrightarrow, f_S, f_T)$ then is a maximal behaviour which extends the original behaviour. This is easy to see if the original behaviour was finite. For infinite behaviours we may use a suitable finite subbehaviour in the proof. ◀

83 Definition OPERATIONAL SEMANTICS OF A PROCESS
The *operational semantics of a process* $\mathcal{P}$ is given by its set of maximal behaviours. ◀

It is straightforward to show the following theorem:

84 Theorem NON-DETERMINISM
If a process is *non-deterministic*, then[15] there exists an initial state for which there is *more than one*[16] *maximal behaviour*. ◀

It may depend on the initial state, whether non-determinism actually appears or not.

3.4 Examples of Branching Structures

In this section, we will provide *some examples* to familiarize ourselves with the branching structure of distributed processes:

85 Example PARALLEL SYSTEM
Let a process $\mathcal{P}$ be given by

$$S = \{A, B, C, D\} \qquad T = \{b, d\} \qquad L = \{\texttt{left, right}\}$$

$$A \stackrel{b}{\longrightarrow} B \qquad C \stackrel{d}{\longrightarrow} D \qquad \mu(a) = \texttt{left} \qquad \mu(b) = \texttt{right}$$

[14] As usual, the bracket notation $[(\mathcal{B}, X)]$ denotes the equivalence class of the element $(\mathcal{B}, X)$ contained in the bracket.

[15] Example 103 provides a *counterexample* for the *converse* of this theorem. There is *no* equivalency of these two properties as it was the case for ordinary transition systems; compare Theorem 35.

[16] Again, isomorphic behaviours only count once.

The initial state shall be $A \cdot B$. The following defines the only maximal behaviour:

$$S_b = \{0,1,2,3\} \qquad T_b = \{l,r\} \qquad 0 \xrightarrow{l}_b 1 \qquad 2 \xrightarrow{r}_b 3$$

$$f_S(0) = A \qquad f_S(1) = B \qquad f_S(2) = C \qquad f_S(3) = D$$

$$f_T(l) = b \qquad f_T(r) = d$$

The other behaviours of $\mathcal{P}$ are the initial behaviour, the behaviour where only the transition labeled `left` has executed, and the behaviour where only the transition labeled `right` has executed.

Figure 3.8 shows all these behaviours and the branching structure of $\mathcal{P}$. In this labeled HASSE-diagram the vertices are labeled by graphical illustrations of the individual behaviours. An edge from a behaviour $\mathcal{B}_1$ to a behaviour $\mathcal{B}_2$ denotes that $\mathcal{B}_2$ is an extension of $\mathcal{B}_1$, hence $\mathcal{B}_1 \trianglerighteq \mathcal{B}_2$, but that there exists no behaviour $\mathcal{B}$ "between"[17] $\mathcal{B}_1$ and $\mathcal{B}_2$. The order relation on the behaviours is therefore the reflexive, transitive hull of the relation which is depicted by directed edges. The directed edge from $\mathcal{B}_1$ to $\mathcal{B}_2$ is labeled with that transition[18] of the process, whose execution is responsible for the fact that $\mathcal{B}_2$ is an (immediate) extension of $\mathcal{B}_1$. The maximal behaviours are contained in a heavy solid line.

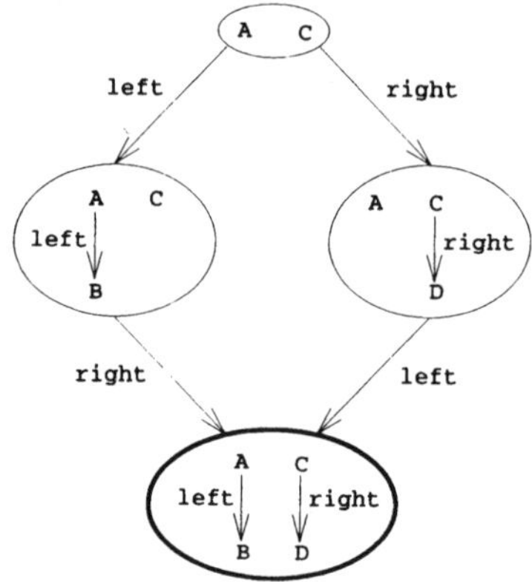

Figure 3.8: Branching Structure of a Parallel System.

Special attention should be paid to the ***parallelogram***. It is characteristic for the branching structure of parallel systems. ◀

86 Example NON-DETERMINISTIC SYSTEM
A non-deterministic system is given in shorthand notation by:

$$S = \{B, X, Y\} \qquad T = \{x, y\} \qquad B \xrightarrow{x} X \qquad B \xrightarrow{y} Y$$

[17] In Definition 89 this concept will be formalized by the notion of an *immediate extension*.

[18] In Lemma 91 below we will show that such a transition always can be found in a systematic manner. For the moment, geometric intuition shall suffice.

Let the initial state be B. This system has three behaviours: The ***initial behaviour*** B_0 given by

$$S_0 = \{0\} \qquad T_0 = \{\} \qquad f_S^0(0) = B \qquad f_T : \emptyset \to T$$

a ***first maximal behaviour*** B_1 given by

$$S_1 = \{0, 1\} \qquad T_1 = \{t\} \qquad 0 \xrightarrow{t}_1 1 \qquad f_S^1(0) = B \qquad f_S^1(1) = X \qquad f_T^1(t) = x$$

and a ***second maximal behaviour*** B_2 given by

$$S_2 = \{0, 1\} \qquad T_2 = \{t\} \qquad 0 \xrightarrow{t}_2 1 \qquad f_S^2(0) = B \qquad f_S^2(1) = Y \qquad f_T^1(t) = y$$

These behaviours and their branching structure are given in Figure 3.9.

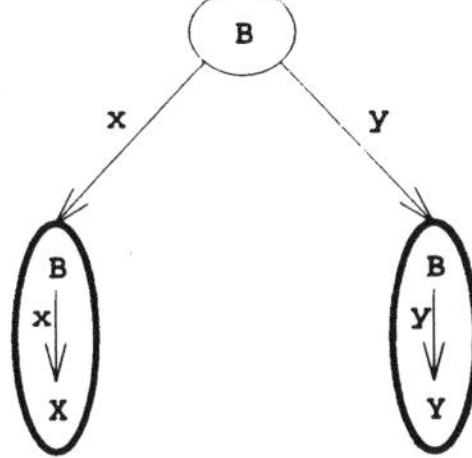

Figure 3.9: Branching Structure of a Non-Deterministic System.

Next we will study the initial state $B \cdot B$. Here we have three maximal behaviours and a total of six behaviours, with a branching structure illustrated in Figure 3.10.

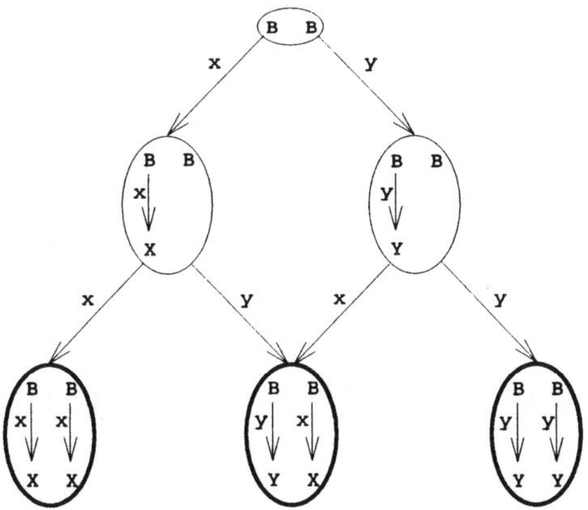

Figure 3.10: Branching Structure of a Non-Deterministic System.

For the initial state $B \cdot X$ we have two maximal behaviours and the branching structure of Figure 3.11. ◀

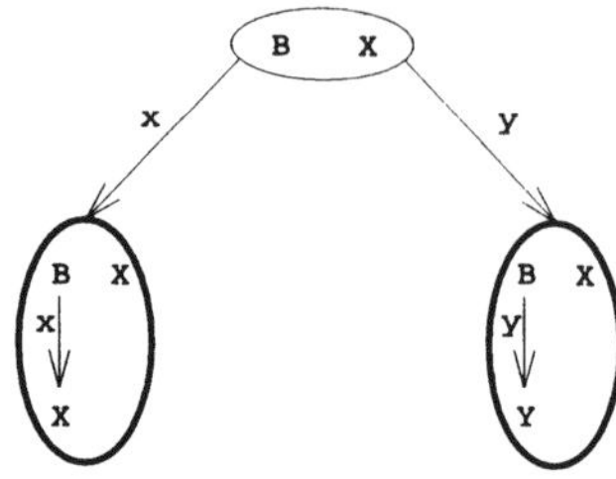

Figure 3.11: Branching Structure of a Non-Deterministic System.

87 Example FORK AND JOIN TRANSITIONS
A process is defined by

$$S = \{A, B, C, D, E, F\} \qquad T = \{\texttt{t}, \texttt{fork}, \texttt{join}, \texttt{u}\}$$

$$A \xrightarrow{\texttt{t}} B \qquad B \xrightarrow{\texttt{fork}} C \cdot D \qquad C \xrightarrow{\texttt{u}} E \qquad D \cdot E \xrightarrow{\texttt{join}} F$$

and initialized in state A.

It has only one maximal behaviour, and it may be interpreted as follows: Initially there is only one computing agent which is in state A. It performs transition `t` and then is in state B. Now transition `fork` takes place, resulting in two computing agents, the first of which is in state C, the second in state D. The first one transitions from C to E via `u`. Then a `join` transition takes place, only one computing agent remains and it is in state F. The maximal behaviour is shown in Figure 3.12. ◀

88 Example SYNCHRONIZATION TRANSITIONS
The system given by

$$S = \{A, B, C, D\} \qquad L = \{\texttt{synch}\} \qquad A \cdot B \xrightarrow{\texttt{synch}} C \cdot D$$

and initialized in state $A \cdot B$ models two computing agents in states A and B which perform a single `synch` transition, comparable to a *synchronization*, and finally assume the states C and D. The maximal behaviour is illustrated in Figure 3.13. ◀

3.5 Branching Structure and Constructions

The branching structure provides geometric information on the phenomena which may take place in a distributed system. In this section we introduce *additional concepts and constructions* which allow us to study it in more detail. Figure 3.33 demonstrates that even a process with a very small number of states and transitions may have a rather complicated branching structure. Its analysis should not be left to geometric intuition but a formal apparatus should rather be used.

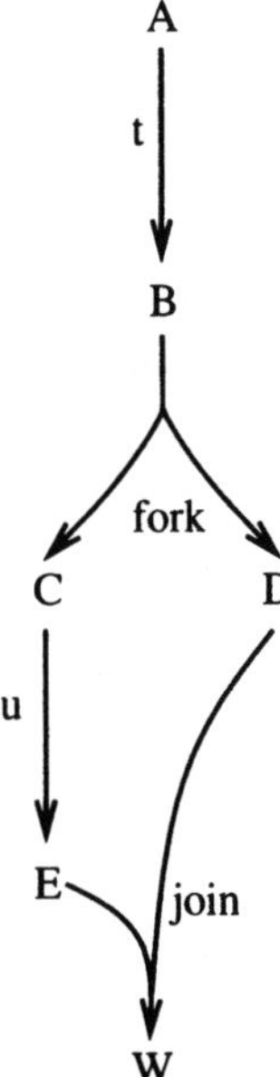

Figure 3.12: Fork and Join Behaviour.

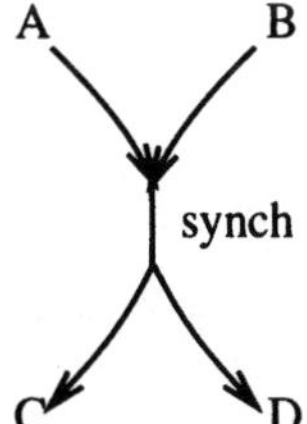

Figure 3.13: Synchronization Behaviour.

The relation of *immediate extension* which we already encountered on an informal basis, tells us, when a behaviour $\mathcal{B}_1$ may be extended into a behaviour $\mathcal{B}_2$ by the *execution of a single transition*:

89 Definition IMMEDIATE EXTENSION
Let $\mathcal{B}_1$ and $\mathcal{B}_2$ be two behaviours of a process $\mathcal{P}$. $\mathcal{B}_2$ is called an *immediate extension of* $\mathcal{B}_1$, iff it is an extension, $\mathcal{B}_1 \trianglerighteq \mathcal{B}_2$ and iff there is no behaviour $\mathcal{B}$, such that $\mathcal{B} \neq \mathcal{B}_1$, $\mathcal{B} \neq \mathcal{B}_2$ so that $\mathcal{B}_1 \trianglerighteq \mathcal{B} \trianglerighteq \mathcal{B}_2$.

We write $\mathcal{B}_1 \vec{\trianglerighteq} \mathcal{B}_2$, iff $\mathcal{B}_2$ is an immediate extension of $\mathcal{B}_1$. ◀

The relation of immediate extension may be carried over to equivalence classes of behaviours, similarly as this was done for the extension relation.

The following example shows that the transition and the states which distinguish a behaviour $\mathcal{B}_1$ from an immediate extension $\mathcal{B}_2$, are not always uniquely defined:

90 Example IMMEDIATE EXTENSION
Let a process $\mathcal{P}$ be given by

$$S = \{A\} \qquad T = \{t\} \qquad A \xrightarrow{t} A$$

a behaviour $\mathcal{B}_1$ by

$$S_1 = \{1,2,3,4,5,6,7\} \qquad T_1 = \{a,b,c,d\}$$

$$A \xrightarrow{a}_1 2 \qquad 2 \xrightarrow{b}_1 3 \qquad 4 \xrightarrow{c}_1 5 \qquad 6 \xrightarrow{c}_1 7$$

$$\forall X \in S_1 : \ f_S^1(X) = A \qquad \forall u \in T_1 : \ f_T^1(u) = t$$

and a behaviour $\mathcal{B}_2$ by

$$S_2 = \{C,D,E,F,G,H,J,K\} \qquad T_2 = \{e,f,g,h,j\}$$

$$C \xrightarrow{e}_2 D \qquad D \xrightarrow{f}_2 E \qquad F \xrightarrow{g}_2 G \qquad H \xrightarrow{h}_2 J \qquad J \xrightarrow{j}_2 K$$

$$\forall X \in S_2 : \ f_S^1(X) = A \qquad \forall u \in T_2 : \ f_T^1(u) = t$$

The two behaviours are shown in Figure 3.14. In contrast to the usual form of labeling in this figure the vertex labels are states of the behaviour (instead of the process) and the edge labels are transitions of the behaviour (instead of labels of the process). This helps understanding the situation.

$\mathcal{B}_2$ is an extension of $\mathcal{B}_1$ in two possible ways which are represented by the injections $i_S : S_1 \to S_2$ and $j_S : S_1 \to S_2$, respectively[19]. They are given by

$$i_S(1) = C \quad i_S(2) = D \quad i_S(3) = E \quad i_S(4) = F \quad i_S(5) = G \quad i_S(6) = H \quad i_S(7) = J$$

$$j_S(1) = H \quad j_S(2) = J \quad j_S(3) = K \quad j_S(4) = F \quad j_S(5) = G \quad j_S(6) = D \quad j_S(7) = D$$

i_S suggests that the behaviours $\mathcal{B}_1$ and $\mathcal{B}_2$ differ in the transition j and the state K, which are present in $\mathcal{B}_2$, but for which there is no analogon in $\mathcal{B}_1$. j_S suggests that the reason is the transition f and the state E. ◀

[19]The injections i_T and j_T of the sets of transitions may be constructed accordingly.

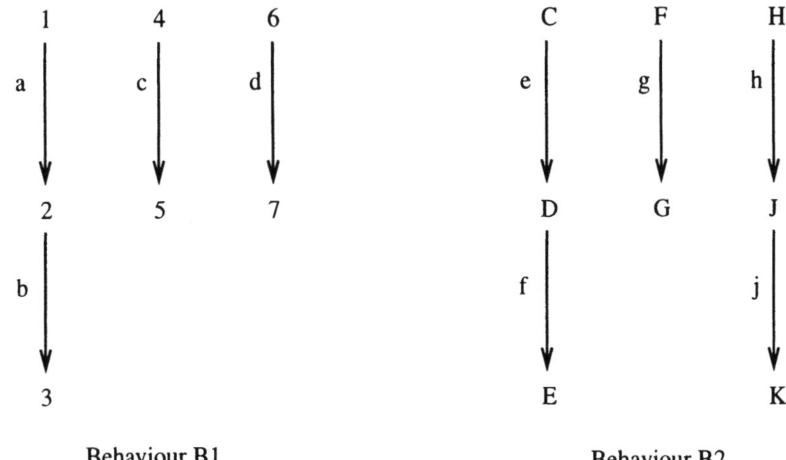

Figure 3.14: Two Behaviours.

Although the transition and the states which separate a behaviour $\mathcal{B}_1$ from an immediate extension $\mathcal{B}_2$ may not be uniquely identified, the observation functions of the behaviour map these transitions and states always to the same transitions and states of the process:

91 Lemma IMMEDIATE EXTENSION
Let $\mathcal{B}_1 = (S_1, T_1, \longrightarrow_1, f_S^1, f_T^1)$ and $\mathcal{B}_2 = (S_2, T_2, \longrightarrow_2, f_S^2, f_T^2)$ be two behaviours of a process $\mathcal{P} = (S, T, \longrightarrow, L, \mu)$, and let $\mathcal{B}_2$ be an immediate extension of $\mathcal{B}_1$.

For every pair of injections $i_S : S_1 \rightarrow S_2$ and $i_T : T_1 \rightarrow T_2$ which show this immediate extension according to Definition 89, there exists a unique transition $t \in T_2$ with $t \notin Im(i_T)$ and unique states $X_1, X_2, \ldots, X_k \in S_2$ with $X_1, X_2, \ldots, X_k \notin Im(i_S)$.

The index, rank, and number[20] of these states $X_1, X_2, \ldots, X_k \in S_2$, as well as their images $f_S^2(X_1), f_S(X_2), \ldots, f_S(X_k) \in S$ under the observation function f_S, and the image $f_T^2(t) \in T$ of the transition $t \in T_2$ are not dependent[21] on the choice of the injection.

PROOF:
The existence of such a t and its uniqueness follows from the definition of the immediate extension. From the typing condition we obtain the X_i. From the Invariance Lemma 79 we obtain the uniqueness of index and rank. For the remainder of the proof, we may disregard all states and transitions of higher index and rank. By the Finite Parallelism Lemma 78, this leaves a finite number of states and transitions to be considered. We now may count how often a certain state or transition of the process $\mathcal{P}$ is observed by the behaviours $\mathcal{B}_1$ and $\mathcal{B}_2$. There will be exactly one transition of the process which will be observed more frequently by $\mathcal{B}_2$ than by $\mathcal{B}_1$. This fact may be established with the help of the injections i_S and i_T, but the numbers of the respective observations will only depend on the behaviours and not on the injections. The number k of postconditions may be deduced from this unique transition. ◀

[20] i.e. k.

[21] Example 90 shows that there may be several injections and that t and the X_i may depend on them. This dependency vanishes only *after* the application of the observation functions .

We may recognize the relation of *immediate extension* as the *combinatorial generator* of the extension relation.

The Immediate Extension Lemma allows us to *annotate the branching structure*:

92 Remark BRANCHING STRUCTURE ANNOTATION
The branching structure is the order structure which the extension order relation $\trianglerighteq$ imposes on the set $\mathbb{B}(\mathcal{P})$ of isomorphism classes of behaviours of a process $\mathcal{P}$. This order structure may be illustrated by its HASSE-diagram and allows several annotations and labelings:

(1) The *vertices* of the HASSE-diagram are isomorphism classes of behaviours. They may be represented by a black dot or by a graph of a behaviour in the respective isomorphism class.

(2) The *vertices* may be *labeled* by the observed final states of the behaviours of the isomorphism class.

(3) The *edges* of the HASSE-diagram illustrate the relation of *immediate extension*. There is a directed edge from behaviour $\mathcal{B}_1$ to behaviour $\mathcal{B}_2$, iff the relation $\mathcal{B}_1 \vec{\trianglerighteq} \mathcal{B}_2$ holds.

(4) The *edges* may be *labeled* with that transition of the process which uniquely is characterized by the Immediate Extension Lemma 91.

◀

This annotated branching structure illustrates many important properties of a distributed process. It must be interpreted correctly:

93 Remark SNAPSHOT INTERPRETATION OF THE BRANCHING STRUCTURE
The semantics of a process is given by its set of *maximal behaviours*. It contains information about the states, the transitions, and the order of their execution.

A behaviour in the branching structure which is not maximal cannot be observed and is an *artificial construction*. A priori, it has not very much to do with the distributed process. To see this, let us conduct a *gedankenexperiment*[22] by which we might obtain such behaviours.

For this purpose, we install a *timer* with every computing agent. We then start our process. After some time, all our timers have elapsed and have stopped all computing agents. We then visit all agents and take notes about all previously executed transitions and the present states of these agents.

As we discussed at length in Sections 2.2 and 3.1, we are not able to setup this gedankenexperiment so that all timers elapse at the same time: In a distributed system there is *no global time* or notion of *synchrony*. We therefore must not interprete a non-maximal behaviour or the states recorded in the gedankenexperiment as an observable entity or as the global state of our distributed system at a certain time T.

Nevertheless, the non-maximal behaviours and the recorded states provide a *mechanism* which can help us in our *understanding of complex processes*. Despite the fact that "they have nothing to do with the 'real' process" we will continue to use them.

This remark is *not only of purely philosophical value*. Take a look at a branching structure of a parallel process as given in Figure 3.8. The semantics of this process is provided by the

[22] In real distributed systems, especially in data bases, such a gedankenexperiment is actually programmed to provide a *snapshot* of the state. These techniques are important for recovery in case of failures. See [TK87] for further information.

maximal behaviour. We may immediately recognize that we are looking at a parallel process: There are two transitions between which there is no relation on the order of execution.

Now let us take a distant look at the branching structure: It contains two non-maximal behaviours where only one of the parallel transitions has executed. Furthermore, from the initial behaviour to the maxial behaviour there are two paths which could be interpreted as two different, sequential executions of the parallel transitions. Therefore, if we take the branching structure as the semantics of a process, we are lead to an interleaving view of parallelism where two parallel transitions are identified with a non-deterministic choice between two sequential executions. ***Only the maximal behaviours*** provide a semantics for a theory which does not reduce parallelism to non-determinism. ◀

94 Definition FINAL STATES OF BEHAVIOURS
Let $\mathcal{P} = (S, T, \longrightarrow, L, \mu)$ be a process and $\mathcal{B} = (S_b, T_b, \longrightarrow_b)$ a not necessarily maximal behaviour.

An elementary state $X \in S_b$ is called a ***final state*** of $\mathcal{B}$, iff there exists no state Y for which X is an immediate cause.

The ***final state of behaviour*** $\mathcal{B}$ is the composite state $X_1 \cdot X_2 \cdot \ldots \cdot X_k$ which contains all finals states X_i of $\mathcal{B}$.

The ***observed final state*** of behaviour $\mathcal{B}$ is the image $f_S^{\circ}(X_1 \cdot X_2 \cdot \ldots \cdot X_k)$ of its final state under the state observation function f_S. ◀

The observed final state is the composite state of the process, which a behaviour observes in its own final state.

A behaviour may contain ***infinitely long branches*** which may be neglected in the determination of final states. Furthermore, there are infinite behaviours without final states.

The final states of a behaviour may be generalized by the notion of a cut.

95 Definition CUT OF A BEHAVIOUR
Let $\mathcal{B} = (S_b, T_b, \longrightarrow_b, f_S, f_T)$ be the behaviour of a process $\mathcal{P} = (S, T, \longrightarrow, L, \mu)$.

A set $C \subseteq S_b$ of states is called a ***cut*** or a ***maximal cut*** of $\mathcal{B}$, iff it is a cut[23] or a maximal cut of the causality relation $\succeq_b \ \subseteq S_b \times S_b$.

Let $\mathfrak{M}_{\mathcal{B}}$ denote the ***set of all maximal cuts*** of behaviour $\mathcal{B}$ and let $\geqslant \ \subseteq \mathfrak{M}_{\mathcal{B}} \times \mathfrak{M}_{\mathcal{B}}$ denote the ***cut order***[24] on $\mathfrak{M}_{\mathcal{B}}$. ◀

A set $C = \{X_1, X_2, \ldots, X_k\}$ of elementary states $X_i \in S$ may be transformed into a bag of states, or, what is the same, a composite state $X_1 \cdot X_2 \cdot \ldots \cdot X_k$. A composite state $X_1 \cdot X_2 \cdot \ldots \cdot X_k \in S_b^{\circ}$ of a behaviour may be transformed into an observed composite state $f_S(X_1) \cdot f_S(X_2) \cdot \ldots \cdot f_S(X_k) \in S^{\circ}$ of the behaviour. If we talk of the ***observed state of a cut***, we assume that these two transformations have been applied to the cut.

96 Lemma CUT LEMMA

(1) The set of ***initial causes*** of behaviour is a ***maximal cut***.

(2) The set of ***final causes*** of a ***finite*** behaviour is a ***maximal cut***.

(3) The set of ***final causes*** of a behaviour is a ***cut*** which is not necessarily maximal.

[23] See Definition 232.
[24] See Definition 233.

PROOF:
(1) is straightforward. In (2) and (3) it is easy to see that the set of final causes is a cut. We show that the set C of final causes of a finite behaviour is a *maximal* cut: Assume by indirect proof that C is a cut, but not maximal. Then there exists a state X which is not element of C and which is no final cause, and which has the property that X cannot be compared with any element of C. Since X is no final cause, there exists a state Y, for which X is an immediate cause. Now, there are two cases: First, if Y is a final cause, then $Y \in C$, and we obtain a contradiction to the property that X cannot be compared with any element of C. Second, if Y is no final cause, we may carry on with this process until we finally find a final cause Z in C, for which X is a cause ($X \succeq Z$). Then we get a similar contradiction as above with Y. Since our behaviour is finite we will find such a final cause Z after a finite number of steps. Example 97 provides a counterexample for the second part of (3). ◀

97 Example CUTS AND MAXIMAL CUTS
In Figure 3.15 we see a behaviour and four maximal cuts $C_1 = \{A, C, D, H\}$, $C_2 = \{B, E, H\}$, $C_3 = \{F, G\}$, and $C_4 = \{F, K\}$. They are related by the following non-trivial[25] cut order relations $C_1 \gtrdot C_2$, $C_1 \gtrdot C_4$, $C_2 \gtrdot C_4$, and $C_3 \gtrdot C_4$.

The set $\{A, C, E, G\}$ is no cut, since it contains a state and one of its causes. The set $\{A, C, D\}$ is a cut, but it is not maximal, since it is a true subset of the cut C_1.

Figure 3.16 shows an infinite behaviour where the set of final causes is no maximal cut.

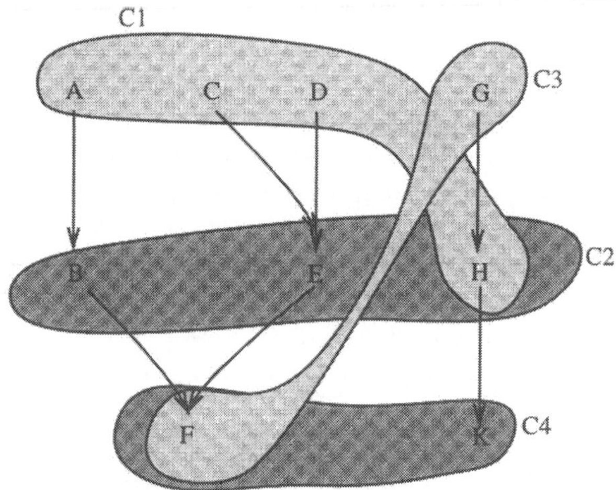

Figure 3.15: Maximal Cuts of a Behaviour.

◀

The observed states of a maximal cut correspond to the *snapshots* of our gedankenexperiment. If for two cuts C_1 and C_2 the relation $C_1 \gtrdot C_2$ holds, then this means that the cut or snapshot C_1 may develop into C_2 if we wait for some time. The direction of time in $\gtrdot$ is, similar as in $\trianglerighteq$, $\vec{\trianglerighteq}$, $\succeq$ and $\gg$, always from the left hand side to the right hand side of the relation.

In the observed state of a behaviour of a cut, certain transitions may be *enabled*:

[25]For every cut C there is the trivial relation $C \gtrdot C$.

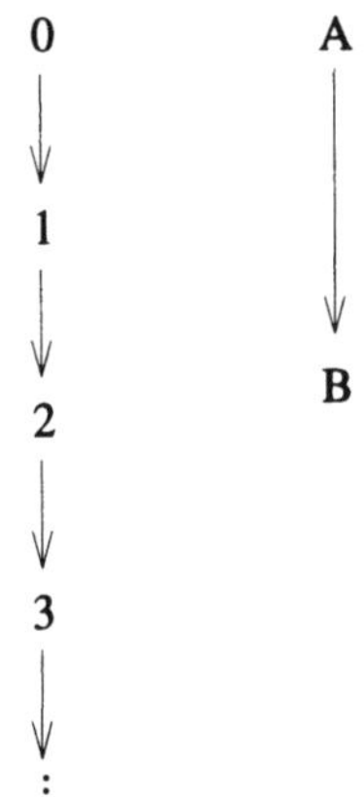

Figure 3.16: Final Causes Do Not Form Maximal Cut.

98 Definition ENABLED
Let $\mathcal{P} = (S, T, \longrightarrow, L, \mu)$ be a process and $X \in S^\circ$ a composite state.

A transition t of type (U, V) is called *enabled by* X, iff there exists a state $R \in X^\circ$, such that $X = U \cdot R$.

A bag $\xi = t_1 \cdot t_2 \cdot \ldots \cdot t_n \in T^\circ$ of transitions[26] $t_i \in T$ of type (U_i, V_i) is called *concurrently enabled by* X, iff there exists a state $R \in X^\circ$, such that $X = U_1 \cdot U_2 \cdot \ldots \cdot U_n \cdot R$.

Now let $\xi, \eta \in T^\circ$ be two bags of behaviours. Then ξ is said to be *concurrent to* η *under* X, iff $\xi \cdot \eta$ is concurrently enabled by X.

A bag $\xi \in T^\circ$ of transitions is called a *maximal enabled bag with respect to* X, iff it is enabled and iff there is no bag $\eta \in T^\circ$ different from the empty bag ε, such that $\xi \cdot \eta$ is enabled by X. ◀

Being *concurrently enabled* under a state X is a relation on the set T° of bags of transitions. This relation generally is *neither reflexive nor transitive*.

If in a state X a transition t is enabled, then we may execute t and obtain a new state X'. The (partially defined) function which *generates this new state* is called the *successor function*:

99 Definition SUCCESSOR FUNCTION
Let $\mathcal{P} = (S, T, \longrightarrow, L, \mu)$ be a process. The *successor function* of this process is the partially defined function

$$\begin{array}{rcl} \omega : \; T \times S^\circ & \rightarrow & S^\circ \\ (t, X) & \mapsto & \begin{cases} Y \cdot V & \Leftrightarrow \quad U \xrightarrow{t} V \wedge \exists Y \in S^\circ : \; X = Y \cdot U \\ \text{undefined} & \Leftrightarrow \quad \text{else} \end{cases} \end{array}$$

◀

[26]The t_i are not required to be different or pairwise different. A bag may contain two "copies" of a transition. This number of copies also affects the question whether a bag of transitions is enabled.

The following proposition states that the order in which two concurrently enabled transitions are executed, has no influence on the state which finally is obtained.

100 Proposition COMMUTING OPERATION
Let $\mathcal{P}$ be a process, X a composite state, and t_1 and t_2 two transitions which are concurrently enabled by X. Then the actions of these transitions on the state X commute:

$$\omega(t_1, \omega(t_2, X)) = \omega(t_2, \omega(t_1, X))$$

◀

3.6 Branching Structure and Geometry

In this section, we will analyze the geometric properties of the branching structure.

Sequential behaviours were already studied in the previous chapter. Since the free commutative monoid S° contains its set of generators, i.e. $S \subseteq S^\circ$, every sequential process is also a distributed process.

Deterministic sequential processes were characterized by *linear branching structures*[27]. Nondeterministic sequential processes could have *tree-like* branching structures. Only parallel processes have a more complex extension order structure. The following process is no sequential process and cannot be described by an ordinary transition system. Nevertheless, its branching structure "looks like" one of a sequential process:

101 Example QUASI-SEQUENTIAL STRUCTURE
Let a process be given by

$$S = \{A, B, C, D, E, F, G, H\} \qquad T = \{r, s, t\}$$

$$A \cdot B \xrightarrow{r} C \qquad C \cdot D \xrightarrow{s} E \cdot F \qquad E \cdot F \xrightarrow{t} G \cdot H$$

The initial state shall be $A \cdot B \cdot D$. The branching structure of this process is shown in Figure 3.17.

◀

Formally we may define:

102 Definition SEQUENTIAL AND QUASI-SEQUENTIAL STRUCTURE
A distributed process $\mathcal{P} = (S, T, \longrightarrow, L, \mu)$ is

of *sequential structure* iff its transition relation $\longrightarrow \ \subseteq S^\circ \times T \times S^\circ$ is a subset of $S \times T \times S$.

of *quasi-sequential structure* iff its branching structure has the following property[28]: For every maximal behaviour $\mathcal{B}_M$, the set $\{\mathcal{B} \mid \mathcal{B} \trianglerighteq \mathcal{B}_M\}$ of all behaviours which may be extended to this maximal behaviour, forms a chain[29].

---

[27] Compare Theorem 29.

[28] Of course we will *identify isomorphic behaviours*. The condition therefore should read: For every maximal isomorphism class of behaviours, the set of all isomorphism classes of behaviours which may be extended to this maximal isomorphism class forms a chain. Until now this should be quite obvious, and in the remaining sections we will relax formal precision in favour of a more concise and readable version.

[29] See Definition 229.

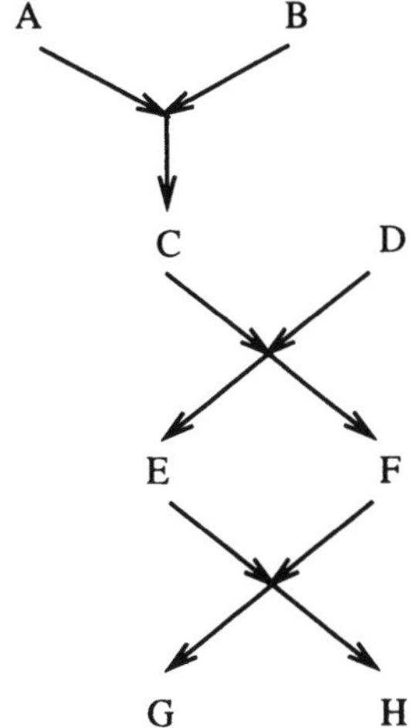

Figure 3.17: Quasi-Sequential Process.

◀

See Example 220 for an explanation of ***interaction graphs*** which may be used to illustrate the behaviour of quasi-sequential processes which are interacting with other processes.

In the following we will present a counterexample to the converse of Theorem 84: A rooted deterministic distributed process may have ***more than one maximal behaviour***:

103 Example ALIASING
A process is defined by shorthand notation as

$$S = \{A, B, C, X\} \qquad T = \{c, j\} \qquad A \cdot B \xrightarrow{j} X \qquad C \xrightarrow{c} A$$

Let the initial state be $A \cdot B \cdot C$.

This process has two maximal behaviours, as we can see in Figure 3.18, but it is not non-deterministic. There is ***no conflict*** between transition j and transition c, but we have ***two different possibilities of attaching*** j, since there are two copies of state A. These maximal behaviours are essentially different from each other: One exhibits parallelism, since there is no order on the execution of j and c, the other looks like[30] a sequential process, since c must execute before j. ◀

We may define:

104 Definition ALIASING
A process is said to ***allow aliasing*** iff it is deterministic, and there exists an initial state for which there is more than one maximal behaviour.

Two non-isomorphic behaviours are called ***aliases of each other*** if they are immediate exten-

[30] In Definition 102 a more precise distinction between sequential and quasi-sequential behaviours will be presented. In this terminology this behaviour is quasi-sequential.

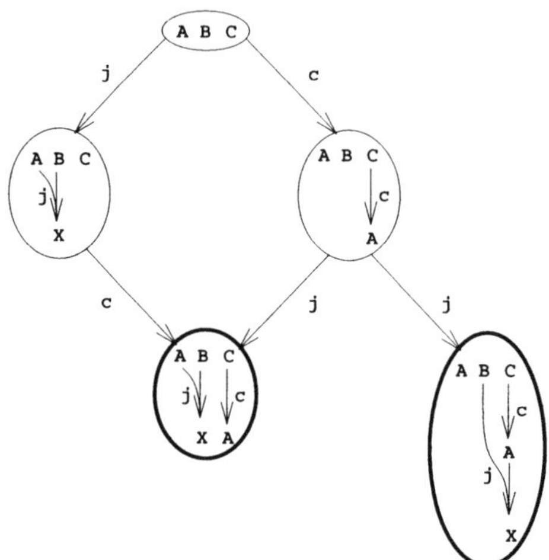

Figure 3.18: Aliasing.

sions of the same behaviour, with the extension process involving[31] the same transition of the process. ◀

Aliasing is only possible iff there is a behaviour which offers more than one choice of *attaching* a transition. This is possible only if the observed final state of this behaviour contains multiple copies of an elementary state. The futures of two behaviours which are aliases of each other are identical, since they also have identical observed final states. Their difference which is due to this aliasing effect will be conserved across future extensions.

We therefore may define:

105 Definition SINGLE-VALUED
A rooted process is called ***single-valued*** iff for every initial state and behaviour the observed final state does not contain multiple copies of an elementary state. ◀

106 Theorem NON-DETERMINISM OF SINGLE-VALUED PROCESSES
A ***single-valued*** process is ***non-deterministic***, if and only if there exists an initial state for which there exists more than one[32] maximal behaviour. A single-valued process does not allow aliasing. ◀

If a process does not allow aliasing, it is not necessarily single-valued. If the observed final state of a behaviour contains multiple copies of an elementary state this does not mean that there are transitions which may be attached in two different ways. It is possible that in such a state no transition is enabled.

107 Definition GEOMETRY OF THE BRANCHING STRUCTURE
Let $\mathcal{P}$ be a process.

We say that a behaviour $\mathcal{B}$ ***splits*** into the behaviours $\mathcal{B}_1, \mathcal{B}_2, \ldots, \mathcal{B}_k$, iff these behaviours are pairwise non-isomorphic immediate extensions of $\mathcal{B}$.

[31] In Example 103 this is the transition j. In Lemma 91 below, this is made more precise.
[32] Isomorphic behaviours only count once.

Let $t_1, t_2, \ldots, t_k$ be the transitions of the process, belonging to these immediate extensions according to the immediate extension Lemma 91.

We say that a behaviour $\mathcal{B}$ ***splits by conflicting transitions***, iff these transitions t_i of the process are enabled, but are not concurrently enabled by the observed final state of $\mathcal{B}$.

We say that a behaviour $\mathcal{B}$ ***splits by aliasing transitions***, iff these transitions t_i all are equal.

We say that a behaviour $\mathcal{B}$ ***splits by concurrent transitions***, iff the transitions t_i of the process are concurrently enabled by the observed final state of $\mathcal{B}$.

We say that two behaviours $\mathcal{B}_1$ and $\mathcal{B}_2$ *join* in a behaviour $\mathcal{B}$, iff $\mathcal{B}_1$ and $\mathcal{B}_2$ both are immediate extensions of $\mathcal{B}$.

Figure 3.19 illustrates split and join situations. ◀

Figure 3.19: Split and Join Geometry.

108 Lemma SPLIT LEMMATA
If a behaviour ***splits by two conflicting transitions***, then the resulting behaviours ***will never join***[33] at any later time.

If a behaviour ***splits by two concurrent transitions***, then the resulting behaviours ***join in the next transition***.

These claims may be formulated in a more precise but less intuitive manner. Assume $\mathcal{B} \mathrel{\vec{\trianglerighteq}} \mathcal{B}_1$ and $\mathcal{B} \mathrel{\vec{\trianglerighteq}} \mathcal{B}_2$, and let t_1 and t_2 be transitions of the process, as mentioned by the immediate extension Lemma 91.

If t_1 and t_2 are enabled but not concurrently enabled by the observed final states, then there exists no behaviour $\mathcal{B}'$ such that $\mathcal{B}_1 \trianglerighteq \mathcal{B}'$ and $\mathcal{B}_2 \trianglerighteq \mathcal{B}'$.

If t_1 and t_2 are concurrently enabled, then there exists a behaviour $\mathcal{B}'$ such that $\mathcal{B}_1 \mathrel{\vec{\trianglerighteq}} \mathcal{B}'$ and $\mathcal{B}_2 \mathrel{\vec{\trianglerighteq}} \mathcal{B}'$. The transitions of the process which belong to these immediate extensions are the transitions t_2 and t_1. ◀

The simple split lemmata provide a geometric characterization and distinction for parallelism and non-determinism involving two transitions:

Assume that in the observed final state of a behaviour $\mathcal{B}$ two transitions t_1 and t_2 are concurrently enabled and hence may execute in ***parallel***. Then there are two immediate extensions $\mathcal{B}_1$ and $\mathcal{B}_2$ of $\mathcal{B}$, each of which corresponds to a snapshot, where just one of the two transitions has executed. Both behaviours may join in the next transitions by executing

[33] Here ***never*** refers to the fact that for the resulting behaviours $\mathcal{B}_1$ and $\mathcal{B}_2$ there is not only no common ***immediate*** extension, but no common extension at all.

the remaining parallel transition. This produces the so called ***concurrency parallelogram***, which is shown in Figure 3.20 in a concrete and abstract concurrency parallelogram.

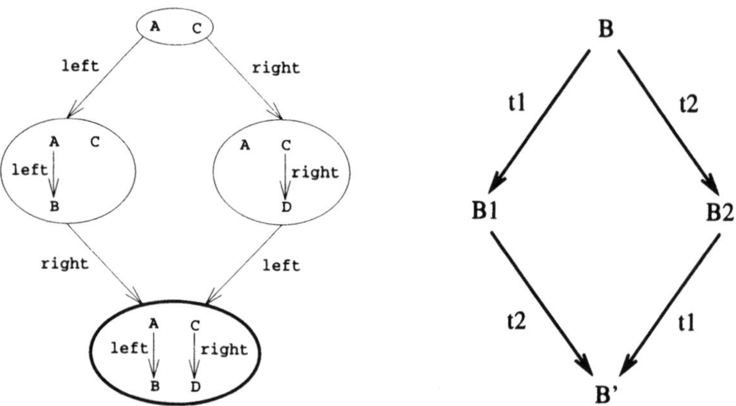

Figure 3.20: Concurrency Parallelogram.

If the states and transitions which are associated with the concurrency parallelogram are equal, then this parallelogram may collapse. This is shown in Figure 3.21, where the branching structure of the process

$$S = \{A, B\} \qquad T = \{t\} \qquad A \xrightarrow{t} B$$

with initialization $A \cdot A$ is shown.

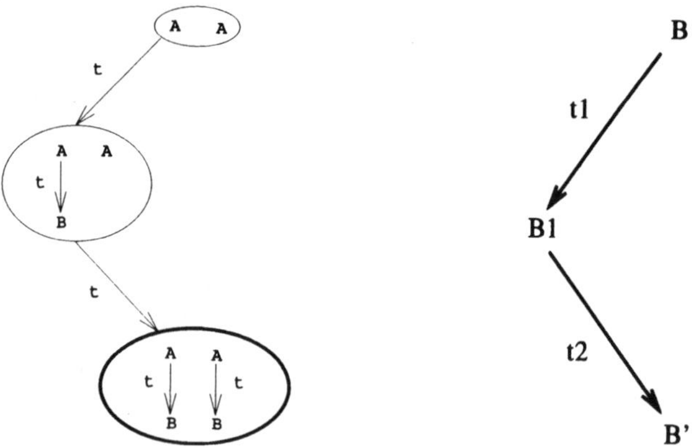

Figure 3.21: Collapsed Concurrency Parallelogram.

Now assume that in the observed final state of $\mathcal{B}$ two transitions t_1 and t_2 are enabled but not concurrently enabled. This requires a non-deterministic choice of either t_1 or t_2. After one of the two transitions is executed, it becomes an irreversible part of the resulting behaviours. Therefore, the produced behaviours $\mathcal{B}_1$ and $\mathcal{B}_2$ will not join at any later time and are extended to different maximal behaviours. This is illustrated in Figure 3.22, where the initial behaviour $\mathcal{B}$ splits into the two maximal behaviours $\mathcal{B}_1$ and $\mathcal{B}_2$.

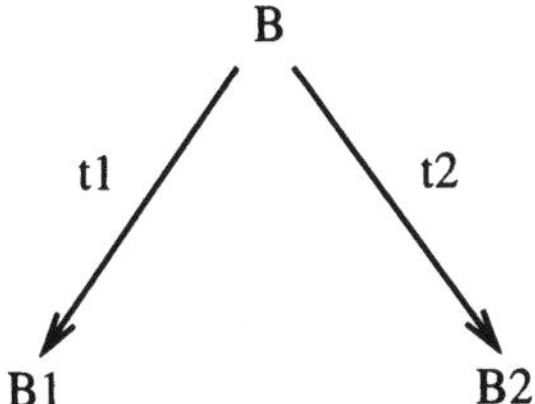

Figure 3.22: Non-Deterministic Split.

109 Lemma UNIQUE JOIN LEMMA
If two behaviours *join*, then they *join in a unique behaviour*:

If $\mathcal{B}_1 \vec{\trianglerighteq} \mathcal{B}$, $\mathcal{B}_2 \vec{\trianglerighteq} \mathcal{B}$, and $\mathcal{B}_1 \vec{\trianglerighteq} \mathcal{B}'$, $\mathcal{B}_2 \vec{\trianglerighteq} \mathcal{B}'$, then $\mathcal{B}$ and $\mathcal{B}'$ are isomorphic. ◄

110 Lemma BACKWARD EXTENSION PROPERTY
Two behaviours *join*, if and only if they are the immediate result of a *concurrent split*:

Let $\mathcal{B}_1$ and $\mathcal{B}_2$ be two non-isomorphic behaviours. If $\mathcal{B}_1 \vec{\triangleright} \mathcal{B}$ and $\mathcal{B}_2 \vec{\triangleright} \mathcal{B}$, then there exists a behaviour $\mathcal{B}'$, such that $\mathcal{B}' \vec{\triangleright} \mathcal{B}_1$ and $\mathcal{B}' \vec{\triangleright} \mathcal{B}_2$. ◄

111 Lemma HYPERCUBE LEMMA
If a behaviour $\mathcal{B}$ *splits by pairwsie different concurrent transitions* $t_1, t_2, \ldots, t_n$, then the resulting behaviours *join* in the form of an *n-dimensional concurrency hypercube*, if no two transitions t_i and t_k which are attached to $\mathcal{B}$ are attached using the same state of $\mathcal{B}$. ◄

In Figure 3.23 a *concurrency hypercube* is shown for three concurrently enabled transitions. The behaviours are represented by their observed final states.

A concurrency hypercube can *collapse*: If in the behaviour of Figure 3.23 the states B and C, and the transitions 2 and 3 are equal, then the hypercube collapses, as indicated in Figure 3.24. The picture of the collpased cube is also obtained if a non-collapsed cube is projected onto a plane from such a projection point where the edges corresponding to the identified transitions are projected onto the same line in the plane.

Figure 3.25 illustrates a more complex collapse. There are two cubes which correspond to two maximally enabled bags of transitions $t_1 \cdot t_2 \cdot t_3$ and $t_2 \cdot t_3 \cdot t_4$. These cubes penetrate each other, sharing the shaded parallelogram. This parallelogram corresponds to the two transitions t_2 and t_3 which form the concurrently (but not maximally) enabled bag $t_2 \cdot t_3$. If t_2 is made equal to t_3, then this intersection parallelogram collapses, producing the branching structure illustrated in Figure 3.26.

Several behaviours may be "attached" to a concurrency hypercube. This may be due to *aliasing phenomena* as illustrated in Figure 3.18, or due to *confusion* which is discussed in Section 3.7.

112 Remark PROOFS OF ABOVE LEMMATA
If one has developed some intuitive understanding of the geometry of the branching structure,

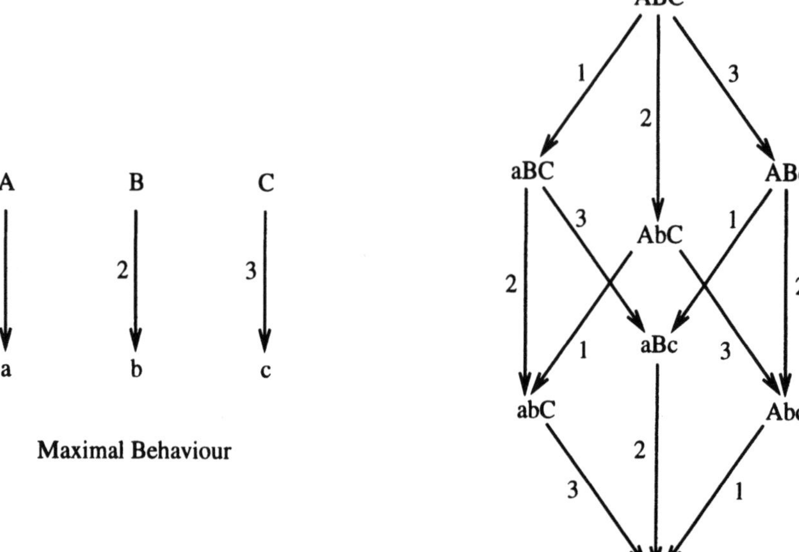

Maximal Behaviour

Branching Structure

Figure 3.23: Concurrency Hypercube.

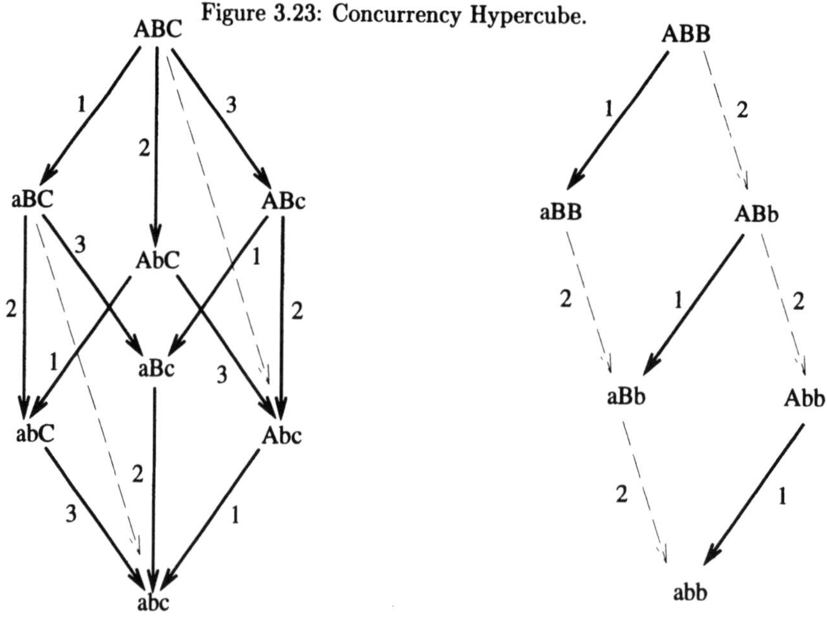

Collapse Indicated

Collapse Performed

Figure 3.24: Collapsed Concurrency Hypercube.

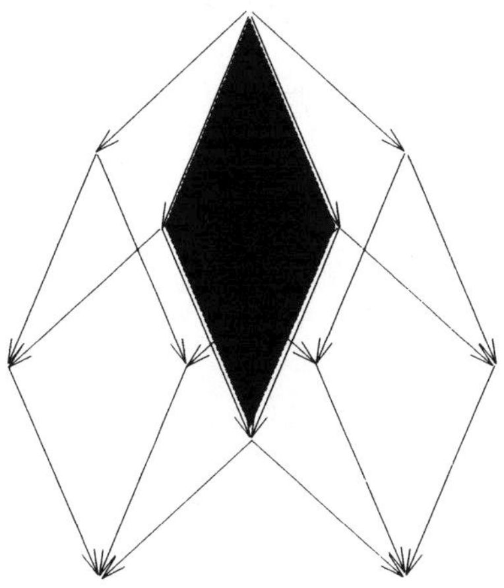

Figure 3.25: Illustration of Hypercube Theorem (Not Collapsed).

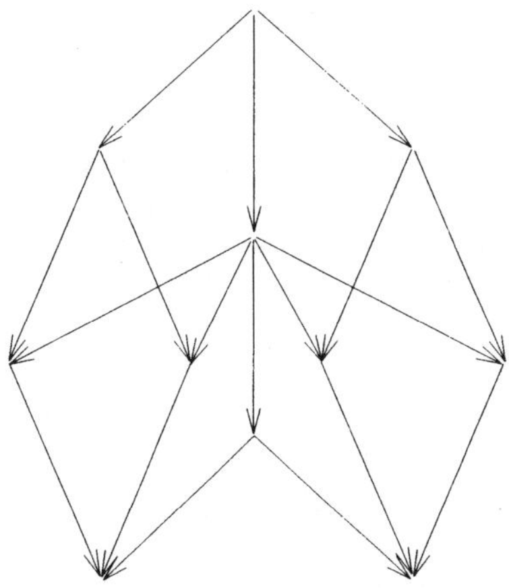

Figure 3.26: Illustration of Hypercube Theorem (Collapsed).

it is quite easy to "see" that the above lemmata hold. Of course, this does not constitute a strict proof. In the following, we will provide a complete proof for the backward extension lemma, to demonstrate the techniques of these proofs. For the remaining lemmata we will only provide a general outline of the proof idea and will skip the details.

Let us analyze a *join situation*: Let $\mathcal{P} = (S_P, T_P, \longrightarrow_P, L, \mu)$ be a process, and let $\mathcal{B}_1 = (S_1, T_1, \longrightarrow_1, f_S^1, f_T^1)$ and $\mathcal{B}_2 = (S_2, T_2, \longrightarrow_2, f_S^2, f_T^2)$ be behaviours which both extend to $\mathcal{B}' = (S', T', \longrightarrow', f_S', f_T')$: Then there exist injections

$$i_S^1 : S_1 \to S' \qquad i_T^1 : T_1 \to T'$$

$$i_S^2 : S_2 \to S' \qquad i_T^2 : T_2 \to T'$$

To make this more readable we write

$$i^1 : \mathcal{B}_1 \to \mathcal{B}' \qquad i^2 : \mathcal{B}_2 \to \mathcal{B}'$$

and skip the distinction between the injections responsible for the states and the transitions, respectively. Since these functions are injections, they provide bijections from their domain to their image. The image then is mapped by a simple inclusion into the codomain of the bijections:

$$\mathcal{B}_1 \xrightarrow{i^1} Im(i^1) \hookrightarrow \mathcal{B}' \qquad \mathcal{B}_2 \xrightarrow{i^2} Im(i^2) \hookrightarrow \mathcal{B}'$$

It is easy to derive that these injections i^1 and i^2 provide an isomorphism of behaviours: The image of $\mathcal{B}_1$ under i^1 can be made a behaviour of $\mathcal{P}$ in a canonic fashion. This behaviour then is isomorphic to $\mathcal{B}_1$ and i^1 is the isomorphism. This is similar for $\mathcal{B}_2$. Since the extension relations are invariant under isomorphism, we may replace $\mathcal{B}_1$ and $\mathcal{B}_2$ by their isomorphic images.

After this replacement the sets of states and transitions of $\mathcal{B}_1$ and $\mathcal{B}_2$ are subsets of the respective sets of $\mathcal{B}'$. The injections which show that $\mathcal{B}_1 \trianglerighteq \mathcal{B}'$ and $\mathcal{B}_2 \trianglerighteq \mathcal{B}'$ hold both are inclusions of subsets. This allows the definition of an intersection behaviour $\mathcal{B}_I := \mathcal{B}_1 \cap \mathcal{B}_2$. For this, the intersection of the sets of states and transitions must be formed. The construction of the transition relation and of the observation functions is straightforward, too, by forming suitable restrictions.

These observations allow a rather direct proof of the *backward extension lemma*. The fact that the i^1 and i^2 are isomorphisms onto their image and hence may be inverted, may be used to provide the required isomorphism in the *unique join lemma*.

The *split lemma*, the *hypercube lemma*, and the *hypercube theorem* follow from a tedious but straightforward analysis of the possible cases. ◀

If we are not interested in the non-maximal behaviours of a process but only in their final states, we may study the *reachability structure*:

113 Definition REACHABILITY STRUCTURE
Let $\mathcal{P} = (S, T, \longrightarrow, L, \mu)$ be a process with initial state $R \in S^\circ$.

The *reachability structure* of $\mathcal{P} = (S, T, \longrightarrow, L, \mu)$ is a directed, edge-labeled multigraph. The set of *vertices* is the set S° of composite states of the process, There is a *directed edge* from state X to state Y if and only if there exist a transition t and states N, X', Y' such that $X' \xrightarrow{t} Y'$, $X = X' \cdot N$, and $Y = Y' \cdot N$. This edge is labeled with t. ◀

3.7 Confusion

In this section, we will discuss confusions and provide a large number of examples for different forms of non-determinism.

A *confusion* is a situation where *non-determinism and parallelism* interact in a peculiar[34] combined form: Assume that two transitions t_1 and t_2 are concurrently enabled in a state X and hence could execute in parallel. If t_1 executes *before* t_2, then this may change the "amount" of non-determinism which is associated with the second, not yet executed transition t_2. It is even possible that transition t_2 will not execute at all.

The notion that transition t_1 is executed *before* transition t_2 does not fully make sense for the case of two *parallel* transitions t_1 and t_2. Again, this concept only represents an *artificial construction* which helps to understand the confusion. It must be interpreted in a similar way as the cuts in Remark 93.

We must specify how the "amount" of non-determinism is measured. For this purpose we use the *conflict set* of a transition t with respect to a state X. It is defined as the set of all transitions which could be executed in state X, but which must not be executed concurrently with t. Hence, there is a non-deterministic choice between t and the elements of the conflict set:

114 Definition CONFLICT SET
Let $\mathcal{P} = (S, T, \longrightarrow, L, \mu)$ be a process, $X \in S^\circ$ a state, and $t \in T$ a transition which is enabled by X.

The *conflict set* of transition t in state X is defined by

$$\mathcal{K}(t, X) := \{u \in T \mid u \text{ enabled by } X \text{ and } t \cdot u \text{ not enabled by } X\}$$

◀

The following definition of confusion is analogous to the PETRI-net version[35] by ROZENBERG and THIAGARAJAN [Rei87], but the subsequent analysis and classification goes beyond that.

In a confusion (t_1, t_2, X) the conflict set of the first transition (t_1) is altered by the execution of the second one (t_2):

115 Definition CONFUSION
Let $\mathcal{P} = (S, T, \longrightarrow, L, \mu)$ be a process.

A *confusion* (t_1, t_2, X) of $\mathcal{P}$ consists of two concurrently enabled transitions $t_1, t_2 \in T$ and a state $X \in S^\circ$, such that

$$\mathcal{K}(t_1, X) \neq \mathcal{K}(t_1, \omega(t_2, X))$$

[34] Or, as the word indicates, in a *confusing* form. This notion is borrowed from PETRI-net theory, where a very similar phenomenon is known.

[35] Several different formal definitions and informal notions of a confusion may be found in PETRI-net literature.

A confusion (t_1, t_2, X) is called *symmetric*, iff also (t_2, t_1, X) is a confusion; otherwise it is called *asymmetric*.

A confusion (t_1, t_2, X) is called a

conflict introduction, iff $\mathcal{K}(t_1, X) \subsetneq \mathcal{K}(t_1, \omega(t_2, X))$,

conflict removal, iff $\mathcal{K}(t_1, X) \supsetneq \mathcal{K}(t_1, \omega(t_2, X))$,

conflict modification, iff it is neither a conflict introduction nor a conflict removal.

◀

Conflict modifications may be understood as the removal of some conflicts and the introduction of other conflicts, and hence may be reduced to introduction and removal.

Assume a transition t is enabled. For a conflict of a transition c with a transition t in a state X, two conditions must be satisfied:

(1) *Enabeling condition:*
Transition c is enabled in state X.

(2) *Concurrency condition:*
Transitions c and t are not concurrently enabled in state X.

A confusion can introduce or remove a conflict by influencing one or the other of these two conditions, or both. In a confusion, several transitions may change their conflict status and the reasons for this change may vary from transition to transition. Furthermore, in symmetric confusions one confusion may be caused by influencing the enabeling condition and the other by influencing the concurrency condition.

In the following we will give several examples of various elementary forms of confusions. They shall illustrate the variety of phenomena in processes which at first sight look extremely simple. This collection does not constitute a complete classification of confusion.

116 Example ASYMMETRIC CONFLICT INTRO, ENABELING CONDITION
We will study the system defined by

$$S = \{A, B, C, D, Z\} \qquad T = \{u, v, w\}$$

$$A \xrightarrow{u} C \qquad B \xrightarrow{v} D \qquad B \cdot C \xrightarrow{w} Z$$

which will be initialized in the state $A \cdot B$.

$(v, u, A \cdot B)$ is a confusion, but $(u, v, A \cdot B)$ is not: If u executes before v, then this increases the conflict set of v.

u and v are concurrently enabled in state $A \cdot B$

$$\mathcal{K}(v, A \cdot B) = \{v\} \qquad \mathcal{K}(v, \omega(u, A \cdot B)) = \mathcal{K}(v, B \cdot C) = \{v, w\}$$

$$\mathcal{K}(u, A \cdot B) = \{u\} \qquad \mathcal{K}(u, \omega(v, A \cdot B)) = \mathcal{K}(u, A \cdot D) = \{u\}$$

Discussion of the branching geometry in Figure 3.27:
The branching structure exhibits a concurrency parallelogram for the transitions u and v. On its left side, an additional maximal behaviour is attached which is lacking on the right side. This situation represents the asymmetrically introduced conflict. This conflict is introduced via the enabeling condition, since in the initial state transition w is not enabled. ◀

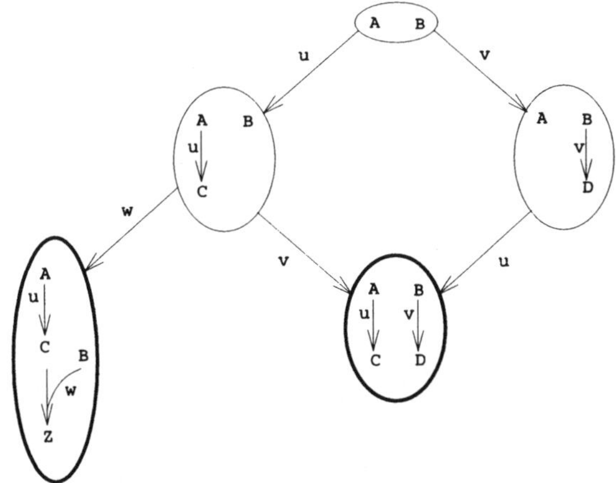

Figure 3.27: Asymmetric Conflict Intro, Enabeling Condition.

117 Example SYMMETRIC CONFLICT INTRO, ENABELING CONDITION
We will study the system defined by

$$S = \{X, Y, V, W, U, Z\} \qquad T = \{r, s, u, v\}$$

$$X \xrightarrow{r} V \qquad Y \xrightarrow{s} W \qquad X \cdot W \xrightarrow{u} U \qquad V \cdot Y \xrightarrow{v} Z$$

which will be initialized in the state $X \cdot Y$.

Both $(r, s, X \cdot Y)$ and $(s, r, X \cdot Y)$ are confusions. If one of the transitions r and s executes before the other, this changes the conflict set of the remaining transition.

r and s are concurrently enabled in state $X \cdot Y$

$$\mathcal{K}(r, X \cdot Y) = \{r\} \qquad \mathcal{K}(r, \omega(s, X \cdot Y)) = \mathcal{K}(r, X \cdot W) = \{r, u\}$$

$$\mathcal{K}(s, X \cdot Y) = \{s\} \qquad \mathcal{K}(s, \omega(r, X \cdot Y)) = \mathcal{K}(s, V \cdot Y) = \{s, v\}$$

Discussion of the branching geometry in Figure 3.28:
The branching structure shows the concurrency parallelogram for r and s. On both sides of the parallelogram additional maximal behaviours are attached belonging to the two symmetric confusions. ◀

118 Example SYMMETRIC CONFLICT INTRO, CONCURRENCY CONDITION
We will study the system defined by

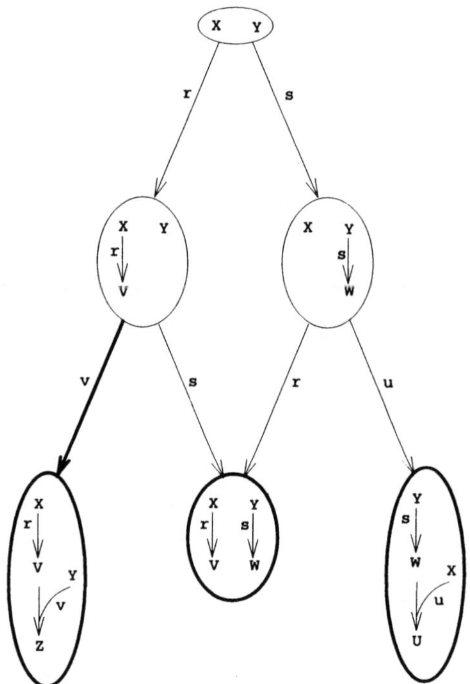

Figure 3.28: Symmetric Conflict Intro, Enabeling Condition.

$$S = \{A, U, V, W\} \qquad T = \{u, v, w\}$$

$$A \xrightarrow{v} V \qquad A \xrightarrow{u} U \qquad A \xrightarrow{w} W$$

which will be initialized in the state $A \cdot A$.

Both $(u, v, A \cdot A)$ and $(v, u, A \cdot A)$ are confusions. Due to the symmetry of the process under arbitrary permutations of transitions and states, also $(u, w, A \cdot A)$, $(w, u, A \cdot A)$, $(v, w, A \cdot A)$, and $(w, v, A \cdot A)$ are confusions. We only prove one confusion:

u and v are concurrently enabled in state $A \cdot A$

$$\mathcal{K}(u, A \cdot A) = \{\} \qquad \mathcal{K}(u, \omega(v, A \cdot A)) = \mathcal{K}(u, A \cdot V) = \{u, v, w\}$$

Discussion of the branching geometry in Figure 3.29:
The branching structure shows three concurrency parallelograms for v and u, v and w, and u and w. There is no concurrency cube for all three transitions u, v, and w. In state $A \cdot A$ there is no conflict between any two transitions, however, the execution of an arbitrary transition introduces a conflict. The conflicts are introduced via the concurrency condition: In the initial state, u is not in conflict with w, but u and w are concurrently enabled. After the exectuion of v, u and w are no longer concurrently enabled. ◀

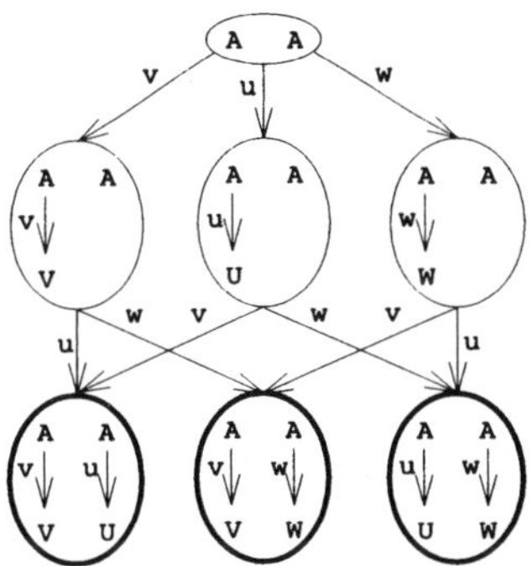

Figure 3.29: Symmetric Conflict Intro, Concurrency Condition.

119 Example ASYMMETRIC CONFLICT INTRO, CONCURRENCY CONDITION
We will study the system[36] defined by

$$S = \{A, U, X, D, C, M, L\} \qquad T = \{u, v, w\}$$

[36]This is the most simple example for this type of confusion, which I have found. It raises the interesting question, how many states are required for the realization of a certain branching type. This number may also be used to measure the complexity of branching structures.

$$U \cdot A \xrightarrow{u} M \qquad D \cdot X \cdot A \xrightarrow{v} X \qquad C \cdot X \cdot A \xrightarrow{w} L$$

which will be initialized in the state $U \cdot C \cdot X \cdot D \cdot A \cdot A$.

$(u, v, U \cdot C \cdot X \cdot D \cdot A \cdot A)$ is a confusion, but $(v, u, U \cdot C \cdot X \cdot D \cdot A \cdot A)$ is not: If u executes before v, then this increases the conflict set of v.

u and v are concurrently enabled in state $U \cdot C \cdot X \cdot D \cdot A \cdot A$

$$\mathcal{K}(u, U \cdot C \cdot X \cdot D \cdot A \cdot A) = \{u\}$$

$$\mathcal{K}(u, \omega(v, U \cdot C \cdot X \cdot D \cdot A \cdot A)) = \mathcal{K}(u, U \cdot C \cdot X \cdot A) = \{u, w\}$$

$$\mathcal{K}(v, U \cdot C \cdot X \cdot D \cdot A \cdot A) = \{v, w\}$$

$$\mathcal{K}(v, \omega(u, U \cdot C \cdot X \cdot D \cdot A \cdot A)) = \mathcal{K}(v, C \cdot X \cdot D \cdot A \cdot M) = \{v, w\}$$

Discussion of the branching geometry in Figure 3.30:
The branching structure shows the concurrency parallelogram of v and u. In the initial state there is no conflict between u and w, since they are concurrently enabled. After the execution of v there is such a conflict. ◀

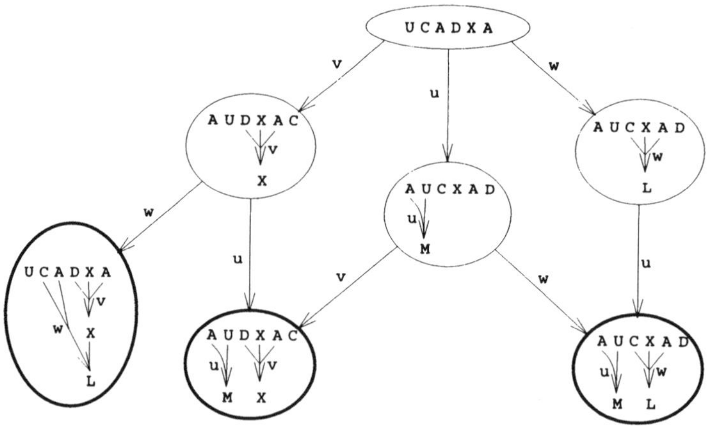

Figure 3.30: Asymmetric Conflict Intro, Concurrency Condition.

120 Example ASYMMETRIC CONFLICT REMOVAL, ENABELING CONDITION
Let our process be defined by

$$S = \{A, B, X, Y, Z\} \qquad T = \{u, v, w\}$$

$$X \xrightarrow{u} Z \qquad A \cdot B \xrightarrow{v} B \qquad B \cdot X \xrightarrow{w} Y$$

and initialized in state $A \cdot B \cdot X$.

Then $(v, u, A \cdot B \cdot X)$ is a confusion, but $(u, v, A \cdot B \cdot X)$ is not:

u and v are concurrently enabled in state $A \cdot B \cdot X$

$$\mathcal{K}(v, A \cdot B \cdot X) = \{v, w\} \qquad \mathcal{K}(v, \omega(u, A \cdot B \cdot X)) = \mathcal{K}(v, A \cdot B \cdot Z) = \{v\}$$

$$\mathcal{K}(u, A \cdot B \cdot X) = \{u, w\} \qquad \mathcal{K}(u, \omega(v, A \cdot B \cdot X) = \mathcal{K}(u, B \cdot X) = \{u, w\}$$

Disussion of the branching geometry in Figure 3.31:
The process is non-deterministic with three maximal behaviours. In the initial state there is a choice between either w or the concurrency parallelogram of u and v. If u executes before v, then this conflict is removed by disabeling the conflicting transition. ◀

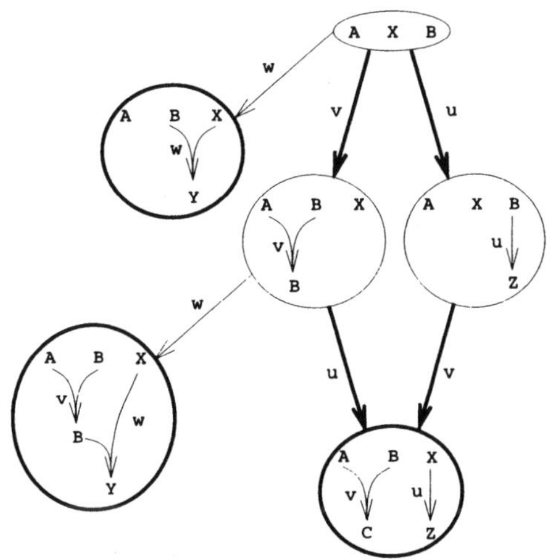

Figure 3.31: Asymmetric Conflict Removal, Enabeling Condition.

121 Example SYMMETRIC CONFLICT REMOVAL, ENABELING CONDITION
Let our process be defined by

$$S = \{A, B, C, D, X, Y\} \qquad T = \{u, v, w\}$$

$$A \cdot X \xrightarrow{u} C \qquad B \xrightarrow{v} D \qquad X \cdot B \xrightarrow{w} Y$$

and initialized in state $A \cdot X \cdot B$.

Then both $(u, v, A \cdot X \cdot B)$ and $(v, u, A \cdot X \cdot B)$ are confusions.

u and v are concurrently enabled in state $A \cdot B \cdot X$

$$\mathcal{K}(u, A \cdot X \cdot B) = \{u, w\} \qquad \mathcal{K}(u, \omega(v, A \cdot X \cdot B)) = \mathcal{K}(u, A \cdot X \cdot D) = \{u\}$$

$$\mathcal{K}(v, A \cdot X \cdot B) = \{v, w\} \qquad \mathcal{K}(v, \omega(u, A \cdot X \cdot B)) = \mathcal{K}(v, C \cdot B) = \{v\}$$

Discussion of the branching geometry in Figure 3.32:
The process is non-deterministic and has two maximal behaviours. One is reached via the concurrency parallelogram of v and u, the other one consists of a single transition. Both parallel transitions are in conflict with w and the execution of any one disables w and symmetrically removes the conflict. ◀

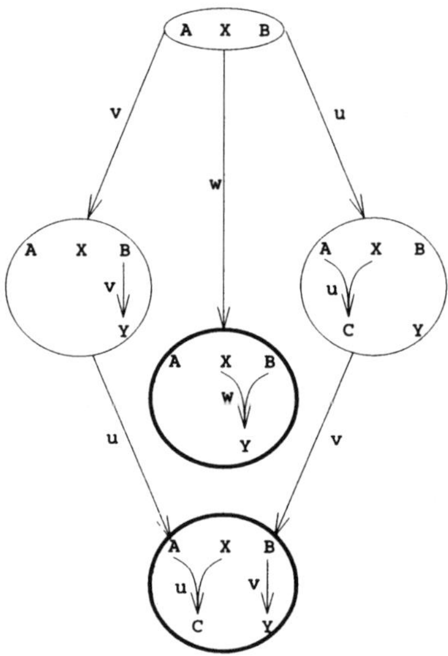

Figure 3.32: Symmetric Conflict Removal, Enabeling Condition.

122 Example ASYMMETRIC CONFLICT REMOVAL, CONCURRENCY COND.
A process is given by

$$S = \{A, B, D\} \qquad T = \{1, 2, c\}$$

$$A \xrightarrow{1} D \qquad A \xrightarrow{2} D \qquad B \xrightarrow{c} A \cdot A$$

and initialized in state $A \cdot B$.

Then $(1, c, A \cdot B)$ is a confusion, but $(c, 1, A \cdot B)$ is not:

1 and c are concurrently enabled in state $A \cdot B$

$$\mathcal{K}(1, A \cdot B) = \{1, 2\} \qquad \mathcal{K}(1, \omega(c, A \cdot B)) = \mathcal{K}(1, A \cdot A \cdot A) = \emptyset$$

$$\mathcal{K}(c, A \cdot B) = \{c\} \qquad \mathcal{K}(c, \omega(1, A \cdot B)) = \mathcal{K}(c, D \cdot B) = \{c\}$$

Discussion of the branching geometry:
It is astonishing that a system consisting of three states and transitions can have such a complex branching structure as it is shown in Figure 3.33. In the initial state, transition 1 is in conflict with transition 2. After the execution of transition c, the transitions 1 and 2 are concurrently enabled. Hence, this conflict removal occurs by influencing the concurrency condition.

The branching structure of this process also may be used for illustrating the lemmata of Section 3.6. For example, the initial behaviour *splits* by the conflicting transitions 1 and 2. The resulting behaviours will *never join* at a later time. If we execute c in the initial condition, we obtain a behaviour in which the bag $1 \cdot 1 \cdot 2$ of transitions is concurrently enabled. In the figure, the respective *concurrency cube* is shown in thicker lines. If we choose the other transition 2 as illustrated by the broken line, we see the effect of *multiple values*: In this situation there is *no concurrency cube*; the behaviours which are obtained by the leftmost transition 1 and the transition 2 with the broken line never join. This illustrates the additional requirement in the split lemma that parallel transitions must not be *attached using the same states* of the behaviour. ◀

Confusion also allows us to understand *race conditions*, a phenomenon very much feared in VLSI design:

123 Remark RACE CONDITIONS
Intuitively, *race conditions* are situations where "something goes wrong due to the relative execution speeds of two processes". Such phenomena are of importance in VLSI technology where the length of an electric connection influences the delay of a signal and may be responsible for unwanted behaviour of the designed chip. They also play a role in the parallel programming of shared memory machines where it is possible that a process executes much faster than another and reaches a certain statement earlier than the programmer thought.

Confusion was the phenomenon where the de facto execution order of transitions which in principle could execute in parallel had an influence on the arising conflicts and hence on the future behaviour of a process. In a physical realization of a process, transitions will be executed by computing agents. If two transitions may be executed in parallel, then this will require two computing agents. The question whether one transition of two parallel transitions was executed before the other may then be translated into the question whether one of the two agents has a higher computing speed than the other. *Confusions*, *race conditions*, and *differing relative computing speeds* hence constitute different view points and notions of the same phenomenon. ◀

3.8 First Order Transition Systems

In Section 2.5 we introduced the *sequential composition* of transitions. For distributed systems we will define an operation which may be interpreted as *parallel composition* of transitions. We will motivate the relevant design decisions in a number of illustrative examples:

124 Definition COMPOSITIONAL DISTRIBUTED TRANSITION SYSTEM
A *compositional distributed transition system* $(S, \mathfrak{T}, \circ)$ consists of

(1) a non-empty set S of *elementary states* which generate the free commutative monoid

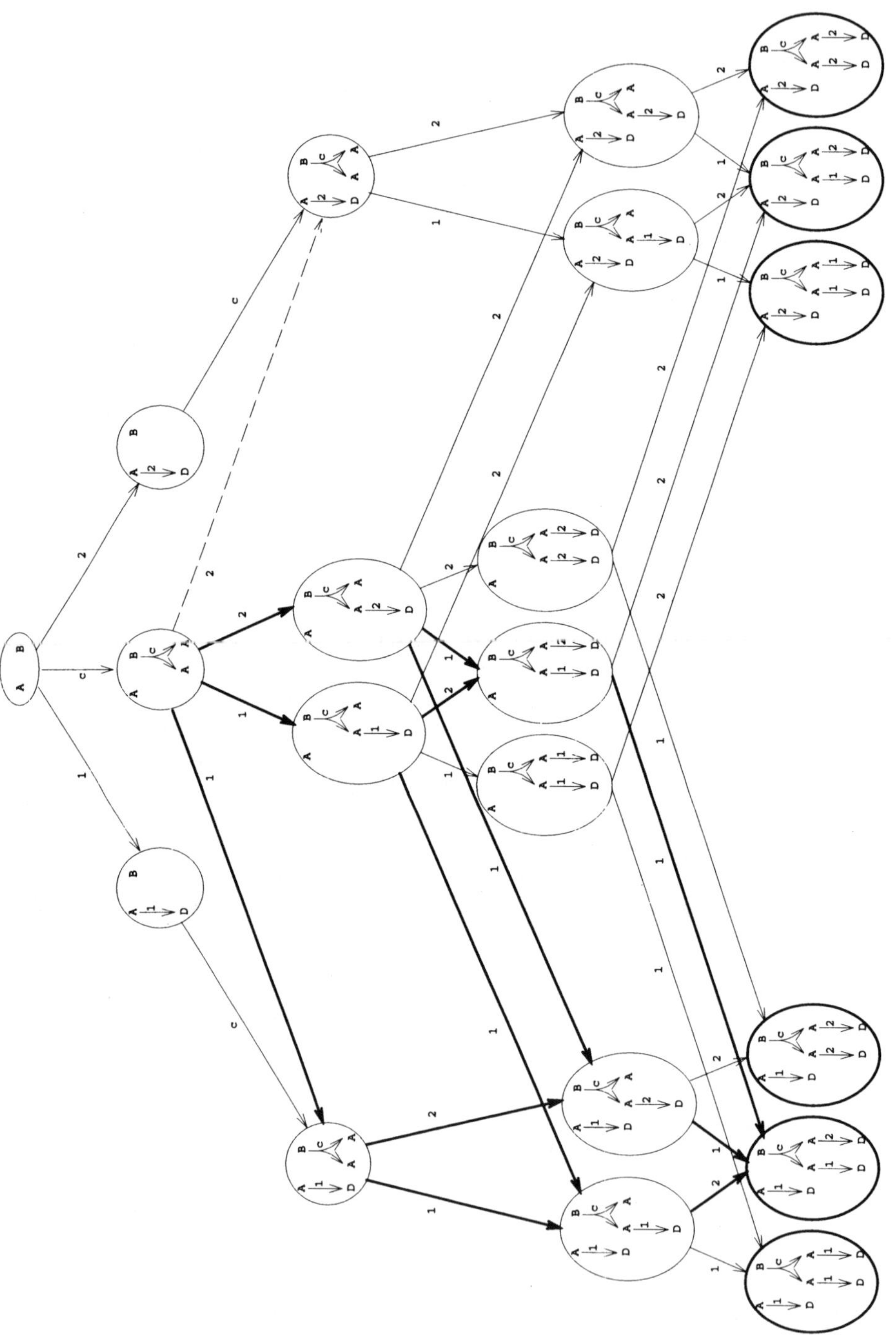

Figure 3.33: Asymmetric Conflict Removal, Concurrency Condition.

$(S^\circ, \cdot, \varepsilon)$ of *composite states*,

(2) for every pair $(X, Y) \in S^\circ \times S^\circ$ of composite states a set $\mathfrak{T}(X, Y)$ of *transitions*, and

(3) for all states $X, Y, Z \in S^\circ$ a *binary operation* $\circ : \mathfrak{T}(Y, Z) \times \mathfrak{T}(X, Y) \to \mathfrak{T}(X, Z)$ on the transitions, the so called *(sequential) composition*,

such that the following conditions are satisfied:

(dis) *Pairwise disjoint*:
The sets $\mathfrak{T}(X, Y)$ of transitions are pairwise disjoint.

(id) *Existence of identity transitions*:
For every state $X \in S^\circ$ there exists an *identity transition* $id_X \in \mathfrak{T}(X, X)$, such that for all $t \in \mathfrak{T}(X, Y)$ and $u \in \mathfrak{T}(Y, X)$ the following equations hold:

$$t \circ id_X = t \qquad id_X \circ u = u$$

(ass) *Associativity of (sequential) composition*:
For all $v \in \mathfrak{T}(C, D)$, $u \in \mathfrak{T}(B, C)$, and $t \in \mathfrak{T}(A, B)$ the following equation holds:

$$v \circ (u \circ t) = (v \circ u) \circ t$$

◀

In the following example we will see that compositional systems do not reflect all the structural information of distributed systems in which we are interested:

125 Example WEAKNESS OF COMPOSITIONAL DISTRIBUTED SYSTEMS
Let us look at the following distributed transition system:

$$S = \{A, B, C, D\} \qquad T := \{r, s\} \qquad A \xrightarrow{r} B \qquad C \xrightarrow{s} D$$

If this system is initialized in state A, it consumes A, performs r, and generates state B. This is similar, if the system is initialized in state C. However, if the system is initialized in state $A \cdot C$, then it may perform the transitions r and s *in parallel*.

Compositional systems allow us to combine transitions which *may* execute in sequence, to a single, more complex transition. In above example this is not possible, since the required typing condition for the formation of the sequential composition $\circ$ is not fulfilled. However, from the transitions

$$A \xrightarrow{r} B \qquad\qquad C \xrightarrow{s} D$$

which could execute in parallel in state $A \cdot C$ we would like to form the transition

$$A \cdot C \xrightarrow{r \odot s} B \cdot D$$

The transition $r \odot s$ describes the state change which takes place if the (parallel) transitions r and s execute. The parallel composition[37] of the transitions r and s to the transition $r \odot s$ must not be interpreted as information on the order of execution of r and s.

[37] In Section 2.5, this was already discussed for the sequential composition.

The new transition $r \odot s$ is called the ***parallel composition*** of the transitions r and s. In the categorical interpretation of higher order processes in Section 7.2, it corresponds to the product functor in the monoidal category. This has motivated the choice of the operator symbol $\odot$. ◀

This example poses a number of questions, all of which shall be at first dealt with in examples, before making a formal definition.

126 Example COMMUTATIVITY AND ASSOCIATIVITY
Whenever we have two transitions

$$A \xrightarrow{f} B \qquad C \xrightarrow{g} D$$

we may form the two parallel compositions

$$A \cdot C \xrightarrow{f \odot g} B \cdot D \qquad C \cdot A \xrightarrow{g \odot f} D \cdot B$$

They are defined on the states $A \cdot C$ and $C \cdot A$ respectively. Syntactically these terms are different, but since the states form a commutative monoid, the values which they represent, are equal. Also from the interpretation of the parallel composition we expect no differences between $f \odot g$ and $g \odot f$ and therefore we expect the following equation

$$f \odot g = g \odot f$$

There is a similar situation with the three transitions

$$A \xrightarrow{f} B \qquad C \xrightarrow{g} D \qquad E \xrightarrow{h} F$$

from which we may form the parallel compositions

$$A \cdot (C \cdot E) \xrightarrow{f \odot (g \odot h)} B \cdot (D \cdot F)$$

and

$$(A \cdot C) \cdot E \xrightarrow{(f \odot g) \odot h} (B \cdot D) \cdot F$$

Again the types of both compositions are equal since $\cdot$ is associative, and there is no reason to distinguish the two transitions. Hence, we expect the equation

$$f \odot (g \odot h) = (f \odot g) \odot h$$

◀

In the following example we will analyze, how the identity transition relates to the parallel composition:

127 Example IDENTITY TRANSITIONS AND PARALLEL COMPOSITION
Assume we have a transition system with

$$A \xrightarrow{f} B$$

and suppose we have another state C. For all states there are identity transitions, hence also for C:

$$C \xrightarrow{id_C} C$$

We therefore may form the following parallel compositions

$$A \cdot C \xrightarrow{f \odot id_C} B \cdot C \qquad C \cdot A \xrightarrow{id_C \odot f} C \cdot B$$

Hence, if there is a transition t from state A to state B, then there is also a transition "of this kind", when in addition to state A also another state C is present: State A still changes to state B via transition t and state C is left unmodified. ◀

128 Example SEQUENTIAL AND PARALLEL COMPOSITION
How does the parallel composition relate to the sequential one? Assume we have a transition system with the following situation:

$$A_1 \xrightarrow{f_1} B_1 \qquad B_1 \xrightarrow{g_1} C_1$$

$$A_2 \xrightarrow{f_2} B_2 \qquad B_2 \xrightarrow{g_2} C_2$$

Then we may form both parallel and sequential compositions: We get the two parallel compositions

$$A_1 \cdot A_2 \xrightarrow{f_1 \odot f_2} B_1 \cdot B_2 \xrightarrow{g_1 \odot g_2} C_1 \cdot C_2$$

from which we may form the sequential composition

$$A_1 \cdot A_2 \xrightarrow{(g_1 \odot g_2) \circ (f_1 \odot f_2)} C_1 \cdot C_2$$

On the other hand we may form the sequential compositions

$$A_1 \xrightarrow{g_1 \circ f_1} C_1 \qquad A_2 \xrightarrow{g_2 \circ f_2} C_2$$

from which we may form the parallel composition

$$A_1 \cdot A_2 \xrightarrow{(g_1 \circ f_1) \odot (g_2 \circ f_2)} C_1 \cdot C_2$$

We thereby obtain two transitions $(g_1 \odot g_2) \circ (f_1 \odot f_2)$ and $(g_1 \circ f_1) \odot (g_2 \circ f_2)$ of the same type $A_1 \cdot A_2 \longrightarrow C_1 \cdot C_2$ which contain the same transitions, f_1, f_2, g_1, and g_2. We therefore expect the equation

$$(g_1 \odot g_2) \circ (f_1 \odot f_2) = (g_1 \circ f_1) \odot (g_2 \circ f_2)$$

Sequential and parallel composition are *a posteriori* structures: They may be constructed from a given process, but they do not imply any order of execution on the involved transitions.

In this sense, it would be *wrong* to make the following statement concerning the above equation: The *left side* indicates that g_1 and g_2 may execute in parallel, and similarly f_1 and f_2 may execute in parallel; furthermore g_1 and g_2 both may execute only after both f_1 and f_2 have executed. The *right side* indicates that g_1 must execute after f_1, and that g_2 must execute after f_2, but there is no restriction on the order of execution between an index-1 transition and an index-2 transition. This interpretation would rather indicate that this equation should not hold.

The *correct interpretation* is that the *effects* of the transitions on the left and on the right side of the equation are the same.

If we apply this equation, we must take care of the partiality of the sequential composition $\circ$. By using the commutativity of $\odot$ we could naively deduce the following equations:

$$(g_1 \circ f_1) \odot (g_2 \circ f_2) = \qquad \text{(by above equation)}$$

$$(g_1 \odot g_2) \circ (f_1 \odot f_2) = \qquad \text{(by commutativity)}$$

$$(g_2 \odot g_1) \circ (f_1 \odot f_2) = \qquad \text{(by above equation)}$$

$$(g_2 \circ f_1) \odot (g_1 \circ f_2)$$

This is *not correct*, since in the last expression the composition $g_2 \circ f_1$ cannot be formed for type reaons. These equations only hold, if the terms on both sides are *existent*. ◀

We may form the parallel composition of two identity transitions:

129 Example PARALLEL COMPOSITION OF IDENTITY TRANSITIONS
Let

$$A \xrightarrow{id_A} A \qquad B \xrightarrow{id_B} B$$

be two identity transitions. Then we may form their parallel compositions

$$A \cdot B \xrightarrow{id_A \odot id_B} A \cdot B$$

The identity transition does not perform any state change, and also the parallel composition of two identity transitions does not perform any work. However, the *unique* transition which does not perform any work on state $A \cdot B$ is the identity transition on $A \cdot B$. Therefore we obtain the equation

$$id_A \odot id_B = id_{A \cdot B}$$

◀

130 Example IDENTITY TRANSITION OF NEUTRAL ELEMENT
We may form the parallel composition of a transition t of the type $A \xrightarrow{t} B$ with the identity transition on the empty state ε:

$$A \cdot \varepsilon \xrightarrow{t \odot id_\varepsilon} B \cdot \varepsilon$$

Since $A \cdot \varepsilon = A$ and $B \cdot \varepsilon = B$, this transition is of the same type as transiion t. Furthermore, transition id_ε does not perform any work, and even the state, on which it operates on, is not visible. Therefore it seems reasonable to require the equation $f \odot id_\varepsilon = id_\varepsilon \odot f = f$. ◀

All the above examples motivate the following definition:

131 Definition FIRST ORDER TRANSITION SYSTEM
A *first order transition system* $(S, \mathfrak{T}, \circ, \odot)$ consists of

(1) a non-empty set S of *elementary states* which generate the free commutative monoid $(S^\circ, \cdot, \varepsilon)$ of *composite states*,

(2) for every pair $(X, Y) \in S^\circ \times S^\circ$ of composite states a set $\mathfrak{T}(X, Y)$ of *transitions*, and

(3) for all states $X, Y, Z \in S^\circ$ a *binary operation* $\circ : \mathfrak{T}(Y, Z) \times \mathfrak{T}(X, Y) \to \mathfrak{T}(X, Z)$ on the transitions, called the *sequential composition*,

(4) for all states $A, B, C, D \in S^\circ$ a *binary operation* $\odot : \mathfrak{T}_{A,B} \times \mathfrak{T}_{C,D} \to \mathfrak{T}_{A\cdot C, B\cdot D}$, called the *parallel composition*,

such that the following conditions are satisfied:

(co) *Compositional System*:
$(S, \mathfrak{T}, \circ)$ is a compositional distributed transition system.

(eqn) For all states and transitions, for which the terms on both sides of the equal sign exist, the following *equations* must hold:

$$f \odot (g \odot h) = (f \odot g) \odot h$$

$$f \odot g = g \odot f$$

$$f \odot id_\varepsilon = id_\varepsilon \odot f = f$$

$$(f \odot g) \circ (h \odot k) = (f \circ h) \odot (g \circ k)$$

$$id_{A\cdot B} = id_A \odot id_B$$

◀

Every distributed transition system *generates* a first order transition system. For this, we have to add identity transitions, and sequential and parallel compositions to the distributed transition system, and we must identify those transitions, whose equality may be deduced from the above equations.

132 Definition FREELY GENERATED FIRST ORDER TRANSITION SYSTEM
Let $\mathcal{D} = (S, T, \longrightarrow)$ be a typed, distributed transition system.

For every pair (X, Y) of states in S°, we define a set $\bar{\mathfrak{T}}(X, Y)$ of expressions as the smallest[38] set of expressions which satisfies the following conditions:

[38] It is necessary but easy to show that such a smallest set indeed exists.

(1) $\bar{\mathfrak{T}}$ is an *extension* of T:
$\bar{\mathfrak{T}}(X, Y)$ contains all transitions from $\mathcal{D}$ which are of type (X, Y).

(2) $\bar{\mathfrak{T}}$ contains the *identity transitions*:
For every state X the set $\bar{\mathfrak{T}}(X, X)$ contains the expression id_X.

(3) $\bar{\mathfrak{T}}$ is *closed under sequential composition*:
If $f \in \bar{\mathfrak{T}}(A, B)$ and $g \in \bar{\mathfrak{T}}(B, C)$, then the expression[39] $(g \circ f)$ is in $\bar{\mathfrak{T}}(A, C)$.

(4) $\bar{\mathfrak{T}}$ is *closed under parallel composition*:
If $f \in \bar{\mathfrak{T}}(A, B)$ and $g \in \bar{\mathfrak{T}}(C, D)$, then the expression $(f \odot g)$ is in $\bar{\mathfrak{T}}(A \cdot C, B \cdot D)$.

Next, define for every pair (X, Y) of states in S° a binary relation $\sim_{X,Y}$ on the set $\bar{\mathfrak{T}}(X, Y)$ as the smallest relation which satisfies the following conditions:

(1) $\sim_{X,Y}$ is an *equivalence relation* on $\bar{\mathfrak{T}}(X, Y)$.

(2) $\sim_{X,Y}$ is a *congruence* with respect to $\circ$ and $\odot$:[40] For all transitions f_1, f_2, k in an arbitrary $\bar{\mathfrak{T}}(A, B)$, with $f_1 \sim_{X,Y} f_2$ the relations $(f_1 \odot k) \sim (f_2 \odot k)$, and $(k \odot f_1) \sim (k \odot f_2)$ hold[41]. If furthermore g and h are in such $\bar{\mathfrak{T}}(A, B)$ sets that the typing condition for forming the composition is fulfilled, then also $(f_1 \circ g) \sim (f_2 \circ g)$, $(h \circ f_1) \sim (h \circ f_2)$.

(3) $\sim_{X,Y}$ contains the *equations* which are required for first order transition systems by Definition 131: If the variables f, g, h, k are instantiated by expressions from $\bar{\mathfrak{T}}(A, B)$ sets, if L and R represent the instantiated left and right sides of an equation which must hold for a first order transition system, and if in all occuring terms the typing condition for forming the composition is fulfilled, then $L \sim R$.

For every pair (X, Y) of states define the set $\mathfrak{T}(X, Y)$ as the set of all equivalence classes of $\bar{\mathfrak{T}}(X, Y)$ with respect to $\sim_{X,Y}$. Since $\sim_{X,Y}$ is a congruence with respect to $\circ$ and $\odot$, we may define $\circ$ and $\odot$ in a canonical[42] way, e.g. $[t] \odot [u] := [(t \odot u)]$.

The resulting first order transition system $(S, \mathfrak{T}, \circ, \odot)$ is called the *first order transition system freely generated by the distributed transition system* $(S, T, \longrightarrow)$. ◂

In first order systems, a transition may be decomposed in more than one way. We therefore have to adapt Definition 40.

133 Definition DECOMPOSABLE AND ATOMIC TRANSITIONS
Let $\mathcal{F} = (S, \mathfrak{T}, \circ, \odot)$ be a first order transition system. A transition $t \in T$ is called *decomposable*, iff there exist $t_1, t_2 \in T$ which both are different from identity transitions, such that $t = t_2 \circ t_1$ or $t = t_1 \odot t_2$. A transition $t \in T$ which is not decomposable, is called *atomic*. ◂

The distributed transition system which freely generates a first order transition system, may be reconstructed easily:

[39] In the sense of a text string.

[40] Since $\circ$ and $\odot$ are not operations on the set $\sim_{X,Y}$ and since $\circ$ is only partially defined, we will provide the details for this somewhat non-standard congruence relation.

[41] $\sim$ must be supplied with indices of the respective types.

[42] This is a standard construction in algebraic specification – compare the reading guide in Section 1.5 for the references.

134 Definition RECONSTRUCTION OF GENERATOR
Let $\mathcal{F} = (S, \mathfrak{T}, \circ, \odot)$ be a first order transition system. The generator of $\mathcal{F}$ is the distributed transition system $\mathcal{G} = (S, T, \longrightarrow)$ obtained from $\mathcal{F}$ by restricting the sets $\mathfrak{T}$ of transitions to atomic transitions. ◀

3.9 Fairness

In distributed transition systems, fairness is a similar phenomenon as in ordinary transition systems. Since the choices are based on local states and can be made in parallel, a more elaborate approach is required. It will turn out that also the relative speed in which the various components of a distributed system are operating plays an important role in the analysis of fairness.

135 Example COOPERATION
An agent may cooperate with two clients A and B. If a client is not serviced for some time, he may temporarily enter a suspended state. The agent offers his cooperation alternatively to A and B, but it may only serve a client once.

The system has the following states:

$$S = \{\texttt{may\_a},\ \texttt{may\_b},\ \texttt{do\_a},\ \texttt{do\_b},\ \texttt{sus\_a},\ \texttt{sus\_b},\ \texttt{did\_a},\ \texttt{did\_b}\}$$

These states have the following significance:

State	Interpretation
may_a	Agent is ready to cooperate with A
do_a	Client A requesting the agent
sus_a	Client A is suspended
did_a	Cooperation with A was performed

For client B this is analogous. The following transitions govern the behaviour of our system.

$$\texttt{do\_a} \cdot \texttt{may\_a} \longrightarrow \texttt{did\_a} \qquad \texttt{do\_b} \cdot \texttt{may\_b} \longrightarrow \texttt{did\_b}$$

represents a successfull cooperation between a client and the agent.

$$\texttt{may\_a} \longrightarrow \texttt{may\_b} \qquad \texttt{may\_b} \longrightarrow \texttt{may\_a}$$

describes that the agent alternatively offers its cooperation to both clients.

$$\texttt{do\_a} \longrightarrow \texttt{sus\_a} \qquad \texttt{do\_b} \longrightarrow \texttt{sus\_b}$$

models the fact that the client may enter a suspension or waiting state, and

$$\texttt{sus\_a} \longrightarrow \texttt{do\_a} \qquad \texttt{sus\_b} \longrightarrow \texttt{do\_b}$$

represents the return from the suspension state into the requesting state.

The process shall be initialized in state $\texttt{do\_a} \cdot \texttt{may\_b} \cdot \texttt{sus\_b}$.

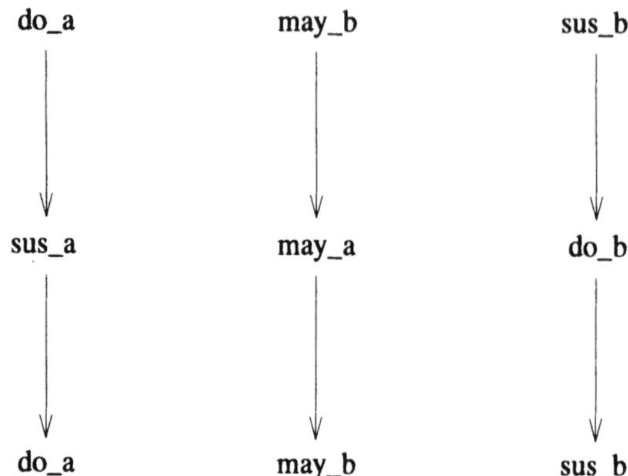

Figure 3.34: Agent and Clients.

This system *seems* to be rather simple, but already shows a large variety of conflicts, confusion and parallelism. The reader is invited to conduct an extensive analysis of the maximal behaviours and confusions of this process.

A very interesting behaviour is represented in Figure 3.34, where no cooperation occurs. Two mechanisms, ***unfair choice*** and ***differing computing speeds***, can be made responsible for this. For the following discussion these mechanisms are ***a posteriori explanations***, which *we* may take as reasons for this behaviour. Of course we are unable to say, what "really" did cause this outcome.

If we claim ***unfair choice*** as the reason, we have the following scenario in mind: From the initial state `do_a·may_b·sus_b` client B transitions from suspension (`sus_b`) to request (`do_b`). In the resulting state `do_a · may_b · do_b` there exists a conflict[43] between the cooperation of the agent and the client B, and between a change of the offer of the agent from client A to client B. If this conflict is resolved in favour of the change of the agent, then the next state is `do_a·may_a·do_b`. Now a similar situation is present, this time for client A. Again assume that the conflict is resolved in favour of a non-cooperating transition. For example, client A may go into suspension.

Conclusion of the ***unfair choice explanation:*** In the maximal behaviour depicted in Figure 3.34 no cooperation took place, because there repeatedly was a non-deterministic choice which offered cooperation as one alternative, but which was always resolved in favour of the non-cooperating transition. Therefore the reason is unfair choice.

If we claim ***differing computing speeds*** or ***race conditions*** as the reason, we have the following scenario in mind: From the initial state `do_a · may_b · sus_b` the system transitioned through the following sequence of states:

[43]There is a third conflicting transition, by which client B changes to suspension again and there is another enabled transition for client A. We will forget about them for a moment, since the purpose of this example is not the *extremely* tedious complete analysis of this process but a motivation for what follows.

`do_a · may_b · sus_b`

`sus_a · may_b · sus_b`

`sus_a · may_a · sus_b`

`sus_a · may_a · do_b`

`sus_a · may_a · sus_b`

`sus_a · may_b · sus_b`

`do_a · may_b · sus_b`

Then this is repeated from the beginning. In all these states there *never* was a *conflict*. Therefore fairness[44] cannot be the reason that no cooperation takes place.

Conclusion of the *differing computing speeds explanation:* The computing speeds of the agents responsible for the three vertical "threads of control" in Figure 3.34 has the effect that a cooperation transition never is enabled.

By taking three fingers and running them down along the three vertical "lines" in Figure 3.34 at the *adequate speed*, the reader is able to reconstruct both scenarios.

In above discussion we always worked with the notion of a global state. Strictly speaking, this approach and interpretation of distributed systems is not correct, but it may serve as an *a posteriori explanation*. ◀

Intuitively, we want to call the above behaviour, in which no cooperation took place, an unfair behaviour: Something that could happen, did not happen. It is not important, whether this was due to the "wrong" choice made between transitions or due to the "wrong" choice of computing speeds.

136 Definition WEAK AND STRONG FAIRNESS
Let $\mathcal{P}$ be a process and $\mathcal{B}$ a behaviour.

Behaviour $\mathcal{B}$ is called

> *weakly fair with respect to transition t*, iff there exists no sequence $C_1, C_2, C_3, \ldots$ of maximal cuts of $\mathcal{B}$ with C_{i+1} being an immediate successor of C_i with respect to the cut order so that t is enabled by all the cuts C_i but is not contained in the transitions executing between C_1 and C_i.
>
> *strongly fair with respect to transition t*, iff there exists no sequence $C_1, C_2, C_3, \ldots$ of maximal cuts of $\mathcal{B}$ for which C_{i+1} is a successor of C_i with respect to the cut order so that t is enabled by all the cuts C_i but t is not contained in the transitions executing between C_1 and C_i. ◀

[44] or rather the lack of it

The following example demonstrates the role of the execution speeds of the individual components of a distributed system.

137 Example CLOCK
Let a system consist of a clock which can execute the transitions `tick` and `setzero`, and of a second processor which non-deterministically manipulates the input pin which decides on the operating mode of the clock. The clock is deterministic and can be modeled by the following transitions which exist for $\mathtt{n} \in \mathbb{N}_0$:

$$\mathtt{clock(n)} \cdot \mathtt{run} \xrightarrow{\mathtt{tick}} \mathtt{clock(n+1)} \cdot \mathtt{run}$$

$$\mathtt{clock(n)} \cdot \mathtt{reset} \xrightarrow{\mathtt{setzero}} \mathtt{clock(0)} \cdot \mathtt{reset}$$

The second processor is non-deterministic and is modeled by the below transitions:

$$\mathtt{run} \xrightarrow{\mathtt{remain1}} \mathtt{run}$$

$$\mathtt{reset} \xrightarrow{\mathtt{remain2}} \mathtt{reset}$$

$$\mathtt{run} \xrightarrow{\mathtt{switch1}} \mathtt{reset}$$

$$\mathtt{reset} \xrightarrow{\mathtt{switch2}} \mathtt{run}$$

The system is initialized in the state $\mathtt{clock(0)} \star \mathtt{run}$.

In the described distributed transition system, there is always a non-deterministic conflict which is caused by the states `run` and `reset`. The system is quasi-sequential and must choose between executing a transition of the first or of the second processor. It is reasonable to assume that every processor repeatedly executes a transition. This can be achieved by considering only maximal behaviours which are fair with respect to the set {`remain1`, `remain2`, `switch1`, `switch2`} and the set {`tick`, `setzero`} of transitions. Thus, the second processor still is allowed to perform an unfair choice among its transitions, but it is guaranteed that eventually both processors execute a transition. ◀

Fairness with respect to sets of transitions, as it was used in the above example, is related to fairness of individual transitions in a similar way as for ordinary transition systems. Hence, Definitions 51 and 52 may easily be generalized to distributed transition systems, thus formalizing the above example.

Due to the lack of a global time, the notions of a *fairness delay* or of *uniform fairness* generally cannot be formulated for distributed systems.

The question whether a transition was enabled during a behaviour could be interpreted as being dependent on the relative execution speeds of the computing agents. Strictly speaking, this question is not well-defined: In an ordinary transition system, a transition is enabled if at a certain moment its precondition state is present. In a distributed system there is no global time according to which this condition may be verified.

In ordinary transition systems, fairness was interpreted as follows: If a transition repeatedly is enabled it will be executed eventually. In a distributed system, this interpretation is more complicated: If it is repeatedly possible to take snapshots of a system in whose observed states a transition is enabled then this transition will be executed eventually.

A more general intuitive interpretation of fairness could read as follows: Something that repeatedly may happen eventually happens. Above, the phrase *something may happen* was translated by *a single transition is enabled*[45]. In the following notion of *global fairness*, this phrase will be translated by *the distributed system has the possibility of behaving in a way that this transition executes*.

Following the tradition[46] of the grand master, this example will be veiled in allegory:

138 Example THE DISCUSSING PHILOSOPHERS
Two philosophers, they shall be called 1 and 2, are locked up in a monastery for the purpose of contemplation. They must obey the following rules:

- The philosophers may sleep, think, or wait for a discussion with the collegue.
- The philosophers may arbitrarily change between these three states.
- If both philosophers are ready for a discussion they may decide to speak to each other for a while after which they return into the state of being ready for a discussion.
- The philosophers must choose between the mentioned options in a strongly fair manner.
- The philosophers are sleeping when they are brought into the monastery.

This story can be described by the following transition system[47]:

```
1_sleep ⟶ 1_think     2_sleep ⟶ 2_think
1_sleep ⟶ 1_ready     2_sleep ⟶ 2_ready
1_ready ⟶ 1_sleep     2_ready ⟶ 2_sleep
1_ready ⟶ 1_think     2_ready ⟶ 2_think
1_think ⟶ 1_sleep     2_think ⟶ 2_sleep
1_think ⟶ 1_ready     2_think ⟶ 2_ready
```

The possibility of discussion is described by the following transitions:

```
1_ready · 2_ready ⟶ talking     talking ⟶ 1_ready · 2_ready
```

The system is initialized in the state `1_sleep · 2_sleep`.

It is easy to see that the philosophers eventually must talk to each other if they obey the above rules. Due to the fairness condition the execution thread of every individual philosopher must contain infinitely many states of all three kinds. Thus, it contains infinitely many cuts in

---

[45]In ordinary transition systems this was easy to define. In distributed transition systems this required the use of *snapshots*, i.e. cuts.

[46]Deadlocks were illustrated by E. W. DIJKSTRA using the famous example of the "dining philosophers". In the mean time a large number of similar amusing stories have been developed, for example the "drinking philosophers", the "cheating husbands", or the "Byzantine generals", in order to illustrate distributed algorithms.

[47]The names of the transitions are omitted for simplicity.

which a transition to talking is possible. Figure 3.35 illustrates an unfair maximal behaviour. Depending on the relative execution speeds of the philosophers an external observer may or may not see the transition to the talking state enabled. As a result of his observations he will pick one of the two possible explanations for unfair behaviour as outlined above.

After a while the philosophers are bored by their monotonous life style and they ask their supervisor for a televison set. Their request is granted but, for reasons of discipline, a number of rules are imposed:

- The philosophers obtain one television set and they may only watch TV together.
- If the philosophers watch TV then they must watch TV all day and they must not think or be ready for philosophical discussions on the same day.
- The philosophers must still choose between all their options in a strongly fair manner.

These additional rules may be implemented by the following transitions:

$$\texttt{1\_sleep} \cdot \texttt{2\_sleep} \longrightarrow \texttt{watch\_tv} \qquad \texttt{watch\_tv} \longrightarrow \texttt{1\_sleep} \cdot \texttt{2\_sleep}$$

As a result of this change the philosophers no longer must engage in discussions. Figure 3.36 illustrates a maximal behaviour which is strongly fair with respect to all transitions but in which the philosophers do not engage in discussions.

The reason for this is that the synchronization of the philosophers in the state `watch_tv` removes the possibility to form maximal cuts in which a transition to the talking state would be enabled. ◀

Although the principle idea of *global fairness* is rather simple, a formal definition is difficult to provide:

A behaviour $\mathcal{B}$ is *weakly globally fair* with respect to a transition t, if there is no sequence of immediate successors of maximal cuts of the behaviour so that the following property holds: Every maximal cut may be extended by a behaviour which contains the transition t but the behaviour $\mathcal{B}$ does not contain transition t in the future of any element of the sequence of maximal cuts.

Modal logic can be used to provide a better definition.

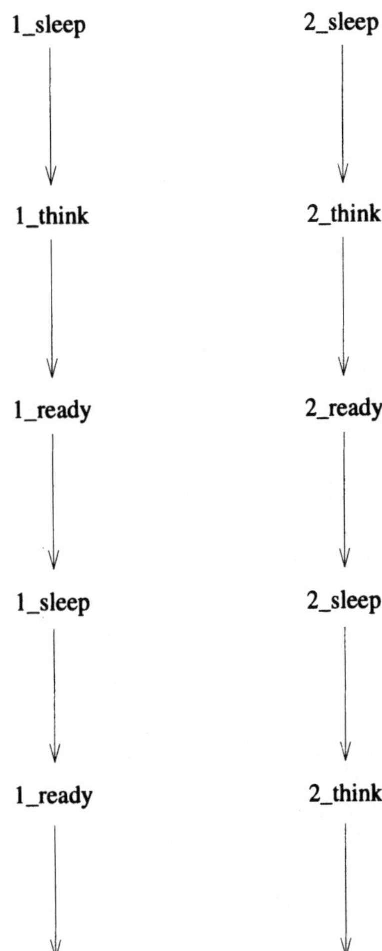

Figure 3.35: The Discussing Philosophers, Unfair Behaviour.

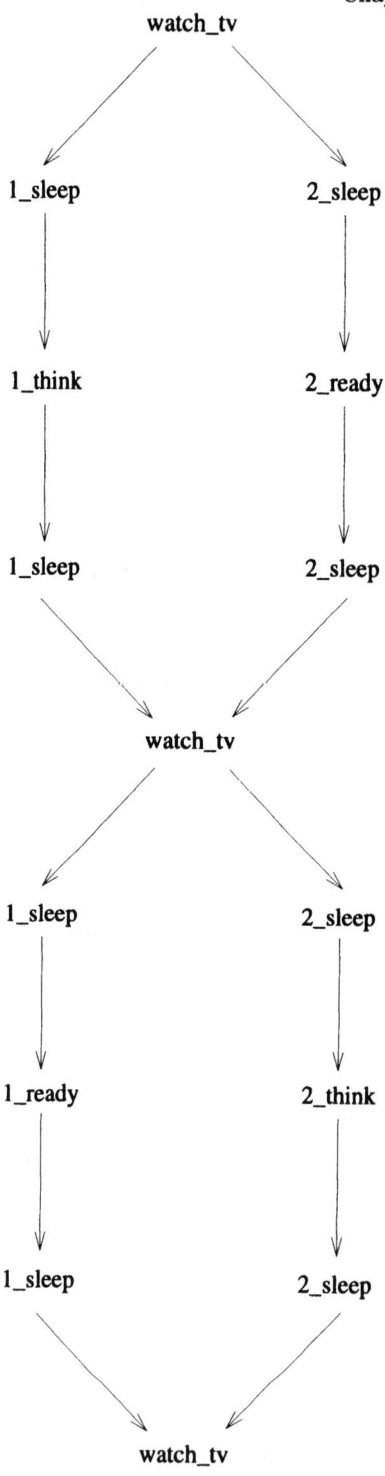

Figure 3.36: The Discussing Philosophers, Fair Behaviour.

Chapter 4

Higher Order Transition Systems

Beauty is the first test; there is no permanent place in the world for ugly mathematics.

G. H. HARDY,
A Mathematician's Apology.

Our theory of distributed processes, as it was presented in the previous chapters, is a *first* order theory: We can define states and state changes, but we cannot define *changes of state changes*. For this purpose, a *higher* order theory is required.

In some formalisms, the addition of higher order concepts increases their *power*. For example, the property that a graph is connected, or the property that it is planar, may only be defined in the framework of second order logic [Cou90a]. Similarly, the property that a program terminates, requires second order HOARE logic [Apt81]. Other formalisms do not gain in power, but in *expressiveness*. For example, a simple imperative programming language contains no higher order features at all, yet it can specify the same set of functions as a functional programming language which does offer many higher order constructions. In both cases, higher order concepts imply additional possibilities.

In this chapter, we will extend our process theory by *adding higher order constructs*.

In Section 1, the *motivations* and *design decisions* for higher order transition systems are presented.

Section 2 contains the *formal theory* of higher order transition systems.

In Section 3, higher order transitions which do not do "real work" but which *only raise and lower the degree of abstraction* are presented. Their job is some kind of book-keeping for the higher order mechanism.

Section 4 outlines how higher order transition systems may be used for *specifying processes*.

4.1 Motivation

In programming languages, the following features have some kind of higher order "touch":

(1) The possibility of passing ***pointers to functions*** as parameters to other functions.

(2) The concepts of ***lazy*** and ***delayed evaluation*** [FH80].

(3) The possibility of writing ***recursive specifications*** [Odi92].

(4) The possibility to ***compose*** programs, to ***store*** them as symbolic expressions and to ***apply*** them to data and/or other programs – ***inside*** of another program.

(5) The possibility of specifying a ***universal*** TURING ***machine*** and to write a ***partial evaluator*** [Coh87], [EL92], [Gom92].

(6) The possibility of writing an ***interpreter*** for a language in this language.

(7) The possibility to write ***self-modifying programs***.

Structurally, a ***first*** order calculus deals with ***objects*** and ***transformations*** of objects. ***Seconder*** order calculi also contain ***transformations of transformations*** of objects.

For example, let the objects be natural numbers and the transformations be functions of natural numbers.

139 Example HIGHER ORDER FUNCTIONS
The ***composition*** of two functions takes two functions $f : \mathbb{N}_0 \to \mathbb{N}_0$ and $g : \mathbb{N}_0 \to \mathbb{N}_0$ as arguments and returns their composition $g \circ f : \mathbb{N}_0 \to \mathbb{N}_0$. Hence, the composition ***transforms transformations*** of objects: It is a higher order construction.

Similarly, the ***partial evaluation*** of a function of two variables is a higher order construction. It takes as arguments a natural number n and a function $h : \mathbb{N}_0 \times \mathbb{N}_0 \to \mathbb{N}_0$ and returns as result the function $h(n, \bullet) : \mathbb{N}_0 \to \mathbb{N}_0$ which is defined by $m \mapsto [h(n, \bullet)](m) := h(n, m)$. ◀

In the case of our ***process calculus***, the objects are states and their transformations are transitions. The following ideas may not only be applied to process calculi but also to similar formalisms which deal with objects and transformations. For example, applying them to natural numbers and functions of natural numbers yields the concepts of higher order functional programming, or, depending on the employed level of abstraction, the theory of typed lambda calculi, or, even more abstract, the theory of cartesian closed categories.[1]

If we want to turn a ***first*** order calculus into a ***second*** order calculus, we must add ***transformations of transformations of objects*** as a third concept. For obtaining a ***third*** order calculus we now must add ***transformations of transformations of transformations of objects*** which leads to a fourth concept. This process continues, the higher the order of our calculus. Unfortunately, this approach will not provide us with a calculus of ***arbitrary*** higher order, since every order requires additional extensions which must be defined separately.

We therefore need a different approach: We will only use the two layers of abstraction, ***objects*** and ***transformations*** of objects, which already are present in the first order calculus.

[1] This theory is presented in Section C.8 of the Appendix.

We will add additional constructions, which may be used to *raise* and to *lower* the degree of abstraction by one.

140 Remark PROGRAMS, HIGHER ORDER STATES, AND EVALUATION
First, we enlarge our set of states by adding a *binary operation* $\twoheadrightarrow$. If A and B are states, then $A \twoheadrightarrow B$ denotes a *higher order state* which may be interpreted as a *program*. If this program is executed in state A, it consumes state A and produces the new state B. In this sense, applying a higher order state or program $A \twoheadrightarrow B$ to a state A has the same effect on A as executing a transition t with $A \xrightarrow{t} B$. The higher order state $A \twoheadrightarrow B$ may only be applied to state A, similarly as transition $A \xrightarrow{t} B$ only may execute, if state A is present.

In a transition system, this application of the higher order state $A \twoheadrightarrow B$ to the state A constitutes a *state change* and therefore will be described by a *transition*. Hence we require that for all states A and B there is an *evaluation transition ev*:

$$(A \twoheadrightarrow B) \cdot A \overset{ev_{A,B}}{\longrightarrow} B$$

Applying a higher order state $A \twoheadrightarrow B$ to a state A has the same effect on A as executing an *arbitrary* transition t with $A \xrightarrow{t} B$, but the state $A \twoheadrightarrow B$ lacks information on a specific transition. In this sense, $A \twoheadrightarrow B$ is an *internal representation* of the *entire set* of transitions $\mathfrak{T}(A, B)$ by a state of the system. ◀

In order to make some formulae more readable we will from now on also write $t : A \longrightarrow B$ to denote $A \xrightarrow{t} B$, which means that t has type (A, B) and hence is in $\mathfrak{T}(A, B)$.

141 Remark PARTIAL APPLICATION
A system is in state A and contains the transition $A \cdot B \xrightarrow{t} C$. In state A we may not apply transition t, since state B is missing. It is reasonable to assume that a *partial application* of t in state A yields the higher order state $B \twoheadrightarrow C$. When at some later time the missing state B is supplied, the evaluation transition consumes state B and generates state C, thereby completing transition t. If the missing state B is not supplied at any later time, then we will never get rid of the higher order state $B \twoheadrightarrow C$ again, indicating that the execution of transition t could not be completed. If t denotes the entire transition $A \cdot B \longrightarrow C$, then the partial transition which leads from state A to the higher order state $B \twoheadrightarrow C$ is denoted by $\Lambda_{A,B,C}(t)$.

Since $A \cdot B = B \cdot A$, t is a transition of type $A \cdot B \longrightarrow C$ as well as of type $B \cdot A \longrightarrow C$. Therefore, partial application may lead to the transition $\Lambda_{A,B,C}(t) : A \longrightarrow (B \twoheadrightarrow C)$ as well as to the transition $\Lambda_{B,A,C} : B \longrightarrow (A \twoheadrightarrow C)$. Hence, the *indices* of the abstraction function Λ are *relevant* and may be dropped only if they may be derived from the context.

$\Lambda_{A,B,C}$ is a function which takes as its argument a transition $t \in \mathfrak{T}(A \cdot B, C)$, whose execution requires the two states A and B. It returns as its value the transition $\Lambda_{A,B,C}(t) \in \mathfrak{T}(A, B \twoheadrightarrow C)$ which already may execute in state A. For all states A, B, C, $\Lambda_{A,B,C}$ is a function on transitions: $\Lambda_{A,B,C} : \mathfrak{T}(A \cdot B, C) \rightarrow \mathfrak{T}(A, B \twoheadrightarrow C)$. ◀

142 Remark EQUATION (η)
If a system is in state $A \cdot B$ and contains the transition $A \xrightarrow{u} (B \twoheadrightarrow C)$, then the transition

$$A \cdot B \overset{u \odot id_B}{\longrightarrow} (B \twoheadrightarrow C) \cdot B$$

may be formed. Appending the evaluation transition $ev_{B,C}$ yields

$$A \cdot B \overset{u \odot id_B}{\longrightarrow} (B \twoheadrightarrow C) \cdot B \overset{ev_{B,C}}{\longrightarrow} C$$

Therefore, transition u which was defined as a transition $A \longrightarrow (B \twoheadrightarrow C)$, gave rise to a state change $A \cdot B \longrightarrow C$. The associated transition was

$$A \cdot B \overset{ev_{B,C} \circ (u \odot id_B)}{\longrightarrow} C$$

Applying the function Λ to $ev_{B,C} \circ (u \odot id_B)$ yields the transition

$$A \overset{\Lambda(ev_{B,C} \circ (u \odot id_B))}{\longrightarrow} (B \twoheadrightarrow C)$$

Hence, the mechanism of partial application, technically described by Λ, generated a second transition $A \longrightarrow (B \twoheadrightarrow C)$ besides u. Therefore, in state A we would have a non-deterministic conflict between the two transitions u and $\Lambda(ev_{B,C} \circ (u \odot id_B))$. Since we do not want to introduce non-determinism by adding higher order constructions, we require that these two transitions are identical by imposing the equation (η):

$$\Lambda_{A,B,C}(ev_{B,C} \circ (u \odot id_B)) = u$$

◀

143 Remark EQUATION (β)
Similarly, if a system is in state $A \cdot B$ and contains the transition

$$A \cdot B \overset{t}{\longrightarrow} C$$

then we may also form its abstraction

$$A \overset{\Lambda(t)}{\longrightarrow} (B \twoheadrightarrow C)$$

and hence

$$A \cdot B \overset{\Lambda(t) \odot id_B}{\longrightarrow} (B \twoheadrightarrow C) \cdot B \overset{ev_{B,C}}{\longrightarrow} C$$

Therefore, we have obtained $ev_{B,C} \circ (\Lambda(t) \odot id_B)$ as a second transition of type $A \cdot B \longrightarrow C$. By a similar reasoning as above, we postulate the following equation (β):

$$ev_{B,C} \circ (\Lambda_{A,B,C}(t) \odot id_B) = t$$

◀

144 Remark t AND $\Lambda(t)$
If we specialize A to the neutral element ε of $\cdot$ we obtain $\Lambda : \mathfrak{T}(B,C) \to \mathfrak{T}(\varepsilon, B \twoheadrightarrow C)$. If t is a transition with $B \overset{t}{\longrightarrow} C$, then $\Lambda(t)$ is a transition so that

$$\varepsilon \overset{\Lambda_{\varepsilon,B,C}(t)}{\longrightarrow} (B \twoheadrightarrow C)$$

Therefore, we obtain

$$B = \varepsilon \cdot B \stackrel{\Lambda(t)\odot id_B}{\longrightarrow} (B \twoheadrightarrow C) \cdot B \stackrel{ev_{B,C}}{\longrightarrow} C$$

In this sense, the transitions t and $\Lambda(t)$ encode the same information. ◀

145 Remark $\twoheadrightarrow$ AND $\rightarrowtail$ FORM A FUNCTOR

In Sections 3.8 and 7.2 it is shown that the operations $\cdot$ and $\odot$ of a first order transition system form a *covariant-covariant bifunctor*[2]. From the transitions $A \stackrel{t}{\longrightarrow} B$ and $C \stackrel{u}{\longrightarrow} D$ we may form the transition $A \cdot C \stackrel{t\odot u}{\longrightarrow} B \cdot D$. The operation $\cdot$ is the *object part* of the functor and the operation $\odot$ is its *morphism part*.

For the new operation $\twoheadrightarrow$ on the states, we similarly would like to have an operation $\rightarrowtail$ on the transitions so that both operations $\twoheadrightarrow$ and $\rightarrowtail$ form another bifunctor. In the following, we will first study how the product bifunctor $(\cdot, \odot)$ may be interpreted. Then, by a similar line of reasoning, we will try to find a bifunctor $(\twoheadrightarrow, \rightarrowtail)$. First, this will yield a function Γ, which *formally* satisfies the transformation laws of a *contravariant-covariant bifunctor*[3], but which is not a bifunctor in the strict sense. Replacing the sets of transitions in the definition of Γ by their internal representation provides a contravariant-covariant bifunctor: This reasoning determines the variance and the interpretation of the exponential functor but does not uniquely define its values. Then we will argue that it is reasonable to require a *naturality condition*, which connects the product bifunctor $(\cdot, \odot)$, the abstraction function Λ, and the exponential bifunctor $(\twoheadrightarrow, \rightarrowtail)$. Finally, this condition will determine the values of $\rightarrowtail$.

Interpretation of the product bifunctor:
If A and B are states, then the state $A \cdot B$ is interpreted as a state which consists both of state A and state B. If $A \stackrel{t}{\longrightarrow} B$ and $C \stackrel{u}{\longrightarrow} D$ are transitions, then transition $t \odot u$ is a transition $A \cdot C \stackrel{t\odot u}{\longrightarrow} B \cdot D$, which operates both on A and C.

Looking for a bifunctor – and finding function Γ:
If $A \stackrel{t}{\longrightarrow} B$ and $C \stackrel{u}{\longrightarrow} D$ are transitions, then there does not necessarily exist a transition from A to D, since there is a gap between B and C. However, if there is a transition v with $B \stackrel{v}{\longrightarrow} C$ which closes the gap from B to C, then we may form the transition $A \stackrel{u\circ v\circ t}{\longrightarrow} D$, which closes the gap between A and D. Hence, the two transitions t and u give rise to a function $\Gamma(t,u) : \mathfrak{T}(B,C) \to \mathfrak{T}(A,D)$, defined by $\Gamma(t,u)(v) := u \circ v \circ t$. The function $\Gamma(t,u)$ tells us how a transition v which closes the gap between B and C may be transformed into a transition $\Gamma(t,u)(v)$ which closes the gap between A and D.

Γ *formally is a contravariant-covariant bifunctor*:
$\Gamma(t,u) : \mathfrak{T}(B,C) \to \mathfrak{T}(A,D)$ is a function on transitions. Γ itself is a function whose arguments are two transitions t and u and whose value is a function on transitions. It is easy to see that Γ depends on t and u in a functor-like way: $\Gamma(id_A, id_A) = id_{\mathfrak{T}(A)}$ and $\Gamma(t_1 \circ u_1, u_2 \circ t_2) = \Gamma(u_1, u_2) \circ \Gamma(t_1, t_2)$. For the latter, observe $\Gamma(t_1 \circ u_1, u_2 \circ t_2)(v) = (u_2 \circ t_2) \circ v \circ (t_1 \circ u_1) = u_2 \circ (t_2 \circ v \circ t_1) \circ u_1 = \Gamma(u_1,u_2)[(\Gamma(t_1,t_2)(v)] = [\Gamma(u_1,u_2) \circ \Gamma(t_1,t_2)](v)$. These laws have the same form as those required in Definition 345 for a contravariant-covariant bifunctor. However, Γ is *not* a bifunctor of the category since its image is a function between sets of transitions.

---

[2]Covariant-covariant bifunctors are introduced in Definition 344 of the appendix.

[3]Contravariant-covariant bifunctors are introduced in Definition 345 of the appendix.

Replacing transition sets by their internal representation:
According to Remark 140, the transition sets $\mathfrak{T}(B,C)$ and $\mathfrak{T}(A,D)$ which occur in the definition of Γ have an internal representation by the higher order states $B \twoheadrightarrow C$ and $A \twoheadrightarrow D$. We therefore are tempted to define $t \rightarrowtail u$ as a transition of the following type:

$$t \rightarrowtail u : (B \twoheadrightarrow C) \longrightarrow (A \twoheadrightarrow D) \qquad \text{or} \qquad B \twoheadrightarrow C \overset{t \rightarrowtail u}{\longrightarrow} A \twoheadrightarrow D$$

Therefore, $\rightarrowtail$ may be looked upon as an operation

$$\rightarrowtail : \mathfrak{T}(A,B) \times \mathfrak{T}(C,D) \to \mathfrak{T}(B \twoheadrightarrow C, A \twoheadrightarrow D)$$

If we furthermore require that the equations

$$id_A \rightarrowtail id_B = id_{A \twoheadrightarrow B}$$

$$(f_1 \circ g_1) \rightarrowtail (g_2 \circ f_2) = (g_1 \rightarrowtail g_2) \circ (f_1 \rightarrowtail f_2)$$

hold, then $(\twoheadrightarrow, \rightarrowtail)$ must be a bifunctor which is contravariant in its first and covariant in its second argument.

Interpretation of $\rightarrowtail$:
Two transitions $A \overset{t}{\longrightarrow} B$ and $C \overset{u}{\longrightarrow} D$ imply a *gap transformation mechanism*: For every transition which closes the gap between B and C, this mechanism provides a transition which closes the gap between A and D. Γ is this gap transformation mechanism for the transitions. Transitions, as we recognized above, are in a certain sense equivalent to higher order states. $\rightarrowtail$ provides this *gap transformation mechanism* represented in the form of higher order states by providing a transition from state $B \twoheadrightarrow C$, which is a state-based description for gap-closing transitions from B to C, to state $A \twoheadrightarrow D$, which is a state-based description for gap-closing transitions from A to D.

Naturality condition:
For the states A, B, and C consider the function $\Lambda_{A,B,C} : \mathfrak{T}(A \cdot B, C) \to \mathfrak{T}(A, B \twoheadrightarrow C)$ and similarly for states A', B', and C' consider the function $\Lambda_{A',B',C'} : \mathfrak{T}(A' \cdot B', C') \to \mathfrak{T}(A', B' \twoheadrightarrow C')$. If there exist transitions between the unprimed states A, B, C, and the primed states A', B' and C', then there are two ways of navigating from the set $\mathfrak{T}(A \cdot B, C)$ to the set $\mathfrak{T}(A', B' \twoheadrightarrow C')$, as outlined in the diagrams in Figure 4.1.

It is now reasonable to assume the following: If we start in above diagram with a transition $t \in \mathfrak{T}(A \cdot B, C)$ in order to obtain a transition $t' \in \mathfrak{T}(A', B' \twoheadrightarrow C')$, then the new transition t' should only depend on the transitions f, g and h, and not on the method which was used to determine t', i.e. the lower-left or the right-upper path in the diagram. This means that the following equation shall hold:

$$\Lambda_{A',B',C'}(h \circ t \circ (f \odot g)) = (g \rightarrowtail h) \circ \Lambda_{A,B,C}(t) \circ f$$

Compare Section C.8 and Figure C.23 for a categorical motivation of this condition. ◀

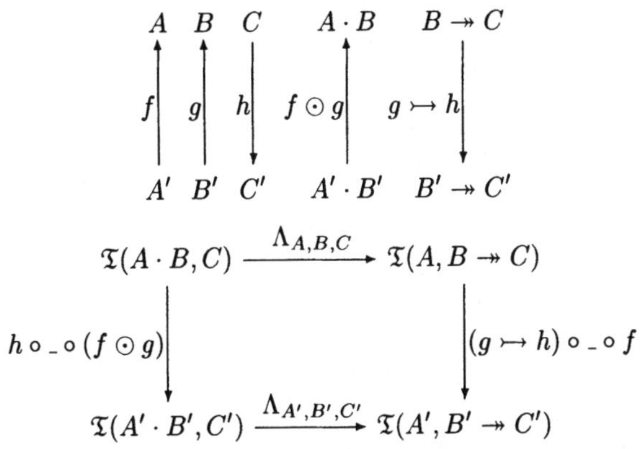

Figure 4.1: Naturality Condition for Higher Order Transition Systems.

4.2 Higher Order Transition Systems

In this section, the motivations and intuitions from the previous section are formalized.

146 Definition ALGEBRA OF HIGHER ORDER STATES
Let S be a set. The *algebra of higher order states of* S is defined as the initial algebra[4] $(\hat{S}, \cdot, \twoheadrightarrow, \varepsilon)$ which contains all elements of S and the element ε as constants, which contains the operations

$$\cdot : \hat{S} \times \hat{S} \to \hat{S}$$

$$\twoheadrightarrow : \hat{S} \times \hat{S} \to \hat{S}$$

and satisfies for all $X, Y, Z \in \hat{S}$ the equations

$$X \cdot \varepsilon = \varepsilon \cdot X = X$$

$$X \cdot Y = Y \cdot X$$

$$X \cdot (Y \cdot Z) = (X \cdot Y) \cdot Z$$

$\cdot$ is called the *product operator* for states and $\twoheadrightarrow$ is called the *exponential operator* for states. ◀

There are no equations for $\twoheadrightarrow$.

147 Definition HIGHER ORDER TRANSITION SYSTEM
A *higher order transition system* $\mathcal{H} = (S, \mathfrak{T}, \circ, \odot, \rightarrowtail, \Lambda)$ consists of

(1) a non-empty set S of *elementary states*, which generates the *algebra* $(\hat{S}, \cdot, \twoheadrightarrow, \varepsilon)$ *of higher order states*,

[4] Consult the reading guide in Section 1.5 for further references on initial algebras.

(2) for every pair $(X,Y) \in \hat{S} \times \hat{S}$ of states a set $\mathfrak{T}(X,Y)$ of *transitions*, and

(3) for all states $A, B, C, D \in \hat{S}$ the *operations*

$$\circ : \mathfrak{T}(B,C) \times \mathfrak{T}(A,B) \to \mathfrak{T}(A,C)$$

$$\odot : \mathfrak{T}(A,B) \times \mathfrak{T}(C,D) \to \mathfrak{T}(A \cdot C, B \cdot D)$$

$$\rightarrowtail : \mathfrak{T}(A,B) \times \mathfrak{T}(C,D) \to \mathfrak{T}(B \twoheadrightarrow C, A \twoheadrightarrow D)$$

$$\Lambda : \mathfrak{T}(A \cdot B, C) \to \mathfrak{T}(A, B \twoheadrightarrow C)$$

such that the following conditions are satisfied:

(fo) *First order transition system*:
$(S, \mathfrak{T}, \circ, \odot)$ is a first order transition system.

(ev) *Existence of evaluation transitions*:
For all states $A, B \in \hat{S}$ there exists an *evaluation transition* $ev_{A,B}$ from state $(A \twoheadrightarrow B) \cdot A$ to state B:

$$(A \twoheadrightarrow B) \cdot A \overset{ev_{A,B}}{\longrightarrow} B$$

(eqn) The following *equations* hold for $f : A \cdot B \longrightarrow C$ and $h : A \longrightarrow (B \twoheadrightarrow C)$:

$$(\beta) \qquad ev_{B,C} \circ (\Lambda_{A,B,C}(f) \odot id_B) = f$$

$$(\eta) \qquad \Lambda_{A,B,C}(ev_{B,C} \circ (h \odot id_B)) = h$$

(nat) Λ and $\rightarrowtail$ satisfy the *naturality condition*:
For every $t : A \cdot B \longrightarrow C$, $f : A' \longrightarrow A$, $g : B' \longrightarrow B$, $h : C \longrightarrow C'$ the following equation holds:

$$\Lambda_{A',B',C'}(h \circ t \circ (f \odot g)) = (g \rightarrowtail h) \circ \Lambda_{A,B,C}(t) \circ f$$

Λ is called the *abstraction function*, and $\rightarrowtail$ is the *exponential operator* for transitions. ◀

The equations (β) and (η) derive their names from lambda calculus, since they correspond to the (β) and (η) *conversion rules*, compare [Bar84]. In the following technical remark some type information for these equations is provided:

148 Remark TYPES FOR THE EQUATIONS (β) AND (η)
For (β) one obtains

$$f : A \cdot B \longrightarrow C$$

$$\Lambda_{A,B,C}(f) : A \longrightarrow (B \twoheadrightarrow C)$$

$$\Lambda_{A,B,C} \odot id_B : A \cdot B \longrightarrow (B \twoheadrightarrow C) \cdot B$$

$$ev_{B,C} \circ (\Lambda_{A,B,C}(f) \odot id_B) : A \cdot B \longrightarrow C$$

and for (η) one obtains

$$h : A \longrightarrow (B \twoheadrightarrow C)$$

$$h \odot id_B : A \cdot B \longrightarrow (B \twoheadrightarrow C) \cdot B$$

$$ev_{B,C} \circ (h \odot id_B) : A \cdot B \longrightarrow C$$

$$\Lambda_{A,B,C}(ev_{B,C} \circ (h \odot id_B)) : A \longrightarrow (B \twoheadrightarrow C)$$

◀

149 Lemma ABSTRACTION IS BIJECTIVE
For all states A, B, C the ***abstraction function*** $\Lambda : \mathfrak{T}(A \cdot B, C) \to \mathfrak{T}(A, B \twoheadrightarrow C)$ of a higher order transition system is a ***bijection***.

PROOF:
This follows immediately from the two equations for Λ. Λ is surjective due to $\Lambda_{A,B,C}(ev_{B,C} \circ (h \odot id_B)) = h$ and injective due to $ev_{B,C} \circ (\Lambda_{A,B,C}(f) \odot id_B) = f$. ◀

The ***abstraction*** of the ***evaluation*** is the ***identity***:

150 Lemma ABSTRACTION OF EVALUATION
In a higher order transition system the following equations hold:

$$\Lambda_{A\twoheadrightarrow B,A,B}(ev_{A,B}) = id_{A\twoheadrightarrow B}$$

$$\Lambda^{-1}_{A\twoheadrightarrow B,A,B}(id_{A\twoheadrightarrow B}) = ev_{A,B}$$

PROOF:
Specializing equation (η) by $h = id_{A\twoheadrightarrow B}$ one obtains

$$\Lambda(ev_{A,B}) = \Lambda(ev_{A,B} \circ id_{(A\twoheadrightarrow B)\cdot A}) =$$

$$= \Lambda(ev_{A,B} \circ (id_{A\twoheadrightarrow B} \odot id_A)) = id_{A\twoheadrightarrow B}$$

◀

The exponential operation $\rightarrowtail$ on the transitions is uniquely determined by the naturality condition.

151 Proposition REPRESENTATION OF $\rightarrowtail$
The exponential operation $\rightarrowtail: \mathfrak{T}(B, A) \times \mathfrak{T}(C, D) \to \mathfrak{T}(A \twoheadrightarrow C, B \twoheadrightarrow D)$ of a higher order transition system is uniquely determined by the following equation which holds for all transitions $g : B \longrightarrow A$ and $h : C \longrightarrow D$:

$$g \rightarrowtail h = \Lambda_{A\twoheadrightarrow C,B,D}(h \circ ev_{A,C} \circ (id_{A\twoheadrightarrow C} \odot g))$$

PROOF:
Immediately, by specializing the naturality condition using $t := ev_{A,C}$ and $f := id_{A \twoheadrightarrow C}$ and by Lemma 150. ◀

In category theory, this result is known in a technically different form as *adjunction with parameters*, compare [Mac88].

It is still necessary to show that $(\twoheadrightarrow, \rightarrowtail)$ is a contra-co bifunctor:

152 Proposition EXPONENTIAL OPERATIONS FORM A BIFUNCTOR
In a higher order transition system $\mathcal{H} = (S, \mathfrak{T}, \circ, \odot, \rightarrowtail, \Lambda)$ the pair $(\twoheadrightarrow, \rightarrowtail)$ is a contra-co bifunctor, i.e. the representation derived in Proposition 151 satisfies the following equations:

$$id_A \rightarrowtail id_B = id_{A \twoheadrightarrow B}$$

$$(f_1 \circ g_1) \rightarrowtail (g_2 \circ f_2) = (g_1 \rightarrowtail g_2) \circ (f_1 \rightarrowtail f_2)$$

PROOF:
By Proposition 151 and equations (β) and (η). ◀

Every distributed transition system *generates* a higher order transition system. To see this, add identity transitions, sequential and parallel compositions, the exponential operators, and the abstraction function to the distributed transition system. Then, identify those transitions, whose equality may be deduced from the above equations.

153 Definition FREELY GENERATED HIGHER ORDER TRANSITION SYSTEM
Let $\mathcal{D} = (S, T, \longrightarrow)$ be a typed, distributed transition system.

For every pair (X, Y) of higher order states in $\hat{S}$, define a set $\bar{\mathfrak{T}}(X, Y)$ of expressions[5] as the smallest set of expressions which satisfy the following conditions:

(1) $\bar{\mathfrak{T}}$ is an *extension* of T:
$\bar{\mathfrak{T}}(X, Y)$ contains all transitions from $\mathcal{D}$ which are of type (X, Y).

(2) $\bar{\mathfrak{T}}$ contains the *identity transitions*:
For every state X the set $\bar{\mathfrak{T}}(X, X)$ contains the expression id_X.

(3) $\bar{\mathfrak{T}}$ contains the *evaluation transitions*:
For every state X and Y the set $\bar{\mathfrak{T}}((X \twoheadrightarrow Y) \cdot Y, Y)$ contains the expression $ev_{X,Y}$.

(4) $\bar{\mathfrak{T}}$ is *closed under sequential composition*:
If $f \in \bar{\mathfrak{T}}(A, B)$ and $g \in \bar{\mathfrak{T}}(B, C)$, then the expression $(g \circ f)$ is in $\bar{\mathfrak{T}}(A, C)$.

(5) $\bar{\mathfrak{T}}$ is *closed under parallel composition*:
If $f \in \bar{\mathfrak{T}}(A, B)$ and $g \in \bar{\mathfrak{T}}(C, D)$, then the expression $(f \odot g)$ is in $\bar{\mathfrak{T}}(A \cdot C, B \cdot D)$.

(6) $\bar{\mathfrak{T}}$ is *closed under abstraction*:
If $f \in \bar{\mathfrak{T}}(A \cdot B, C)$, then the expression $\Lambda_{A,B,C}(t)$ is in $\bar{\mathfrak{T}}(A, B \twoheadrightarrow C)$.

Next, define for every pair (X, Y) of higher order states in $\hat{S}$ a binary relation $\sim_{X,Y}$ on the set $\bar{\mathfrak{T}}(X, Y)$ as the smallest relation which satisfies the following conditions:

[5] In the sense of a text string.

(1) $\sim_{X,Y}$ is an *equivalence relation* on $\bar{\mathfrak{T}}(X,Y)$.

(2) $\sim_{X,Y}$ is a *congruence relation* with respect to $\circ$, $\odot$, and Λ. The conditions for $\circ$ and $\odot$ are the same as in Definition 132. For Λ this means the following: If for $t, t' \in \bar{\mathfrak{T}}(A{\cdot}B, C)$ the relation $t \sim t'$ holds, then also $\Lambda_{A,B,C}(t) \sim \Lambda_{A,B,C}(t')$ holds.

(3) $\sim_{X,Y}$ contains the *equations* which are required for higher order transition systems by Definition 147: If the variables f, g, h, k are instantiated by expressions from the $\bar{\mathfrak{T}}(A,B)$, and if L and R represent the instantiated left and right sides of an equation which must hold[6] for a higher order transition systems, and if in all occuring terms the typing condition for forming the composition is fulfilled, then $L \sim R$ holds.

For every pair (X,Y) of states the set $\mathfrak{T}(X,Y)$ is defined as the set of all equivalence classes of $\bar{\mathfrak{T}}(X,Y)$ with respect to $\sim_{X,Y}$. Since $\sim_{X,Y}$ is a congruence with respect to $\circ$, $\odot$, and Λ it is possible to define $\circ$, $\odot$, and Λ in the canonical[7] way, e.g. $[t] \odot [u] := [(t \odot u)]$.

The operation $\rightarrowtail$ is defined by the representation from Proposition 151. As a consequence, the naturality condition holds.

The resulting higher order transition system $(S, \mathfrak{T}, \circ, \odot, \twoheadrightarrow, \Lambda)$ is called the *higher order transition system freely generated by the distributed transition system* $(S, T, \longrightarrow)$.

By a similar procedure, a higher order transition system $\mathcal{H} = (S, \mathfrak{T}, \circ, \odot, \rightarrowtail, \Lambda)$ can also be *freely generated from a suitable*[8] *relation* $\longrightarrow \ \subseteq \hat{S} \times T \times \hat{S}$. ◀

154 Example A SIMPLE HIGHER ORDER TRANSITION SYSTEM
Let $S = \{A, B\}$ be the set of elementary states of a system which is capable of a state change from A to B. This may be described by the *first order transition system* consisting of the set $T = \{t\}$ of transitions, and a relation $\longrightarrow \ \subseteq S^\circ \times T \times S^\circ$, defined by

$$A \xrightarrow{t} B$$

A *higher order description* of this system is provided by a relation $\longrightarrow \ \subseteq \hat{S} \times T \times \hat{S}$, defined by

$$\varepsilon \xrightarrow{u} (A \twoheadrightarrow B)$$

The resulting higher order transition systems contain the same state changes but the names of the respective transitions are different. ◀

4.3 Reversible and Invertible Transitions

In this section transitions which perform real work are distinguished from transitions which implement the higher order mechanisms and do not produce any further effect.

[6]Compare Definition 131 for first order systems and Definition 147 for higher order systems.

[7]This is a standard construction in algebraic specification – compare the reading guide in Section 1.5 for the references.

[8]For example, typing conditions must be satisfied, the elements of T must be different from the expressions id_A and $ev_{B,C}$, etc.

155 Example REVERSIBLE TRANSITIONS

Let $\mathcal{H}$ be the higher order transition system over the set $S = \{A, B\}$ of states, generated freely by the relation containing only $A \xrightarrow{t} B$ and $B \xrightarrow{u} A$. The compositions of these transitions have the following types:

$$A \xrightarrow{u \circ t} A \qquad B \xrightarrow{t \circ u} B$$

Neither $u \circ t$ nor $t \circ u$ is an identity transition on the respective state. Although the sequential composition $u \circ t$ imposes a state "change" from state A to the same state A, two transitions have executed and real work has been performed. ◀

We define reversible transitions as follows:

156 Definition REVERSIBLE TRANSITIONS

Let $\mathcal{H}$ be a higher order transition system. A transition $t : A \longrightarrow B$ is called *reversible*, iff there exists a transition $u : B \longrightarrow A$.

Two states A and B in a higher order transition system $\mathcal{H}$ are called *connected*, iff there exist transitions $t : A \longrightarrow B$ and $u : B \longrightarrow A$. ◀

Hence, two states are connected, iff there exists a reversible transition from one state to the other.

If two states are connected, then *no state information is lost* by a transition from one state to the other: After the transition $t : A \longrightarrow B$ it is possible to perform the transition $u : B \longrightarrow A$ which *restores the original state*.

For transitions $u : B \longrightarrow A$ and $t : A \longrightarrow B$, it is possible that $u \circ t = id_A$ and $t \circ u = id_B$ hold. In this case not only no state information is lost but also no real work is performed by the transitions since it is possible to perform another transition so that the entire effect equals the empty transition:

157 Definition ISO-TRANSITIONS

Let $\mathcal{H}$ be a higher order transition system. A transition $t : A \longrightarrow B$ is called *invertible* or an *iso-transition*, iff there exists a transition $u : B \longrightarrow A$, such that $u \circ t = id_A$ and $t \circ u = id_B$.

Two states are called *isomorphic*, iff there exists an iso-transition between these states. ◀

In the following it will be analyzed which transitions perform real work and which are invertible and thus do not perform real work.

The *evaluation transition* lowers the level of abstraction by applying a higher order state $A \twoheadrightarrow B$ to its "input state" B. Similarly, the *placeholder transition* raises the level of abstraction by generating a higher order state $B \twoheadrightarrow A \cdot B$ from the state A:

158 Definition PLACEHOLDER TRANSITION

Let $\mathcal{H}$ be a higher order transition system and let A and B be two states. Then the transition

$$pl_{A,B} : A \longrightarrow (B \twoheadrightarrow A \cdot B) \qquad pl_{A,B} = \Lambda_{A,B,A \cdot B}(id_{A \cdot B})$$

is called the *placeholder transition*. ◀

Both evaluation and placeholder perform real work in the sense of the iso-transition concept. Furthermore, every transition in the following sequence

$$A = A \cdot \varepsilon A \stackrel{id_A @ f}{\longrightarrow} A \cdot (A \twoheadrightarrow B) \stackrel{ev_{A,B}}{\longrightarrow} B$$

of transitions performs real work. Above, this work consists of the generation of the higher order state by the transition $f : \varepsilon(A \twoheadrightarrow B)$ and in its application to A. This can be shown more formally:

159 Proposition EVALUATION AND PLACEHOLDER
Let $\mathcal{H}$ be a higher order transition system which is freely generated by a relation.

In $\mathcal{H}$ *neither evaluation nor placeholder* transitions are *invertible*.

If the generating transition is empty, then they are not even reversible.

PROOF:
In Section 5.1, it will be demonstrated that linear logic provides a *correct and complete axiomatization* for higher order transition systems. In linear logic it is possible to show[9] that although the deductions $((A \multimap B) \star A) \vdash B$ and $A \vdash (B \multimap (A \star B))$ hold – they represent the evaluation and placeholder transitions in a sense which will be made more precise in Section 5.1 – the dual deductions $B \vdash ((A \multimap B) \star A)$ and $(B \multimap (A \star B)) \vdash A$ do not hold. Hence, in the higher order transition system which is generated by the empty relation, there are no transitions of the form $B \longrightarrow (A \twoheadrightarrow B) \cdot A$ nor $(B \twoheadrightarrow (A \cdot B)) \longrightarrow A$ and thus these transitions are neither reversible nor invertible.

If the freely generating relation is not empty, then there may exist transitions of this kind and then evaluation and placeholder may be reversible. However, there are no equations which connect evaluation and placeholder with these additional transitions in a way that the composition equals the identity transitions. ◀

160 Theorem ISOMORPHIC STATES
Let $\mathcal{H}$ be a higher order transition system. Then the states $A \cdot B \rightarrowtail C$ and $A \twoheadrightarrow (B \twoheadrightarrow C)$ are isomorphic.

PROOF:
By using the categorical interpretation of higher order processes it follows that Λ induces natural isomorphisms of the following form:

$$\mathfrak{T}(\varepsilon, A \cdot B \twoheadrightarrow C) \to \mathfrak{T}(A \cdot B, C) \to \mathfrak{T}(A, B \twoheadrightarrow C) \to \mathfrak{T}(\varepsilon, A \twoheadrightarrow (B \twoheadrightarrow C))$$

The dual version of the application of the YONEDA Lemma in Theorem 364 immediately provides an isomorphism. ◀

4.4 Forgetting the Higher Order Structure

A process is a distributed transition system. The higher as well as the first order transition systems provide only additional means of understanding processes. Hence, if a higher order transition system is generated by a relation, then this additional structure must be dropped in order to obtain a process.

[9]The proof that the respective deductions do not hold was carried out with the help of the PROLOG decision algorithm in [LS94].

161 Definition FIRST ORDER RESTRICTION
Let $\mathcal{H} = (S, \mathfrak{T}, \circ, \odot, \rightarrowtail, \Lambda)$ be a higher order transition system. The *first order restriction* of $\mathcal{H}$ is the first order transition system $(S, \mathfrak{T}^1, \circ^1, \odot^1)$. It is constructed over the same set of elementary states. All the other components defined by suitable restriction:

For $X, Y \in S^\circ$ the set of transitions is defined by $\mathfrak{T}^1(X,Y) = \mathfrak{T}(X,Y)$. The operations $\circ^1$ and $\odot^1$ are identical to their higher order counterpart except that they are only defined for transitions involving states in S° instead of $\hat{S}$. ◀

162 Definition SHADOW
Let S and T be sets, and let $\longrightarrow\ \subseteq \hat{S} \times T \times \hat{S}$ be a relation. The *shadow* of the relation $\longrightarrow$ is the generator of the first order transition system which was obtained as restriction of the higher order transition system generated freely by $\longrightarrow$. ◀

A shadow can also be formed from the transition relation $\longrightarrow$ of a higher order transition system.

In higher order transition systems, there are the three operations $\circ$, $\odot$, and $\rightarrowtail$ on transitions, all of which could be used to decompose a transition. Here, only the first order decomposition is of interest:

163 Definition FIRST-ORDER DECOMPOSABLE AND ATOMIC TRANSITIONS
Let $\mathcal{H} = (S, \mathfrak{T}, \circ, \odot, \rightarrowtail, \Lambda)$ be a higher order transition system. A transition $t \in T$ is called *first-order decomposable*, iff there exist $t_1, t_2 \in T$ which both are different from identity transitions, such that $t = t_2 \circ t_1$ or $t = t_1 \odot t_2$. A transition $t \in T$ which is not first-order decomposable is called *first-order atomic*. ◀

164 Example DECOMPOSABLE TRANSITIONS
Let $S = \{A, B, X\}$, $T = \{t\}$ and $t : X \longrightarrow (A \twoheadrightarrow B)$ freely generate a higher order transition system $\mathcal{H}$. This system contains, among others, the following transitions:

$$X \cdot A \xrightarrow{\Lambda^{-1}(t)} B$$

$$X \cdot A \xrightarrow{t \odot id_A} (A \twoheadrightarrow B) \cdot A \xrightarrow{ev_{A,B}} B$$

$$\varepsilon \xrightarrow{\Lambda(t)} (X \twoheadrightarrow (A \twoheadrightarrow B))$$

The transition $\Lambda^{-1}(t)$ is first-order decomposable, since due to equation (η) it is equal to the transition $ev_{A,B} \circ (t \odot id)$.

The first order restriction contains, among others, the following transitions:

$$X \cdot A \xrightarrow{\Lambda^{-1}(t)} B$$

$$X \cdot A \xrightarrow{t \odot id_A} (A \twoheadrightarrow B) \cdot A \xrightarrow{ev_{A,B}} B$$

$$X \cdot A \cdot A \xrightarrow{\Lambda^{-1}(t) \odot id_A} B$$

$$X \cdot A \cdot X \cdot A \xrightarrow{\Lambda^{-1}(t) \odot \Lambda^{-1}(t)} B \cdot B$$

The expressions $\Lambda^{-1}(t)$ and $ev_{A,B} \circ (t \odot id)$ still denote the same transition, but this transition is no longer decomposable. Its decomposition in the form $ev_{A,B} \circ (t \odot id)$ can no longer be made within a first order system, since neither of the two components are transitions in the respective first order system.

Finally, the shadow may be obtained by dropping all decomposable transitions. We thus obtain the distributed transition system $S = \{A, B, X\}$, $T = \{\Lambda^{-1}(t)\}$ and $X \cdot A \xrightarrow{\Lambda^{-1}(t)} B$. ◀

In the above example, *two* expressions for transitions from $X \cdot A$ to B denoted the same transition, due to equation (η). Could it still be possible, by making more extensive use of abstraction and evaluation, to construct a very complex expression for a transition from $X \cdot A$ to B which is not equal to $\Lambda^{-1}(t)$? If this were possible, then the higher order construction mechanism would *introduce non-determinism* beyond the non-determinism which already is contained in the generating relation. In order to avoid this, in Remarks 142, 143, and 145 the equations (β) and (η) and the naturality condition were added. Was this sufficient?

The following theorem shows that this was sufficient for state changes, the subsequent remark conjectures that this still is sufficient for transitions[10].

165 Theorem CONSERVATIVITY THEOREM FOR STATE CHANGES
Let $\mathcal{D}$ be a distributed transition system, let $\mathcal{H}$ be the higher order transition system which is freely generated by $\mathcal{D}$, and let $\mathcal{F}$ be the first order transition system which is the restriction of $\mathcal{H}$ to the first order. Then the generator of $\mathcal{F}$ contains a state change from A to B if and only if the distributed transition system $\mathcal{D}$ contains this state change.

PROOF:
Also the proof of this theorem follows directly from technical results which are developed in the subsequent chapter: The completeness and correctness of linear logic in Theorem 170 and the construction of the freely generated higher order transition system can be used to translate this question into linear logic. The logical version claims that the positive fragment of linear logic is a conservative extension of the tensor fragment. This is true by Theorem 261. ◀

We expect a similar result to hold for transitions:

166 Remark CONSERVATIVITY CONJECTURE FOR TRANSITIONS
Let $\mathcal{D}$ be a distributed transition system, let $\mathcal{H}$ be the higher order transition system, which is freely generated by $\mathcal{D}$, and let $\mathcal{F}$ be the first order transition system, which is the restriction of $\mathcal{H}$ to the first order. Then the generator of $\mathcal{F}$ contains a transition $t : A \longrightarrow B$ if and only if the distributed transition system $\mathcal{D}$ contains this transition.

The above Theorem 165 states that the higher order mechanism does not add new state changes. This conjecture states that the higher order mechanism even does not add new transitions between two states between which transitions already exist.

The linear logic Theorem 261 is not sufficient to prove this conjecture since it does not contain statements on transitions. The corresponding results in categorical linear logic which could provide a proof of the conservativity conjecture are known by the name of coherence theorems. Theorem 373 provides an example of such a result. [Jay90] has shown in Theorem

[10] In this statement transitions and state changes are distinguished: A ***transition*** is an element $t \in T$ such that $t : A \longrightarrow B$ holds, whereas a ***state change*** is a pair (A, B) for which a transition $t : A \longrightarrow B$ exists.

374 that the respective categorical coherence theorem does not hold. However, after a small but technically involved modification of Theorem 374, a coherence result can be obtained, compare [Jay90]. Therefore we still hope that this conjecture can be proved in the present or in a modified form. ◀

In the following example the factorial is specified using this higher order approach:

167 Example HIGHER ORDER SPECIFICATION OF FACTORIAL
Let the set of states be $S = \{\mathtt{fac(n)}, \mathtt{is(m)} \mid \mathtt{n}, \mathtt{m} \in \mathbb{N}_0\}$ and let $\mathcal{H}$ be the higher order transition system generated by the transitions $T = \{\mathtt{a}, \mathtt{s(n, m)} \mid \mathtt{n}, \mathtt{m} \in \mathbb{N}_0\}$ with

$$\mathtt{fac(0)} \xrightarrow{\mathtt{a}} \mathtt{is(1)}$$

$$(\mathtt{fac(n)} \twoheadrightarrow \mathtt{is(m)}) \xrightarrow{\mathtt{s(n,m)}} (\mathtt{fac(n+1)} \twoheadrightarrow \mathtt{is((n+1)m)}) \qquad \forall \mathtt{n}, \mathtt{m} \in \mathbb{N}_0$$

This transition system contains the following maximal behaviour:

$$\begin{aligned}\mathtt{fac(3)} &\xrightarrow{\Lambda_{\varepsilon,\,\mathtt{fac(0)},\mathtt{is(1)}}(\mathtt{a}) \odot id_{\mathtt{fac(3)}}} (\mathtt{fac(0)} \twoheadrightarrow \mathtt{is(1)}) \cdot \mathtt{fac(3)} \\ &\xrightarrow{\mathtt{s(0,1)} \odot id_{\mathtt{fac(3)}}} (\mathtt{fac(1)} \twoheadrightarrow \mathtt{is(1)}) \cdot \mathtt{fac(3)} \\ &\xrightarrow{\mathtt{s(1,1)} \odot id_{\mathtt{fac(3)}}} (\mathtt{fac(2)} \twoheadrightarrow \mathtt{is(2)}) \cdot \mathtt{fac(3)} \\ &\xrightarrow{\mathtt{s(2,2)} \odot id_{\mathtt{fac(3)}}} (\mathtt{fac(3)} \twoheadrightarrow \mathtt{is(6)}) \cdot \mathtt{fac(3)} \\ &\xrightarrow{ev_{\mathtt{fac(3)},\mathtt{is(6)}}} \mathtt{is(6)}\end{aligned}$$

It also contains the following behaviour:

$$\begin{aligned}\mathtt{fac(3)} &\xrightarrow{\Lambda_{\varepsilon,\,\mathtt{fac(0)},\mathtt{is(1)}}(\mathtt{a}) \odot id_{\mathtt{fac(3)}}} (\mathtt{fac(0)} \twoheadrightarrow \mathtt{is(1)}) \cdot \mathtt{fac(3)} \\ &\xrightarrow{\mathtt{s(0,1)} \odot id_{\mathtt{fac(3)}}} (\mathtt{fac(1)} \twoheadrightarrow \mathtt{is(1)}) \cdot \mathtt{fac(3)} \\ &\xrightarrow{\mathtt{s(1,1)} \odot id_{\mathtt{fac(3)}}} (\mathtt{fac(2)} \twoheadrightarrow \mathtt{is(2)}) \cdot \mathtt{fac(3)} \\ &\xrightarrow{\mathtt{s(2,2)} \odot id_{\mathtt{fac(3)}}} (\mathtt{fac(3)} \twoheadrightarrow \mathtt{is(6)}) \cdot \mathtt{fac(3)} \\ &\xrightarrow{\mathtt{s(3,6)} \odot id_{\mathtt{fac(3)}}} (\mathtt{fac(4)} \twoheadrightarrow \mathtt{is(24)}) \cdot \mathtt{fac(3)} \\ &\xrightarrow{\mathtt{s(4,24)} \odot id_{\mathtt{fac(3)}}} (\mathtt{fac(5)} \twoheadrightarrow \mathtt{is(120)}) \cdot \mathtt{fac(3)} \ \ldots\end{aligned}$$

◀

Chapter 5

Process Specification Formalisms

You can only find truth with logic
if you have already found truth without it.

G. K. CHESTERTON,
The Man who was Orthodox.

In this chapter, transition systems are connected with a *logical framework* and a *specification language*.

Section 1 shows, how a higher order transition system may be interpreted as a *model* for linear logic, and it defines when a formula from linear logic shall be considered *valid* in a higher order transition system. Finally, *correctness* and *completeness* results are established.

In Section 2, linear logic formulae are employed to *specify* higher order transition systems and processes.

In Section 3, some extensions to the present specification formalism are discussed.

Section 4 discusses equivalence notions for processes and illustrates how a concept for abstracting from implementational details can be developed.

This chapter assumes that the reader is familiar with linear logic to the extent of Appendix B.

5.1 Linear Logic and Transition Systems

The connection between linear logic and transition systems is based upon the following correspondence: Every *formula* φ in linear logic may be interpreted as a *state* $\mathcal{I}(\varphi)$ of a higher order transition system, and every *deduction* $\varphi \vdash \psi$ in linear logic may be interpreted as a *transition* from the state $\mathcal{I}(\varphi)$ to the state $\mathcal{I}(\psi)$.

This approach may be used to *specify* transition systems: If a transition system shall contain a transition from a state A to a state B then linear logic formulae α and β must be found such that $A = \mathcal{I}(\alpha)$ and $B = \mathcal{I}(\beta)$. The linear logic formula $\alpha \multimap \beta$ may be used as a non-logical axiom to specify the deduction $\vdash_{\{\alpha \multimap \beta\}}$, as outlined in Definition 299. As a consequence, the deduction $\alpha \vdash_{\{\alpha \multimap \beta\}} \beta$ holds which in turn corresponds to the existence of a transition from state $\mathcal{I}(\alpha)$ to state $\mathcal{I}(\beta)$.

Linear logic formulae are constructed on a set C of constants and transition systems are built around a set S of elementary states. Every function $\mathrm{i} : C \to S$ which interprets the constant symbols $\in C$ of linear logic by elementary states $\in S$ may be extended to a function $\mathcal{I} : \mathcal{L}_p[C] \to \hat{S}$ which interprets ***all*** linear logic formulae over the set C of constants as states over $\hat{S}$:

168 Definition STATE INTERPRETATION OF LINEAR LOGIC FORMULAE
Let C and S be sets, $\mathcal{L}_p[C]$ the set of all formulae of the positive fragment of linear logic[1] over C, and $\hat{S}$ the algebra of higher order states[2] over S. Finally let $\mathrm{i} : C \to S$ be an arbitrary function.

The ***interpretation function generated by*** i is the function $\mathcal{I} : \mathcal{L}_p[C] \to \hat{S}$ which is defined by structural induction over the generators of $\mathcal{L}_p[C]$:

$$\mathcal{I}(\mathbf{1}) = \varepsilon$$

$$\mathcal{I}(c) = \mathrm{i}(c) \qquad \text{for all } c \in C$$

$$\mathcal{I}(\varphi \star \psi) = \mathcal{I}(\varphi) \cdot \mathcal{I}(\psi) \qquad \text{for all } \varphi, \psi \in \mathcal{L}_p[C]$$

$$\mathcal{I}(\varphi \multimap \psi) = \mathcal{I}(\varphi) \twoheadrightarrow \mathcal{I}(\psi) \qquad \text{for all } \varphi, \psi \in \mathcal{L}_p[C]$$

A function $\mathcal{I} : \mathcal{L}_p[C] \to \hat{S}$ is called an ***interpretation function***, iff there exists a function $\mathrm{i} : C \to S$ such that $\mathcal{I}$ is the interpretation function generated by i. ◀

For every function i, there is exactly one interpretation function $\mathcal{I}$ which extends i from C to $\mathcal{L}_p[C]$ and satisfies the above equations.

169 Definition MODEL RELATION
Let C and S be sets, let $\mathrm{i} : C \to S$ be a function, and let $\mathcal{I} : \mathcal{L}_p[C] \to \hat{S}$ be the interpretation function generated by i. A linear logic formula $\varphi \in \mathcal{L}_p[C]$ is called ***valid*** in a higher order transition system $\mathcal{H} = (S, \mathfrak{T}, \circ, \odot, \rightarrowtail, \Lambda)$ with respect to the interpretation $\mathcal{I}$, iff there exists a transition t such that $\varepsilon \xrightarrow{t} \mathcal{I}(\varphi)$.

The ***model relation*** $\models$ is defined by:

$$\mathcal{H} \models \varphi \;\Leftrightarrow\; \exists t : \varepsilon \xrightarrow{t} \mathcal{I}(\varphi)$$

◀

The notion of validity as well as the model relation depend on the choice of the function i. The model relation may be extended to sets $V \subseteq \mathcal{L}_p[C]$ of formulae by

[1] See Definitions 254 and 259.
[2] See Definition 146.

$$\mathcal{H} \models V \;\Leftrightarrow\; \forall \varphi \in V : \mathcal{H} \models \varphi$$

In the following, $\vdash$ denotes the *pure deduction relation of the positive fragment* of linear logic from Definition 279. It is *correct and complete* with respect to the model relation $\models$:

170 Theorem CORRECTNESS AND COMPLETENESS FOR LINEAR LOGIC (1)
Let S and C be sets and let $\mathcal{I} : \mathcal{L}_p[C] \to \hat{S}$ be an interpretation function. Let all higher order transition systems be constructed over the set S and let all linear logic formulae be from the positive fragment constructed over the set C. Then the following statements hold:

(1) *Correctness*:
If a formula φ may be deduced from the unit element then it is valid in all higher order transition systems.

(2) *Completeness*:
If a formula φ is valid in *every* higher order transition system then it may be deduced from the unit element.

Both properties are expressed by the following formula:

$$(\forall \mathcal{H} : \mathcal{H} \models \varphi) \;\Leftrightarrow\; \mathbf{1} \vdash \varphi$$

PROOF:
First it is shown that

$$(\forall \mathcal{H} : \mathcal{H} \models \varphi) \;\Leftrightarrow\; \mathbf{1} \vdash \varphi$$

is equivalent to

$$(\forall \mathcal{H} : \mathcal{H} \models (\varphi \multimap \psi)) \;\Leftrightarrow\; \varphi \vdash \psi$$

Then this equivalent version will be proven.

Equivalence of claims:
$\Rightarrow$ direction: Specializing the assumption yields $(\forall \mathcal{H} : \mathcal{H} \models (\varphi \multimap \psi)) \Leftrightarrow \mathbf{1} \vdash \varphi \multimap \psi$. The deduction $\mathbf{1} \vdash \varphi \multimap \psi$ is equivalent to $\varphi \vdash \psi$ due to the deduction rules $(\multimap I)$ and $(\multimap I\,\mathrm{inv})$, and the neutrality of $\mathbf{1}$.

$\Leftarrow$ direction: Specializing the assumption yields $(\forall \mathcal{H} : \mathcal{H} \models (\mathbf{1} \multimap \psi)) \Leftrightarrow \mathbf{1} \vdash \psi$. By the definition of the model relation, $\mathcal{H} \models (\mathbf{1} \multimap \psi)$ holds, iff there exists a transition $t : \varepsilon \longrightarrow (\varepsilon \twoheadrightarrow \mathcal{I}(\psi))$. With the help of Λ^{-1} it can be shown that this is equivalent to the existence of a transition $u : \varepsilon \longrightarrow \mathcal{I}(\psi)$. Hence, $\mathcal{H} \models (\mathbf{1} \multimap \psi)$ is equivalent to $\mathcal{H} \models \psi$.

Correctness:
Since $\vdash$ is defined by structural induction it suffices to show the correctness of all induction rules.

The correctness of the *axiom* (Ax) with $\varphi \vdash \varphi$ follows from the fact that higher order transition systems contain identity transitions for every state: $id_{\mathcal{I}(\varphi)} : \mathcal{I}(\varphi) \longrightarrow \mathcal{I}(\varphi)$. Hence, $\Lambda(id_{\mathcal{I}(\varphi)}) : \varepsilon \longrightarrow (\mathcal{I}(\varphi) \twoheadrightarrow \mathcal{I}(\varphi))$, and therefore $\mathcal{H} \models (\varphi \multimap \varphi)$.

Now assume[3] that the (cut) rule is the last rule which was used for proving a deduction of the form $\gamma \star \delta \vdash \psi$, and that (cut) was applied to the deductions $\gamma \star \varphi \vdash \psi$ and $\delta \vdash \varphi$.

Assume by the principle of complete structural induction that the correctness for these deductions has already been shown, i.e. $\mathcal{H} \models (\gamma \star \varphi \multimap \psi)$ and $\mathcal{H} \models (\delta \multimap \varphi)$ for every $\mathcal{H}$. Hence, there exist transitions $\varepsilon \longrightarrow \mathcal{I}(\gamma \star \varphi \multimap \psi)$ and $\varepsilon \longrightarrow \mathcal{I}(\delta \multimap \varphi)$.

With the help of the abstraction function Λ and the equations for $\mathcal{I}$ the existence of transitions $f : \mathcal{I}(\gamma) \cdot \mathcal{I}(\varphi) \longrightarrow \mathcal{I}(\psi)$ and $g : \mathcal{I}(\delta) \longrightarrow \mathcal{I}(\varphi)$ may be derived. Hence, in $\mathcal{H}$ the following transition

$$\mathcal{I}(\gamma) \cdot \mathcal{I}(\delta) \stackrel{id_{\mathcal{I}(\gamma)} \odot g}{\longrightarrow} \mathcal{I}(\gamma) \cdot \mathcal{I}(\varphi) \stackrel{f}{\longrightarrow} \mathcal{I}(\psi)$$

may be constructed. Again, by using the abstraction function and the equations for $\mathcal{I}$, a transition $\varepsilon \longrightarrow \mathcal{I}(\gamma \star \delta \multimap \psi)$ may be constructed. Therefore $\mathcal{H} \models (\gamma \star \delta \multimap \psi)$.

For the other deduction rules, the proofs are similar. For the correctness of (P) the fact that $\hat{S}$ is a commutative and associative monoid is used, for $(1I)$ and $(1L)$ correctness can be shown very easily, for $(\star I)$ it is obtained by an application of the parallel composition $\odot$, for $(\star L)$ it is an immediate consequence of the definition of the pure deduction calculus, for $(\multimap I)$ the abstraction Λ is used. Only the correctness proof for $(\multimap L)$ is more elaborate and will be presented in the following:

Assume that the deductions $\gamma \vdash \varphi$ and $\delta \star \psi \vdash \vartheta$ have been proven correct, i.e. $\mathcal{H} \models (\gamma \multimap \varphi)$ and $\mathcal{H} \models ((\delta \star \psi) \multimap \vartheta)$. This means that there exist transitions $\varepsilon \longrightarrow (\mathcal{I}(\gamma) \twoheadrightarrow \mathcal{I}(\varphi))$ and $\varepsilon \longrightarrow ((\mathcal{I}(\delta) \cdot \mathcal{I}(\psi)) \twoheadrightarrow \mathcal{I}(\vartheta))$. By using the abstraction function, a transition $\mathcal{I}(\gamma) \longrightarrow \mathcal{I}(\varphi)$ can be constructed and by using parallel composition, identity transitions, and evaluation transition a transition

$$\mathcal{I}(\gamma) \cdot (\mathcal{I}(\varphi) \twoheadrightarrow \mathcal{I}(\psi)) \longrightarrow \mathcal{I}(\varphi) \cdot (\mathcal{I}(\varphi) \twoheadrightarrow \mathcal{I}(\psi)) \longrightarrow \mathcal{I}(\psi)$$

can be derived. From the assumptions and by using the abstraction function a transition $\mathcal{I}(\delta) \longrightarrow (\mathcal{I}(\psi) \twoheadrightarrow \mathcal{I}(\vartheta)$ can be derived. Finally, the existence of a transition

$$\mathcal{I}(\gamma) \cdot (\mathcal{I}(\varphi) \twoheadrightarrow \mathcal{I}(\psi)) \cdot \mathcal{I}(\delta) \longrightarrow \mathcal{I}(\psi) \cdot (\mathcal{I}(\psi) \twoheadrightarrow \mathcal{I}(\vartheta)) \longrightarrow \mathcal{I}(\vartheta)$$

may be derived.

Completeness:
Assume that for every higher order transition system $\mathcal{H}$ the relation $\mathcal{H} \models (\varphi \multimap \psi)$ holds, or, equivalently, contains a transition $\mathcal{I}(\varphi) \longrightarrow \mathcal{I}(\psi)$. Then this is true for the higher order transition system freely generated by the empty relation. This higher order transition system contains only those transitions which is required to contain by Definition 153. Therefore it is sufficient to show that every transition t which is generated by the structural induction rules (1) – (6) in Definition 153 gives rise to a corresponding deduction.

It is easy to see that the interpretation function $\mathcal{I} : \mathcal{L}_p[C] \to \hat{S}$ is surjective.

If t is the ***identity transition*** id_X of a state $X = \mathcal{I}(\varphi)$, then the deduction rule (Ax) can be used for proving $\varphi \vdash \varphi$.

[3]In fact, this proof step is superfluous, since the (cut) rule may be eliminated due to the Cut Elimination Theorem 280. We nevertheless present this proof step since it provides the interesting insight to which transitional mechanism the (cut) rule corresponds.

If t is the *evaluation transition* $ev_{X,Y} : (X \twoheadrightarrow Y) \cdot Y \longrightarrow Y$ for two states $X = \mathcal{I}(\varphi)$ and $Y = \mathcal{I}(\psi)$ then the corresponding deduction $(\varphi \multimap \psi) \star \varphi \vdash \psi$ can be derived from the rule (mp) in Theorem 287.

Let a transition $h : X \longrightarrow Z$ be derived by *sequential composition* of the transitions $f : X \longrightarrow Y$ and $g : Y \longrightarrow Z$. Assuming $X = \mathcal{I}(\alpha)$, $Y = \mathcal{I}(\beta)$, and $Z = \mathcal{I}(\gamma)$ for the states, and $\alpha \vdash \beta$ as well as $\beta \vdash \gamma$ as induction hypotheses, the deduction rule $(tra1)$ from Theorem 256 can be used for deriving the corresponding deduction $\alpha \vdash \gamma$.

For the *parallel composition* the deduction rule $(\star I)$ is employed and for the *abstraction* the deduction rule $(\multimap I)$ is used. ◀

More often, correctness and completeness are applied in the following form which is mentioned in the above proof:

171 Corollary CORRECTNESS AND COMPLETENESS, EQUIVALENT VERSION
Under the assumptions of Theorem 170, we obtain:

$$(\forall \mathcal{H} : \ \mathcal{H} \models (\varphi \multimap \psi)) \ \Leftrightarrow \ \varphi \vdash \psi$$

◀

In classical logic, completeness commonly is expressed in the following different form, [Ham78], [BM77]:

172 Remark COMPLETENESS IN CLASSICAL AND LINEAR LOGIC
In classical logic, *completeness* usually is defined as a statement of the following form: Assume that for every model $\mathcal{H}$ the following holds: Whenever $\mathcal{H} \models \varphi$, then $\mathcal{H} \models \psi$. Then it is possible to deduce $\varphi \vdash \psi$. More formally this may be written as:

$$[\forall \mathcal{H} : \ (\mathcal{H} \models \varphi) \Rightarrow (\mathcal{H} \models \psi)] \ \Rightarrow \ (\varphi \vdash \psi)$$

This form of completeness does *not* hold in linear logic, as may be seen from the following *counter-example*:

Let $a, b, c \in C$ be three constants. It is possible to deduce $\mathbf{1} \vdash (((a \multimap b) \star a) \multimap b)$ and $\mathbf{1} \vdash (((c \multimap d) \star c) \multimap d)$. Define $S := C$ and use the identity on this set as the function i which generates the interpretation function $\mathcal{I}$. By correctness, for every higher order transition system $\mathcal{H}$ the relations $\mathcal{H} \models (((a \multimap b) \star a) \multimap b)$ and $\mathcal{H} \models (((c \multimap d) \star c) \multimap d)$ hold. Therefore, both statements

$$\forall \mathcal{H} : \ \mathcal{H} \models (((a \multimap b) \star a) \multimap b) \ \Rightarrow \ \mathcal{H} \models (((c \multimap d) \star c) \multimap d)$$

and

$$\forall \mathcal{H} : \ \mathcal{H} \models (((c \multimap d) \star c) \multimap d) \ \Rightarrow \ \mathcal{H} \models (((a \multimap b) \star a) \multimap b)$$

are true. On the other hand, neither

$$(((a \multimap b) \star a) \multimap b) \vdash (((c \multimap d) \star c) \multimap d)$$

nor

$$(((c \multimap d) \star c) \multimap d) \vdash (((a \multimap b) \star a) \multimap b)$$

holds, as may be shown with the help of a PROLOG implementation of the decision procedure in Theorem 282.

In *classical logic*, both variants of the completeness property are equivalent. Here, the implication in

$$\forall \mathcal{H}: \ (\mathcal{H} \models \varphi) \Rightarrow (\mathcal{H} \models \psi)$$

may be *internalized* from the *meta-level* where it connects the statements $(\mathcal{H} \models \varphi)$ and $(\mathcal{H} \models \psi)$ into the *object-level* where it connects the classical logic formulae φ and ψ. This yields the following equivalence:

$$(\mathcal{H} \models \varphi \Rightarrow \mathcal{H} \models \psi) \text{ if and only if } (\mathcal{H} \models (\varphi \Rightarrow \psi))$$

This observation can be used to prove the equivalency of both cited forms of completeness for classical logic.

In *linear logic*, this internalization is not possible: The implication on the meta-level belongs to *classical logic* whereas the object level consists of *linear logic* which has a different kind of implication. However, the classical implication may be reconstructed in linear logic with the help of the exponential. ◀

Correctness and completeness results also hold for deduction systems with *non-logical axioms*:

173 Theorem CORRECTNESS AND COMPLETENESS, NON-LOGICAL AX.
Let A be a set of non-logical axioms, giving rise to the deduction relation[4] $\vdash_A$. The remaining assumptions are as in Theorem 170. Then the following *correctness* and *completeness* result holds:

$$[\forall \mathcal{H} : (\mathcal{H} \models A \Rightarrow \mathcal{H} \models \varphi)] \Leftrightarrow \mathbf{1} \vdash_A \varphi$$

PROOF:
For the $\Rightarrow$ direction we assume that $\forall \mathcal{H} : \ (\mathcal{H} \models A \ \Rightarrow \ \mathcal{H} \models \varphi)$. Then this holds for the higher order transition system $\mathcal{H}$ which is freely generated by the set $T := A$ of transitions and the transition relation $\longrightarrow \ \subseteq \hat{S} \times T \times \hat{S}$ which is defined by $\longrightarrow := \{(\varepsilon, a, \mathcal{I}(a)) \mid a \in A\}$. For this transition system $\mathcal{H}$ the relation $\mathcal{H} \models A$ holds and by assumption also $\mathcal{H} \models \varphi$. Hence, there exists a transition $\varepsilon \longrightarrow \mathcal{I}(\varphi)$. Since the state changes of $\mathcal{H}$ are obtained by an inductive definition, there is a finite sequence of applications of rules (1) – (6) from Definition 153 which proves the existence of a transition of type $\varepsilon \longrightarrow \mathcal{I}(\varphi)$. In the proof of Theorem 170 every rule was associated with a deduction rule of linear logic and thus this finite sequence of rule-applications may be translated into a proof for the deduction $\mathbf{1} \vdash_A \varphi$.

For the reverse direction $\Leftarrow$ assume that $\mathbf{1} \vdash_A \varphi$ and $\mathcal{H} \models A$. By Theorem 300 there exists a finite subset $\{\gamma_1, \gamma_2, \ldots, \gamma_n\} \subseteq A$ and $m_1, m_2, \ldots, m_n$ so that $\gamma_1^{m_1} \star \gamma_2^{m_2} \star \ldots \star \gamma_n^{m_n} \vdash \varphi$. $\mathcal{H} \models A$ and thus for every i also $\mathcal{H} \models \gamma_i$. As a consequence there exists a transition $\varepsilon \longrightarrow \mathcal{I}(\gamma_i)$ and by parallel composition also transitions $\varepsilon \longrightarrow (\mathcal{I}(\gamma_i))^{m_i}$. Thus, $\mathcal{H} \models \gamma_1^{m_1} \star \gamma_2^{m_2} \star \ldots \star \gamma_n^{m_n}$. By using correctness $\mathcal{H} \models \varphi$ is obtained. ◀

Theorem 300 and Proposition 301 show how $\vdash_A$ may be reduced to $\vdash$.

[4] Compare Definition 299 and Theorem 300.

174 Theorem CORRECTNESS AND COMPLETENESS FOR LINEAR LOGIC (2)
Under the assumptions of Theorem 170 the following correctness and completeness result may be shown:

$$[\forall \mathcal{H}: (\mathcal{H} \models \varphi \Rightarrow \mathcal{H} \models \psi)] \quad \Leftrightarrow \quad !\varphi \vdash \psi$$

In the above statement, $\vdash$ denotes the pure deduction for the *exponential fragment* of linear logic.

PROOF:
This follows directly from Theorems 170 and 300. ◀

As shown in [Ono90] and [Gir87a], the classical implication $\varphi \Rightarrow \psi$ may be *embedded* into linear logic by the construction $!\varphi \multimap \psi$. Therefore, the above theorem also illustrates the linear analogon of the classical internalization technique from remark 172.

Similar as for classical logic, the compactness result for the syntactic level, i.e. the deduction relation, may be transported to the semantic level, i.e. the model relation:

175 Theorem COMPACTNESS OF THE MODEL RELATION
Under the assumptions of Theorem 170 it is possible to deduce the following result:

If $\forall \mathcal{H} : (\mathcal{H} \models A \Rightarrow \mathcal{H} \models \varphi)$, then there exists a finite subset $F \subseteq A$, such that $\forall \mathcal{H}: (\mathcal{H} \models F \Rightarrow \mathcal{H} \models \varphi)$.

PROOF:
This follows from the correctness and completeness Theorem 173 and from the syntactic compactness Theorem 301. The choice of F may depend on φ. ◀

In contrast to classical logic, every set of linear logic formulae is *satisfiable*. Hence, also the concept of *consistency* which is the syntactic analogon of satisfiability is missing in linear logic.

176 Theorem SATISFIABILITY
For every set A of linear logic formulae of the positive fragment, there exists a higher order transition system $\mathcal{H}$, such that $\mathcal{H} \models A$.

PROOF:
Definition 153 shows how such a higher order transition system is constructed. ◀

The formulae in Theorem 174 motivate the following definition:

177 Definition CONSEQUENCE RELATION
Let S, C and $\mathcal{I}$ be given as in Theorem 170.

The *consequence relation* $\Vdash \subseteq \mathcal{L}_p[C] \times \mathcal{L}_p[C]$ is defined by

$$\varphi \Vdash \psi \Leftrightarrow \forall \mathcal{H}: (\mathcal{H} \models \varphi \Rightarrow \mathcal{H} \models \psi)$$

It may be extended to sets of formulae $A \subseteq \mathcal{L}_p[C]$ by

$$A \Vdash \psi \Leftrightarrow \forall \mathcal{H}: \mathcal{H} \models A \Rightarrow \mathcal{H} \models \psi$$

◀

178 Remark COMPLETENESS AND COMPACTNESS VIA CONSEQUENCE
Some of the previous statements may be elegantly reformulated using the consequence relation:

Completeness may be rephrased as

$$\varphi \Vdash \psi \;\Leftrightarrow\; !\varphi \vdash \psi$$

Compactness may be formulated in the following form: If $A \Vdash \psi$, then there exists a finite subset $F \subseteq A$, such that $F \Vdash \psi$. ◀

5.2 Linear Logic Process Specification

A ***process specification*** consists of two parts: In the ***data part***, the elementary states of the transition systems are specified. For this purpose, an algebraic specification is provided whose formal expressions yield the set C of constants of linear logic, and whose initial algebra provides the set S of elementary states. An interpretation function may be derived from the canonical function which interpretes every formal expression as element of the initial algebra.

In the ***process part***, the transitions of the process are specified. For this purpose, a set of labeled linear logic formulae is specified which define the transitions of a higher order transition system according to the above model relation.

As a result, a ***higher order transition system*** is obtained. Finally, the higher order constructions and the parallel and sequential compositions are "forgotten" in order to obtain a ***distributed transition system***. It represents the process which is specified by this process specification.

If a linear logic formula $\varphi \multimap \psi$ is valid in a higher order transition system then it contains ***at least one*** transition from state $\mathcal{I}(\varphi)$ to state $\mathcal{I}(\psi)$.

In order to distinguish several different transitions between the same two states this distinction must be provided already on the level of linear logic. For this purpose labels are added to the logic formulae:

179 Definition LABELED LINEAR LOGIC FORMULAE
Let C and L be sets.

A ***labeled linear logic formula*** is an expression $l : \varphi$, where $l \in L$ is a label and $\varphi \in \mathcal{L}_p[C]$ is a linear logic formula. A set A of labeled linear logic formulae is called ***strictly labeled***, iff there is no element $l \in L$ which is a label for more than one formula. ◀

180 Definition PROCESS SPECIFICATION
A ***process specification*** (Σ, R, E, A) consists of

(1) a ***signature*** $\Sigma = (\mathfrak{S}, \mathfrak{F}, \alpha)$,

(2) a set $R \subseteq \mathfrak{S}$ of sorts which represent the ***elementary states***,

(3) a set E of ***conditional equations*** over the signature Σ, such that the algebraic specification admits an initial model[5], and

(4) a ***strictly labeled*** set $A \subseteq \mathcal{L}_p[\Sigma_R]$ of ***linear logic formulae*** of the positive fragment and over the signature. ◀

[5]Compare Section 4 of [Wir90] on the existence of initial models in the case of conditional equations, and on other details on algebraic specification.

Above, the set Σ_R is the set of all expressions over the signature which are of a sort contained in the set R of sorts. It is used as the set of constants over which the linear logic formulae are constructed.

181 Definition SPECIFIED TRANSITION SYSTEM
The *higher order transition system* $\mathcal{H} = (S, \mathfrak{T}, \circ, \odot, \rightarrowtail, \Lambda)$ which is *specified* by the process specification from Definition 180 is given by the following construction:

(0) The signature Σ and the conditional equations E generate an *initial algebra*: For every sort $s \in \mathfrak{S}$ there is a carrier set $\mathcal{Z}_s$, and for every function symbol $f \in \mathfrak{F}$ there is a function F_f on the carrier sets, so that the equations in E are satisfied. Furthermore, the algebra shall be initial.

(1) The set S of elementary states contains all elements of the initial algebra whose sort is in R:

$$S = \bigcup_{r \in R} \mathcal{Z}_r$$

(2) The function i which is used for the construction of an interpretation $\mathcal{I}$ is the canonical interpretation of expressions over the signature as elements of the algebra:

$$\mathrm{i} : \bigcup_{r \in R} \Sigma_r \to \bigcup_{r \in R} \mathcal{Z}_r$$

(3) Let T be the set of all labels of formulae in A, and let $\longrightarrow\ \subseteq \hat{S} \times T \times \hat{S}$ be the set of all transitions $\varepsilon \xrightarrow{l} \mathcal{I}(\varphi)$, with $l : \varphi$ ranging over all labeled formulae in A. Then $\mathcal{H}$ shall be the higher order transition system freely generated by this relation.

◀

In the higher order transition system $\mathcal{H}$ specified by this process specification, all formulae φ in A are valid[6]. $\mathcal{H}$ contains no transitions with the exception of those which are required by A and by the fact that $\mathcal{H}$ is a higher order transition system. In algebraic specification, this approach is known by the name of *initial semantics*. In algebra, as well as in our process theory, also *lose semantics* could be used. Compare [EGLer] and [Wir90] on the respective algebraic theory.

In the following a *pocket calculator* will be specified which takes as input a state `add(n,m)`, with `n` and `m` natural numbers, and returns as an output state `is(k)`, where the natural number `k` is the sum of `m` and `n`. The state `add(n,n)` serves as a command to the calculator.

The theoretical foundations require a considerable formalism, since for several objects the syntactic and semantic levels must be distinguished. In practical applications, often a *light-weight formalism* is (ab)used in which several *different objects* are denoted by the *same symbol*. Therefore this notation should not be used in proofs, but it may be useful in practical calculations. In the following example, the full formalism will be used.

182 Example ADDITION
The *signature* is given by the set

[6] After removal of their label.

$$\mathfrak{S} = \{\texttt{nat},\ \texttt{input},\ \texttt{output}\}$$

of sorts and by the function symbols

$$\mathfrak{F} = \{\texttt{0},\ \texttt{succ},\ \texttt{add},\ \texttt{is}\}$$

with the following arities:

$$\alpha(\texttt{0}) = (\varepsilon, \texttt{nat})$$

$$\alpha(\texttt{succ}) = (\texttt{nat}, \texttt{nat})$$

$$\alpha(\texttt{add}) = (\texttt{nat} \times \texttt{nat}, \texttt{input})$$

$$\alpha(\texttt{is}) = (\texttt{nat}, \texttt{output})$$

Usually, this is specified by using the following *short-hand notation*:

$$\texttt{0} :\rightarrow \texttt{nat}$$

$$\texttt{succ} : \texttt{nat} \rightarrow \texttt{nat}$$

$$\texttt{add} : \texttt{nat} \times \texttt{nat} \rightarrow \texttt{input}$$

$$\texttt{is} : \texttt{nat} \rightarrow \texttt{output}$$

There shall be no equations.

This specification gives rise to an *initial algebra* with the following carrier sets:

$$\mathcal{Z}_{\texttt{nat}} = \mathbb{N}_0$$

$$\mathcal{Z}_{\texttt{input}} = \{\texttt{ADD(n,m)} \mid \texttt{n}, \texttt{m} \in \mathbb{N}_0\}$$

$$\mathcal{Z}_{\texttt{output}} = \{\texttt{IS(n)} \mid \texttt{n} \in \mathbb{N}_0\}$$

Note the distinction between "is" which denotes an *operator symbol*, and "IS" which is used to *construct the elements of the algebra*. This is essential for theoretical purposes, but for practical applications the symbols may be overloaded.

There are the following functions on the carrier sets:

$$F_0 : \{\bullet\} \rightarrow \mathcal{Z}_{\texttt{nat}} \qquad \texttt{n} \mapsto 0$$

$$F_{\texttt{succ}} : \mathcal{Z}_{\texttt{nat}} \rightarrow \mathcal{Z}_{\texttt{nat}} \qquad \texttt{n} \mapsto \texttt{n} + 1$$

$$F_{\texttt{add}} : \mathcal{Z}_{\texttt{nat}} \times \mathcal{Z}_{\texttt{nat}} \rightarrow \mathcal{Z}_{\texttt{input}} \qquad (\texttt{n}, \texttt{m}) \mapsto \texttt{ADD(n,m)}$$

$$F_{\texttt{is}} : \mathcal{Z}_{\texttt{nat}} \rightarrow \mathcal{Z}_{\texttt{output}} \qquad \texttt{n} \mapsto \texttt{IS(n)}$$

The elementary states shall be represented by the sorts $R = \{\texttt{input}, \texttt{output}\}$. Hence, the set of elementary states for the higher order transition systems $\mathcal{H}$ is $S = \mathcal{Z}_{\texttt{input}} \cup \mathcal{Z}_{\texttt{output}} = \{\texttt{ADD(n,m)}, \texttt{IS(k)} \mid \texttt{n,m,k} \in \mathbb{N}_0\}$.

Let Σ_s denote the set of all expressions of sort s which may be formed in the signature Σ. Examples for expressions in $\Sigma_{\texttt{output}}$ are $\texttt{is(0)}$ or $\texttt{is(succ(0))}$. Let $\Psi_s : \Sigma_s \to \mathcal{Z}_s$ be the *canonical epimorphism*[7] which interprets expressions of sort s as elements of carrier sets. For example

$$\Psi_{\texttt{nat}}(\texttt{succ(succ(succ(0)))}) = 3$$

$$\Psi_{\texttt{output}}(\texttt{is(succ(0))}) = \texttt{IS(1)}$$

This interpretation translates syntax, e.g. the symbol "is" in the above expression, into semantics, i.e. the algebra element denoted by the symbol "IS".

The set over which the linear logic formulae is constructed, is the set $C = \Sigma_{\texttt{input}} \cup \Sigma_{\texttt{output}}$ of all expressions over the signature which are of sort input or of sort output.

The strictly labeled set of linear logic formulae is given by

$$\{\texttt{r(n)} : \texttt{add(0,n)} \multimap \texttt{is(n)} \mid \texttt{n} \in \Sigma_{\texttt{nat}}\} \bigcup$$

$$\{\texttt{b(m,n)} : \texttt{add(succ(m),n)} \multimap \texttt{add(m,succ(n))} \mid \texttt{n,m} \in \Sigma_{\texttt{nat}}\}$$

The interpretation function $\mathcal{I}$ interprets these formulae as states of higher order transition systems. $\mathcal{H}$ satisfies this specification, if it contains the following transitions:

$$\texttt{ADD(0,n)} \overset{\Lambda^{-1}(r(\texttt{n}))}{\longrightarrow} \texttt{IS(n)} \qquad \forall \texttt{n} \in \mathbb{N}_0$$

$$\texttt{ADD(m+1,n)} \overset{\Lambda^{-1}(b(\texttt{m,n}))}{\longrightarrow} \texttt{ADD(m,n+1)} \qquad \forall \texttt{n,m} \in \mathbb{N}_0$$

Furthermore, $\mathcal{H}$ must contain all those transitions which are required by the definition of a higher order transition system. Examples for such transitions are the identity and the evaluation transitions for all states, and transitions which may be obtained by abstraction Λ, by sequential composition $\circ$, and by parallel composition $\odot$. The transitions generated by the exponential operation $\rightarrowtail$ can be expressed with the help of the former operations.

By forgetting the higher and first order structures as outlined in Definitions 161 and 162, a distributed transition system, i.e. a process, is obtained. This process represents the calculator. ◀

In the above example the specified process contains *exactly* the transitions $\Lambda^{-1}(\texttt{r(n)})$ and $\Lambda^{-1}(\texttt{b(m,n)})$ which already were "visible" in the set A of logic formulae: All these formulae were of the form $l : \gamma \multimap \delta$ with γ and δ being ***first order states***, i.e. being elements of S°.

In the next example we demonstrate how the higher order mechanism generates additional transitions, if more complicated logic formulae are provided:

[7]See introductions into algebraic specification as listed in the reading guide in Section 1.5.

183 Example ADDITION WITH HIGHER ORDER MECHANISM
This specification differs from the previous one only in the set A of linear logic formulae. Here, we will use the following set:

$$A = \{\mathtt{r(n)} : \mathtt{add(0,n)} \multimap \mathtt{is(n)} \mid \mathtt{n} \in \Sigma_{\mathtt{nat}}\} \bigcup$$

$$\{\mathtt{d(m,n,k)} : (\mathtt{add(m,n)} \multimap \mathtt{is(k)}) \multimap (\mathtt{add(succ(m),n)} \multimap \mathtt{is(succ(k))}) \mid$$

$$\mathtt{m,n,k} \in \Sigma_{\mathtt{nat}}\}$$

The higher order transition system which is defined by this specification contains, among others, the following transitions:

$$\varepsilon \xrightarrow{\mathtt{r(3)}} (\mathtt{ADD(0,3)} \twoheadrightarrow \mathtt{IS(3)})$$

$$\mathtt{ADD(0,3)} \xrightarrow{\Lambda^{-1}(\mathtt{r(3)})} \mathtt{IS(3)}$$

$$\varepsilon \xrightarrow{\mathtt{d(0,3,3)}} [(\mathtt{ADD(0,3)} \twoheadrightarrow \mathtt{IS(3)}) \twoheadrightarrow (\mathtt{ADD(1,3)} \twoheadrightarrow \mathtt{IS(4)})]$$

$$(\mathtt{ADD(0,3)} \twoheadrightarrow \mathtt{IS(3)}) \xrightarrow{\Lambda^{-1}(\mathtt{d(0,3,3)})} (\mathtt{ADD(1,3)} \twoheadrightarrow \mathtt{IS(4)})$$

$$(\mathtt{ADD(7,3)} \twoheadrightarrow \mathtt{IS(2)}) \xrightarrow{\Lambda^{-1}(\mathtt{d(7,3,2)})} (\mathtt{ADD(8,3)} \twoheadrightarrow \mathtt{IS(3)})$$

As sequential composition of the first and the fourth transition, we obtain

$$\varepsilon \xrightarrow{\mathtt{r(3)}} (\mathtt{ADD(0,3)} \twoheadrightarrow \mathtt{IS(3)}) \xrightarrow{\Lambda^{-1}(\mathtt{d(0,3,3)})} (\mathtt{ADD(1,3)} \twoheadrightarrow \mathtt{IS(4)})$$

and thus

$$\mathtt{ADD(1,3)} \xrightarrow{\Lambda^{-1}(\Lambda^{-1}(\mathtt{d(0,3,3)}) \circ \mathtt{r(3)})} \mathtt{IS(4)}$$

Also the distributed transition system which is obtained after the removal of the higher and first order structures contains those transformations. It was not "visible" in the set A of logic formulae, since it was calculated by the higher and first order mechanisms. In the distributed transition system, the transition $\Lambda^{-1}(\Lambda^{-1}(\mathtt{d}(0,3,3)))$ only has a meaning as a textual expression, since on the level of a process, Λ and $\circ$ have no other meaning than that of a graphical symbol. ◀

5.3 Extensions of the Formalism

In the previous section, *labels* were added to the linear logic formulae for distinguishing several transitions between the same two states. In the higher order transition system which is specified by a set of labeled linear logic formulae according to Definition 181, these labels were used as basis for the construction of transition expressions – according to the rules (1) – (6) from Definition 153. However, they do not appear in the model relation of Definition 169 nor in the deduction system of linear logic.

In the following, a deduction system and a model relation are developed which also deal with these labels. First the syntax of labeled linear logic , restricted to the positive fragment, must be defined:

184 Definition LABELED LINEAR LOGIC
$A := \{\star, \multimap, \mathbf{1}, (,)\}$ is the set of *elementary symbols* of linear logic. Let C be a set of so-called *theory constants* and L a set of so-called *basic labels*.

The language $\mathcal{L}_l[C, L]$ of *well formed formulae of labeled linear logic of the positive fragment over the sets C and L* is generated by the following rules:

First, the following context-free grammar defines the set of *unlabeled formulae*:

$$S \to \mathbf{1} \mid (SBS)$$

$$B \to \star \mid \multimap$$

For every constant $c \in C$, a rule of the form

$$S \to c$$

is added.

Next, the following inductive rules define the set of *label expressions*:

(1) For every unlabeled formula φ, the expression id_φ is a label expression.

(2) For every unlabeled formulae φ, ψ, the expression $ev_{\varphi,\psi}$ is a label expression.

(3) If t and u are label expressions, then also the expressions $(t \circ u)$, $(t \odot u)$, and $(t \rightarrowtail u)$ are label expressions. $\rightarrowtail$ is defined as an abbreviation of the representation given in Proposition 151.

(4) If t is a label expression and φ, ψ, ϑ are unlabeled formulae, then the expressions $\Lambda_{\varphi,\psi,\vartheta}(t)$ and $\Lambda^{-1}_{\varphi,\psi,\vartheta}(t)$ are label expressions.

A *labeled linear logic formula* is an expression $l : \varphi$, where l is a label expression and φ is a linear logic expression.

The set of all *label expressions* will be denoted by $\hat{L}_C$. ◀

Next, a *one-sided deduction calculus* $\vdash^1$ is defined for labeled linear logic. Neglecting technical details, the *motivation* for this deduction calculus is contained in the following correspondence:

The *labels* shall be interpreted as *transitions*. If $\varphi \vdash \psi$ holds in the pure and unlabeled deduction calculus, then there must exist a label expression l such that $\vdash^1 l : (\varphi \multimap \psi)$ holds in the labeled, one-sided calculus. In this sense, the label expressions provide information on the proof of the deduction. Similarly, if $\vdash^1 l : \varphi$ holds, then in the pure deduction calculus $\mathbf{1} \vdash \varphi$ must hold. Additionally, $\vdash^1 l : \varphi$ will correspond to $\varepsilon \xrightarrow{\mathcal{J}(l)} \mathcal{I}(\varphi)$ in the higher order transition system. $\mathcal{J}$ is an interpretation function for transitions.

The labeled one-sided deduction calculus provides a logical framework which is tailored to the higher order transition systems and takes care of the labels and transitions. In the following, this will be made more precise.

185 Definition ONE-SIDED DEDUCTION FOR LABELED LOGIC
The one-sided deduction system for labeled linear logic is the relation $\vdash^1 \subseteq \mathcal{L}[C, L]$ which is defined by the deduction rules of Table 5.1.

If $A \subseteq \mathcal{L}[C, L]$ is a set of labeled linear logic formulae then the deduction relation $\vdash^1_A$ with the set A of non-logical axioms is obtained similarly as in Definition 299. ◀

The label expressions must be interpreted by transitions. This can be done in the following way:

186 Proposition INTERPRETATION FUNCTION
Let C, S, L, and T be sets, $\mathcal{H} = (S, \mathfrak{T}, \circ, \odot, \rightarrowtail, \Lambda)$ a higher order transition system, $T = \cup_{X,Y \in \hat{S}} \mathfrak{T}(X, Y)$, $\mathrm{j} : L \to T$ a function, and $\mathcal{I} : \mathcal{L}[C] \to \hat{S}$ an interpretation. Then there exists a unique partial function $\mathcal{J} : \hat{L}_C \to T$ such that the following equations are satisfied:

$$\mathcal{J}(t) = \mathrm{j}(t) \qquad \text{for all } t \in L$$

$$\mathcal{J}(t \circ u) = \mathcal{J}(t) \circ \mathcal{J}(u) \text{ iff } \mathcal{J}(t) \circ \mathcal{J}(u) \text{ is defined in } \mathcal{H}$$

$$\mathcal{J}(t \circ u) \quad \text{is undefined, iff } \mathcal{J}(t) \circ \mathcal{J}(u) \text{ is not defined in } \mathcal{H}$$

$$\mathcal{J}(t \odot u) = \mathcal{J}(t) \odot \mathcal{J}(u)$$

$$\mathcal{J}(t \rightarrowtail u) = \mathcal{J}(t) \rightarrowtail \mathcal{J}(u)$$

$$\mathcal{J}(\Lambda_{\varphi,\psi,\vartheta}(t)) = \Lambda_{\mathcal{I}(\varphi),\mathcal{I}(\psi),\mathcal{I}(\vartheta)}(\mathcal{J}(t))$$

$$\mathcal{J}(\Lambda^{-1}_{\varphi,\psi,\vartheta}(t)) = \Lambda^{-1}_{\mathcal{I}(\varphi),\mathcal{I}(\psi),\mathcal{I}(\vartheta)}(\mathcal{J}(t))$$

◀

Note the symbol overloading in the above equations: On the left hand side of the equality sign, $\circ$, $\odot$, $\rightarrowtail$, Λ, and Λ^{-1} denote expressions, whereas on the right hand side they denote operations in the higher order transition system $\mathcal{H}$.

The function $\mathcal{J}$ whose existence is shown in Proposition 186 is called the *interpretation function for transitions*.

187 Definition MODEL RELATION FOR LABELED LOGIC
Let C, L and S, T be sets, and let $\mathrm{i} : C \to S$ and $\mathrm{j} : L \to T$ be functions. A labeled linear logic formula $l : \varphi \in \mathcal{L}_p[C, L]$ is called *valid* in a higher order transition system $\mathcal{H} = (S, \mathfrak{T}, \circ, \odot, \rightarrowtail, \Lambda)$ with respect to the interpretation functions $\mathcal{I}$ and $\mathcal{J}$, iff the following holds:

$$\frac{}{\vdash^1 \Lambda_{\varepsilon,\varphi,\varphi}(id_\varphi) : (\varphi \multimap \varphi)}\,(Ax)$$

$$\frac{}{\vdash^1 \Lambda_{\varepsilon,(\varphi \multimap \psi)\star\varphi,\psi}(ev_{\varphi,\psi}) : ((\varphi \multimap \psi) \star \varphi) \multimap \psi}\,(ev)$$

$$\frac{}{\vdash^1 \Lambda_{\varepsilon,\varphi\star(\psi\star\vartheta),(\varphi\star\psi)\star\vartheta}(id_{\varphi\star(\psi\star\vartheta)}) : \varphi \star (\psi \star \vartheta) \multimap (\varphi \star \psi) \star \vartheta}\,(ass)$$

$$\frac{}{\vdash^1 \Lambda_{\varepsilon,(\varphi\star\psi)\star\vartheta,\varphi\star(\psi\star\vartheta)}(id_{(\varphi\star\psi)\star\vartheta}) : (\varphi \star \psi) \star \vartheta \multimap \varphi \star (\psi \star \vartheta)}\,(ass^*)$$

$$\frac{}{\vdash^1 \Lambda_{\varepsilon,\varphi\star\gamma,\gamma\star\varphi}(id_{\varphi\star\gamma}) : \varphi \star \gamma \multimap \gamma \star \varphi}\,(com)$$

$$\frac{\vdash^1\!: l : \varphi}{\vdash^{:} \Lambda_{\varepsilon,\varepsilon,\varphi}(l) : \mathbf{1} \multimap \varphi}\,(add) \qquad \frac{\vdash^1 l : \mathbf{1} \multimap \varphi}{\vdash^1 \Lambda^{-1}_{\varepsilon,\varepsilon,\varphi}(l) : \varphi}\,(rem)$$

$$\frac{}{\vdash^1 \Lambda_{\varepsilon,\varphi\star\mathbf{1},\varphi}(id_\varphi) : \varphi \star \mathbf{1} \multimap \varphi}\,(\mathbf{1}R) \qquad \frac{}{\vdash^1 \Lambda_{\varepsilon,\varphi,\varphi\star\mathbf{1}}(id_{\varphi\star\mathbf{1}}) : \varphi \multimap \varphi \star \mathbf{1}}\,(\mathbf{1}L)$$

$$\frac{\vdash^1 t : \varphi \multimap \psi \qquad \vdash^1 u : \psi \multimap \vartheta}{\vdash^1 \Lambda_{\varepsilon,\varphi,\vartheta}(\Lambda^{-1}_{\varepsilon,\psi,\vartheta}(u) \circ \Lambda^{-1}_{\varepsilon,\varphi,\psi}(t))\varphi \multimap \vartheta}\,(\circ)$$

$$\frac{\vdash^1 t : \gamma \star \varphi \multimap \vartheta}{\vdash^1 \Lambda_{\varepsilon,\gamma,\varphi \multimap \vartheta}(\Lambda_{\gamma,\varphi,\vartheta}(\Lambda^{-1}_{\varepsilon,\gamma\star\varphi,\vartheta}(t)))}\,(\Lambda)$$

$$\frac{\vdash^1 t : \gamma \multimap (\varphi \multimap \vartheta)}{\vdash^1 \Lambda_{\varepsilon,\gamma\star\varphi,\vartheta}(\Lambda^{-1}_{\gamma,\varphi,\vartheta}(\Lambda^{-1}_{\varepsilon,\gamma,\varphi \multimap \vartheta}(t))) : \gamma \star \varphi \multimap \vartheta}\,(\Lambda^{-1})$$

Table 5.1: Labeled, One-Sided Deduction Rules.

$$\varepsilon \xrightarrow{\mathcal{J}(l)} \mathcal{I}(\varphi)$$

Thus, the *model relation* $\models^1$ is defined by:

$$\mathcal{H} \models^1 l : \varphi \;\Leftrightarrow\; \varepsilon \xrightarrow{\mathcal{J}(l)} \mathcal{I}(\varphi)$$

◀

By minor modifications of the above proofs for the unlabeled logic, the following theorems may be obtained:

188 Theorem CORRECTNESS AND COMPLETENESS FOR LABELED LOGIC
Let C, L, S, and T be sets, and $\mathcal{I}$ and $\mathcal{J}$ interpretation functions. Let all higher order transition systems be constructed over the set S and let all linear logic formulae be from the labeled positive fragment constructed over the sets C and L. Then the following statements hold:

$$\mathcal{H} \models^1 l : \varphi \text{ if and only if } \vdash^1 l : \varphi$$

Let A be a set of labeled[8] linear logic formulae.

$$\forall \mathcal{H} :\; \mathcal{H} \models^1 A \Rightarrow \mathcal{H} \models l : \varphi \text{ if and only if } \vdash^1_A l : \varphi$$

◀

Also the other correctness and completeness results from the previous section may be translated to this new setting.

The following observation connects the labeled and unlabeled versions of the formalism:

189 Theorem LABELED AND UNLABELED LOGIC
The pure, unlabeled deduction $\varphi \vdash \psi$ holds if and only if there exists a label expression l such that the one-sided, labeled deduction $\mathbf{1} \vdash^1 l : (\varphi \multimap \psi)$ holds. Furthermore, the deduction $\vdash^1 l : \varphi$ holds if and only if the deduction $\mathbf{1} \vdash \varphi$ holds.

Let A be a set of labeled linear logic formulae and let A' be the set of all formulae in A with the labels removed. Then the pure, unlabeled deduction $\varphi_{A'} \vdash \psi$ holds if and only if there exists a label expression l such that the one-sided, labeled deduction $\mathbf{1} \vdash^1_A l : (\varphi \multimap \psi)$ holds. Furthermore, the deduction $\vdash^1_A l : \varphi$ holds if and only if the deduction $\mathbf{1} \vdash_A \varphi$ holds. ◀

190 Remark QUANTIFIERS AS A SHORT-CUT
The above process specifications contain possibly infinite sets of labeled formulae. In many situations an infinite set can be described by universally quantifying the respective formulae. For example, the set

$$\{\mathtt{r(n)} : \mathtt{add(0,n)} \multimap \mathtt{is(n)} \mid \mathtt{n} \in \Sigma_{\mathtt{nat}}\} \bigcup$$

$$\{\mathtt{b(m,n)} : \mathtt{add(succ(m),n)} \multimap \mathtt{add(m,succ(n))} \mid \mathtt{n},\mathtt{m} \in \Sigma_{\mathtt{nat}}\}$$

[8] In the same sense as in Definition 180, i.e. the labels are elements of the set L and not label expressions. This is necessary to avoid problems as they are caused by the set $\{t : c \multimap d,\ u : e \multimap f,\ t \odot u : c \multimap (e \multimap f)\}$, with $c, d, e, f \in C$. Here, $t \odot u$ is of type $c \multimap (e \multimap f)$ whereas it should be of the type $c \star e \multimap d \star f$. In every higher order transition system the states by which these formulae are interpreted are different.

of labeled linear logic equations from Example 182 can also be represented by the two formulae

$$\forall \mathtt{n}: \ \mathtt{r}(\mathtt{n}) : \mathtt{add}(0,\mathtt{n}) \multimap \mathtt{is}(\mathtt{n})$$

$$\forall \mathtt{n},\mathtt{m}: \ \mathtt{b}(\mathtt{m},\mathtt{n}) : \mathtt{add}(\mathtt{succ}(\mathtt{m}),\mathtt{n}) \multimap \mathtt{add}(\mathtt{m},\mathtt{succ}(\mathtt{n}))$$

The universal quantifier introduces non-determinism since an arbitrary instance of the respective formula can be deduced. ◀

5.4 Equivalence and Abstraction for Processes

For a more abstract point of view, processes and higher order transition systems contain too many ***implementational details***. Definitions 10 and 63 demonstrated the abstraction from the concrete names of transitions and Definition 16 illustrated means of abstracting from the concrete names of states. In this section, these results are applied to higher order transition systems and process specifications. Furthermore, a technique for abstracting from implementational details is presented.

If two process specifications differ only in the employed labels, they give rise to related higher order transition systems:

191 Example DIFFERENT LABELS
Let the algebra consist of one sort, and of the constants a and b. For the first specification, the set of linear logic formulae shall be given by $A_1 = \{s : a \star a, r : a \star a \multimap b\}$, and for the second one by $A_2 = \{u : a \star a,\ r : a \star a \multimap b\}$.

The specifications and the defined higher order transition systems only differ in the concrete names which are used for the transition $\varepsilon \longrightarrow a \cdot a$. ◀

Two higher order transition systems may still be related in a similar way, even if the specifications differ also in the employed logic formulae:

192 Example DIFFERENT FORMULAE
Let the algebra consist again of one sort, and of the constants a and b. For the first specification, the set of linear logic formulae shall be given by $A_1 = \{s : a \star a,\ r : a \star a \multimap b\}$, and for the second by $A_2 = \{s : a \star a,\ t : a \multimap (a \multimap b)\}$.

Both specified higher order transition systems contain the same state changes, although the associated transitions may have different names:

First System	**Second System**
$r : \varepsilon \to (a \cdot a \twoheadrightarrow b)$	$\Lambda_{\varepsilon,a\cdot a,b}(\Lambda^{-1}_{a,a,b}(\Lambda^{-1}_{\varepsilon,a,a\twoheadrightarrow b}(t))) : \varepsilon \to (a \cdot a \twoheadrightarrow b)$
$\Lambda^{-1}_{\varepsilon,a\cdot a,b}(r) : a \cdot a \to b$	$\Lambda^{-1}_{a,a,b}(\Lambda^{-1}_{\varepsilon,a,a\twoheadrightarrow b}(t)) : a \cdot a \to b$
$\Lambda_{a,a,b}(\Lambda^{-1}_{\varepsilon,a\cdot a,b}(r)) : a \to (a \twoheadrightarrow b)$	$\Lambda^{-1}_{\varepsilon,a,a\twoheadrightarrow b}(t) : a \to (a \twoheadrightarrow b)$
$\Lambda_{\varepsilon,a,a\twoheadrightarrow b}(\Lambda_{a,a,b}(\Lambda^{-1}_{\varepsilon,a\cdot a,b}(r))) : \varepsilon \to (a \twoheadrightarrow (a \twoheadrightarrow b))$	$t : \varepsilon \to (a \twoheadrightarrow (a \twoheadrightarrow b))$

Hence, a bijection between the elements of the transition sets of these higher order systems may be constructed:

First System	Second System
r	$\Lambda_{\varepsilon,a\cdot a,b}(\Lambda^{-1}_{a,a,b}(\Lambda^{-1}_{\varepsilon,a,a\twoheadrightarrow b}(t)))$
$\Lambda^{-1}_{\varepsilon,a\cdot a,b}(r)$	$\Lambda^{-1}_{a,a,b}(\Lambda^{-1}_{\varepsilon,a,a\twoheadrightarrow b}(t))$
$\Lambda_{a,a,b}(\Lambda^{-1}_{\varepsilon,a\cdot a,b}(r))$	$\Lambda^{-1}_{\varepsilon,a,a\twoheadrightarrow b}(t)$
$\Lambda_{\varepsilon,a,a\twoheadrightarrow b}(\Lambda_{a,a,b}(\Lambda^{-1}_{\varepsilon,a\cdot a,b}(r)))$	t

◀

The above observations lead to the subsequent definitions:

193 Definition SIMILARITY
Let the sets C and S and the function $\mathrm{i} : C \to S$ be given.

Two higher order transition systems $\mathcal{H}_1$ and $\mathcal{H}_2$ over the same set S of elementary states are called ***similar***, iff the sets of linear logic formulae which are valid in $\mathcal{H}_1$ and $\mathcal{H}_2$, respectively, are equal.

Two process specifications are called ***similar***, iff their specified higher order transition systems are similar. ◀

194 Definition ISOMORPHIC HIGHER ORDER TRANSITION SYSTEMS
Two higher order transition systems $\mathcal{H}_1 = (S_1, \mathfrak{T}_1, \circ_1, \odot_1, \twoheadrightarrow_1, \Lambda_1)$ and $\mathcal{H}_2 = (S_2, \mathfrak{T}_2, \circ_2, \odot_2, \twoheadrightarrow_2, \Lambda_2)$ are called ***isomorphic***, iff they have the same set of states, i.e. $S_1 = S_2$, and iff for all states $X, Y \in \hat{S}_1$, there exists a bijection $\beta_{X,Y} : \mathfrak{T}_1(X,Y) \to \mathfrak{T}_2(X,Y)$ such that for all transitions for which the respective operations are defined the following equations hold[9]:

$$\beta(f \circ_1 g) = \beta(f) \circ_2 \beta(g)$$

$$\beta(f \odot_1 g) = \beta(f) \odot_2 \beta(g)$$

$$\beta(f \rightarrowtail_1 g) = \beta(f) \rightarrowtail_2 \beta(g)$$

$$\beta(\Lambda_1(f)) = \Lambda_2(\beta(f))$$

◀

The collection β of bijections respects the algebraic structure o the transitions.

In the above examples the specified higher order transitions systems had the same structure. In the following example, the specified higher order transition systems are different but the specified processes are equal:

195 Example DIFFERENT HIGHER ORDER TRANSITION SYSTEM
Let the algebra consist of one sort and of the constants a, b, c, and d. For the first specification, the set of linear logic formulae is given by $A_1 = \{s : a \multimap b, t : b \multimap d\}$ and for the second specification by $A_2 = \{s : a \multimap b, u : (a \multimap b) \multimap (c \multimap d)\}$.

[9] The indices for β are omitted for simplicity.

The specified higher order transition systems are different: The PROLOG implementation of the decision algorithm [LS94] and the correctness and completeness of linear logic can be used to show that the first higher order transition system does not contain a transition $(a \twoheadrightarrow b) \longrightarrow (c \twoheadrightarrow d)$ whereas the second one does.

However, the specified processes are equal. For both specifications they consist of the transitions $a \longrightarrow b$ and $c \longrightarrow d$. Hence, these two specifications differ only in their higher order component. ◀

This observation can be formalized in the following way:

196 Definition PROCESS-EQUIVALENT
Two higher order transition systems $\mathcal{H}_1$ and $\mathcal{H}_2$ are called ***process-equivalent***, iff their shadows are S-isomorphic processes. ◀

197 Example S-EQUIVALENCE
Let a process $\mathcal{P}_1$ be given over the set $\{a, b, c, d\}$ of elementary states by

$$a \xrightarrow{t} b \qquad b \xrightarrow{u} c \qquad c \xrightarrow{v} d$$

and let a process $\mathcal{P}_2$ be given over the set $\{a, d\}$ of elementary states by

$$a \xrightarrow{x} d$$

Neglecting the names of the transitions, the second process is equivalent to the first process if one considers only the set $\{a, d\}$ of states. Considering only this set of states, both processes have the same state changes, namely $a \longrightarrow d$.

The process $\mathcal{P}_3$ which is defined over the set $\{a, b, c, d\}$ by

$$a \xrightarrow{t} b \qquad b \xrightarrow{u} c \qquad c \xrightarrow{z} b \qquad c \xrightarrow{v} d$$

is equivalent to $\mathcal{P}_2$ in a weaker sense. In addition to the transition $a \longrightarrow d$, $\mathcal{P}_3$ also allows an infinite behaviour $a \longrightarrow b \longrightarrow c \longrightarrow b \longrightarrow c \longrightarrow b \longrightarrow \ldots$. Although this behaviour does not evolve in the considered state set $\{a, d\}$, it started in this set of states. Hence, an observer who is capable of observing the states $\{a, d\}$ notices that the processes $\mathcal{P}_1$ and $\mathcal{P}_2$ always terminate if they are initialized in state a. This is in contrast to $\mathcal{P}_3$ which after being initialized in state a has the possibility to engage in an infinite behaviour whose states cannot be observed. Such an infinite behaviour whose states cannot be observed due to abstraction is called diverging. ◀

In process algebraic calculi there exists an elaborate theory on abstraction from implementational details and on divergent processes. This theory comprises only *interleaving* parallelism and the notion of abstraction is based on *atomic actions*, i.e. transitions, compare [BW90a]. Our approach deals with *partial order* parallelism and the notion of *abstraction* is constructed based on *states*. A more refined theory of abstraction still must be developed for our process calculus.

The large number of notions for equivalence and abstraction, compare [Hüt94], [BW90a], [BR83], which presently are known for algebraic process theories suggest that first more experience with the present linear logic approach must be gathered before a reasonable approach to equivalence and abstraction can be made.

Chapter 6

Examples

There is nothing more practical than a good theory.

L. I. BREZHNEV,
Nature **270**(1977), 470.

In this chapter, we will demonstrate the use of our theory of processes for *small real world examples*. We will present a *specification language* and a *prototypical implementation* of a subset, and we will apply both to application processes.

This chapter demonstrates that our ideas *can be put to work in practical applications*. No effort has been made to design a modern and user-friendly specification language. Therefore, in the following examples, the user must also work out such low level details which normally are taken care of by macros, libraries or compilers. The presentation of these details may provide for a better understanding of our approach, but it is a nuisance in every day programming. The development of a practical programming environment for large scale software engineering is an important task which still must be carried out. Work of this kind is in progress.

The below examples show that our process theory is suitable for modeling real world systems, that it can be implemented, and that this implementation can be used for protocol prototyping. These observations motivate further research in order to develop equivalence notions for this process theory and to develop complete and correct axiomatizations for these relations.

In Section 1, a *specification language* is presented for use in the remainder of this chapter.

In Sections 2 and 3, the *stack* and the *bag* are specified. Both are well-known abstract data types which will be treated as processes which interact with a user. The purpose of these examples is to demonstrate these *interactions*, to illustrate advantages in the treatment of *errors* and *exceptions*, and to show how *procedural* and *declarative* as well as *first* and *higher order* aspects of the presented specification framework may be transformed into each other.

Section 4 shows how an *interrupt handling mechanism* can be specified in linear logic.

Section 5 discusses *data replication* and specifies the *primary site algorithm*.

In Section 6, a detailed discussion of the *alternating bit protocol* is presented. An *implementation* of a subset of the linear logic specification language is for *prototyping the alternating bit protocol*. This approach is compared to a specification and simulation using process algebra ACP and the related process specification formalism PSF.

In Section 7, a *business process* in a workflow management setting is specified with the help of linear logic.

In Section 8, the FIBONACCI *numbers* are specified.

Finally, Section 9 describes the present *prototypical implementation* of a subset of the specification language.

6.1 A Specification Language

We will *informally* introduce a specification language for use in the subsequent applications. No effort is made to provide a more formal approach for this language. The language contains the following *keywords*:

module `<module name>` **is** denotes the begin of a module definition

endmodule denotes the end of a module definition

sorts defines the sorts of the abstract data type

operators defines the operator symbols

states defines sorts on which the operators of linear logic may operate

init specifies the initial state

variables defines the type of variables

imports *syntactically* includes other modules

daxioms provides the data axioms which are equations

paxioms provides the process axioms which are labeled linear logic expressions

if `<boolean>` **then** `<equation>` denotes conditional equations

The language uses different *name spaces* for modules, sorts, variables, functions and labels.

Sorts and operator definitions provide information on the signature. *Constants* are denoted as in the following specification:

0: $\rightarrow$ nat

Equations may also be *labeled* as in the following example:

popeq: pop(push(d,s)) = s

These labels can be used in the presentation of (machine-assisted) proofs and as a reference. They can include references to variables as in the following equation:

popeq(d,s): pop(push(d,s)) = s

The free variables in process and data axioms are implicitly assumed to be quantified by the *universal quantifier* of the respective sort. To keep the specifications more readable, variable references are often omitted. They may be added by using a suitable preprocessor so that no two formulae obtain the same label and the requirements of Definition 179 are satisfied.

The *semantics* of this specification language is provided by Section 5.2: The equations and labeled linear logic formulae can directly be translated into process specifications.

6.2 The Stack

198 Example THE STACK, INTUITIVELY
Intuitively, a stack is an "object" with the following properties:

(1) There is an "empty" stack.

(2) Given a data object *d*, we can "push" this object "onto" the stack. The resulting stack is a "non-empty" stack.

(3) We can perform a "pop" on a "non-empty" stack which "undoes" the effect of the most recent "push".

(4) We can perform a "top" on a "non-empty" stack which returns the data object of the most recent "push" which was not "undone" by a "pop".

(5) "pop" and "top" on an "empty" stack have no well-defined meaning. ◀

Above description characterizes a stack intuitively. In [BT90] BERGSTRA outlines problems which arise when we want to describe a stack more precisely with algebraic means.

199 Example SPECIFICATION OF STACK ELEMENTS
An algebraic description of the stack also requires a specification of its elements. Since we do not want to care about the elements themselves within the specification of the stack, we will import a suitable module. Two examples are given below: Module "data_1" consists of two data elements "d1" and "d2", module "data_2" specifies the natural numbers:

module data_1 **is**
  sorts data
  operators
    d1: → data
    d2: → data
endmodule

module data_2 **is**
  sorts data
  operators
    0: → data
    S: data → data
endmodule

◀

200 Example STACK AS ALGEBRA
Algebraically, the stack can be described by the following specification:

module stack_alg **is**
  imports data_1
  sorts data, stack
  operators
    empty: → stack

```
    push: data × stack → stack
    pop: stack → stack
    top: stack → data
  variables data d; stack s;
  daxioms
    popeq: pop(push(d,s)) = s
    topeq: top(push(d,s)) = d
endmodule
```

Specification "stack_alg" allows to write down the expressions "pop(empty)" and "top(empty)" which correspond to no "real" stack. Usually this problem is solved with the help of error algebras, as explained in the following remark. There are several other possibilities, the difficulties of which are explained in detail in [BT90]. ◀

201 Remark ERROR ALGEBRAS
The *error algebra construction* is a standard construction [Wir90] which consists of the following steps:

(1) Add an *error constant* to every sort.

(2) Add *error generation equations* as required by the abstract data type. For example, if the expression "pop(empty)" shall have "no well-defined meaning", then add an error generation equation which makes "pop(empty)" equal to the error constant of the sort "stack".

(3) Add *error propagation equations* which guarantee that if an error eventually occurs somewhere during the evaluation of an expression, then this error is propagated and the entire expression finally evaluates to the error constant of the correct sort.

(4) Add *non-error* or *"okay" predicates* for every sort which can determine whether an element is equal to the error constant of the respective sort. If not already present, add the booleans as sorts, since they are required as codomain sort for the predicates. Add equations for the non-error predicates.

(5) Transform the original equations into *conditional equations*, since they only are required to hold when all occuring elements are different from the error elements. For example, the equation $0 * x = 0$ shall not hold, if x is an error element: In this case error propagation must occur: $0 * \mathit{error} = \mathit{error}$.

◀

202 Example STACK AS ERROR ALGEBRA
Applying the above error algebra construction to the algebraic version "stack_alg" of the stack yields the specification "stack_error":

```
module stack_error is
  imports data_1, boolean
  sorts data, stack, boolean
  operators
    empty: → stack
    push: data × stack → stack
```

```
        pop: stack → stack
        top: stack → data
        derror: → data
        serror: → stack
        dokay: data → boolean
        sokay: stack → boolean
    variables data d; stack s;
    daxioms
        popceq: if dokay(d) and sokay(s) then pop(push(d,s)) = s
        topceq: if dokay(d) and sokay(s) then top(push(d,s)) = d
        poperr: pop(empty) = serror
        toperr: top(empty) = derror
        popprop: pop(serror) = serror
        topprop: top(serror) = derror
        pushprop1: push(derror,s) = serror
        pushprop2: push(d,serror) = serror
        sokay1: sokay(empty) = true
        sokay2: sokay(serror) = false
        sokay3: sokay (push(d,s)) = dokay(d) and sokay(s)
        dokay1: dokay(derror) = false
        dokay2: dokay(d1) = true
        dokay3: dokay(d2) = true
endmodule
```

The error construction has a number of problems:

First, it is *inadequately complicated* and significantly enlarges the original specification. In case of the stack example, the error construction added 12 equations to the 2 of the original specification.

Second, the error construction *complicates parametric specifications*. A generic stack construction in the form "stack **of** data" does not leave the sort "data" invariant since it adds an error constant. Hence, also the specification of "data" must be changed. For example, if "data" is the sort of the natural numbers, further equations and predicates must be added in order to deal with integer arithmetic involving this new error constant.

Third, the error construction is an *inadequate modelling* of an error situation: If an error occurs, we would like to abort the computation and obtain an information where the error occured. Maybe we want to activate an exception handler. Algebraic error handling is not able to provide information where the error occured. Every correct implementation would continue calculations beyond an error condition, with the final result being the error constant of the correct sort. ◀

203 Example STACK AS PROCESS

In our theory a stack is modeled as a process which can interact with a user by exchanging the following interaction tokens:

Token	**Issued by**	**Meaning**
push(d)	user	stack, please "push" data element d
pop	user	stack, please "pop" yourself
top	user	stack, please show the "top" element
pushok	stack	I performed a "push" successfully
popok	stack	I performed a "pop" successfully
topis(d)	stack	I performed a "top" successfully and found element d
poperr	stack	I performed a "pop" and an error occurred
toperr	stack	I performed a "top" and an error occurred

The only other state tokens are of sort stack. This sort is built from the two *constructor functions* "empty" and "pushcon". Note the new meaning of "push(d)".

module stack_1 **is**
  imports data_1
  sorts data, stack, user
  states stack, user
  init empty
  operators
    empty: $\rightarrow$ stack
    pushcon: data $\times$ stack $\rightarrow$ stack
    push: data $\rightarrow$ user
    pop: $\rightarrow$ user
    top: $\rightarrow$ user
    pushok: $\rightarrow$ user
    popok: $\rightarrow$ user
    poperr: $\rightarrow$ user
    topis: data $\rightarrow$ user
    toperr: $\rightarrow$ user
  variables data d; stack s;
  paxioms
    pushp: push(d) $\star$ s $\multimap$ pushcon(d,s) $\star$ pushok
    popp: pop $\star$ empty $\multimap$ empty $\star$ poperr
    topp: top $\star$ empty $\multimap$ empty $\star$ toperr
    popc: pop $\star$ pushcon(d,s) $\multimap$ s $\star$ popok
    topc: top $\star$ pushcon(d,s) $\multimap$ pushcon(d,s) $\star$ topis(d)
endmodule

The user of the stack supplies commands to the stack by issuing suitable tokens. As soon as the stack has performed these tasks, it produces an answer token. This *coupling* of the stack and its user process is an example of the usual mechanism of *procedure invocation and return*. ◀

In the following module an example of a user process for a stack is specified. The token "pc" which is parametrized by natural numbers serves as *program counter* and guarantees *sequential execution* of commands.

204 Example STACK USER PROCESS
The specification text is as follows:

```
module stack_user_1 is
  imports data_1, nat
  sorts user, nat, data, prog_counter
  states user, prog_counter
  init pc(0)
  operators
    pc: nat → prog_counter
    push: data → user
    pop: → user
    top: → user
    pushok: → user
    popok: → user
    poperr: → user
    topis: data → user
    toperr: → user
  paxioms
    ex0a: pc(0) ⊸ push(d0) ⋆ pc(1)
    ex0b: pc(0) ⊸ push(d1) ⋆ pc(1)
    ex1: pc(1) ⊸ push(d1) ⋆ pc(2)
    ex2: pc(2) ⊸ top ⋆ pc(3)
    ex3a: pc(3) ⋆ topis(d0) ⊸ push (d0)
    ex3b: pc(3) ⋆ topis(d1) ⊸ push (d1)
endmodule
```

This user process is initialized in state pc(0). Then there is a *non-deterministic choice* between ex0a and ex0b: The state pc(0) is consumed and the states push(d0) (or push(d1)) and pc(1) are produced. Due to this program counter transformation, now only transition ex1 is possible. State pc(1) is consumed, and push(d1) and pc(2) are generated. This is followed by transition ex2 which consumes pc(2) and generates top and pc(3). Then, ex3a and ex3b demonstrate a *conditional statement*. As a result to the prior "top", the stack has supplied a suitable "topis" state. Depending on the reported value of this element (d0 or d1), either ex3a or ex3b execute.

Unfortunately, this user process still suffers from a major flaw: The user supplies commands for the stack, but no mechanism *suspends* the user process until the stack returns its answer. In fact, the user process could perform ex0, ex1 and ex2 without waiting for any reaction of the stack. In this case, the stack faces push(d0) $\star$ push(d1) $\star$ top. In this state the stack may non-deterministically choose in which sequence the commands will be executed. A stack should not exhibit such a behaviour.

To solve this problem, the specification is modified to have the *user process waiting* for the return from the stack. Only now we have correctly modeled the "call-and-return" mechanism of a procedure.

```
module stack_user_2 is
  imports data_1, nat
  sorts user, nat, data, prog_counter
  states user, prog_counter
  init pc(0)
  operators
```

```
      pc: nat → prog_counter
      push: data → user
      pop: → user
      top: → user
      pushok: → user
      popok: → user
      poperr: → user
      topis: data → user
      toperr: → user
   paxioms
      fx0: pc(0) ⊸ push(d0) ⋆ pc(1)
      fx1: pc(1) ⋆ pushok ⊸ push(d1) ⋆ pc(2)
      fx2: pc(2) ⋆ pushok ⊸ top ⋆ pc(3)
      fx3a: pc(3) ⋆ topis(d0) ⊸ push(d0)
      fx3b: pc(3) ⋆ topis(d1) ⊸ push(d1)
endmodule
```
◀

For testing purposes it is now very useful to define a stack user process which *iterates a non-deterministic choice of stack commands*. In this process the state "free" guarantees that the user process waits with issuing new commands until the stack has completed the previous call:

205 Example RANDOMIZED STACK USER PROCESS
The specification reads as follows:

```
module stack_user_random is
   imports data_1
   sorts data, user
   states user
   init free
   operators
      free: → user
      push: data → user
      pop: → user
      top: → user
      pushok: → user
      popok: → user
      poperr: → user
      topis: data → user
      toperr: → user
   variables data d;
   paxioms
      u_pop: free ⊸ pop
      u_push: free ⊸ push(d)
      u_top: free ⊸ top
      u_pushok: pushok ⊸ free
      u_popok: popok ⊸ free
      u_poperr: poperr ⊸ free
      u_topis: topis(d) ⊸ free
```

```
    u_toperr: toperr ⊸ free
endmodule
```

The state "free" is responsible for suspending the user process when the stack process is executing. The universally quantified process axiom

u_push: free ⊸ push(d)

provides a non-deterministic choice of the data element *d*. Since this axiom is universally quantified, an instance with an arbitrary choice of *d* may be deduced, as outlined in Remark 190. ◀

In the following example higher order axioms will be used for the specification of the stack:

206 Example STACK AS PROCESS, HIGHER ORDER
The specification is:

```
module stack_2 is
  imports data_1
  sorts data, stack, user
  states stack, user
  init empty
  operators
    empty: → stack
    pushcon: data × stack → stack
    push: data → user
    pop: → user
    top: → user
    pushok: → user
    popok: → user
    poperr: → user
    topis: data → user
    toperr: → user
  variables data d; stack p, q;
  paxioms
    pushp: push(d) ⋆ p ⊸ pushcon(d,p) ⋆ pushok
    popp: pop ⋆ empty ⊸ empty ⋆ poperr
    tope: top ⋆ empty ⊸ empty ⋆ toperr
    popc2: (push(d) ⋆ p ⊸ q ⋆ pushok) ⊸ (pop ⋆ q ⊸ p ⋆ popok)
    topc2: (push(d) ⋆ p ⊸ q ⋆ pushok) ⊸ (top ⋆ q ⊸ q ⋆ topis(d))
endmodule
```

It is straightforward to see that the process defined by this specification contains all the state changes of the process from Example 203. Presently we can only argue on an informal basis that this process does not contain additional and unwanted transitions, or that it is non-deterministic[1] to a degree which was not intended. ◀

In the following specification of the stack, the process axioms only define the *interaction structure* between the stack and a user process. They manage, when and how interaction tokens of which sort may be exchanged. Their *contents*, namely the choice of the specific interaction token, is determined by algebraic means:

[1] For non-determinism, compare Remark 166.

207 Example STACK AS PROCESS, ALGEBRAICALLY
The specification reads:

```
module stack_3 is
  imports data_1
  sorts data, stack, user
  states stack, user
  init empty
  operators
    empty: → stack
    pushcon: data × stack → stack
    push: data → user
    pop: → user
    top: → user
    pushok: → user
    popok: → user
    poperr: → user
    topis: data → user
    toperr: → user
    topans: stack → user
    popans: stack → user
    popnext: stack → stack
  variables data d; stack s;
  daxioms
    topanse: topans(empty) = toperr
    topansc: topans(pushcon(d,s)) = topis(d)
    popanse: popans(empty) = poperr
    popansc: popans(pushcon(d,s)) = popok
    popnexte: popnext(empty) = empty
    popnextc: popnext(pushcon(d,s)) = s
  paxioms
    pushpx: push(d) ⋆ s ⊸ pushcon(d,s) ⋆ pushok
    poppx: pop ⋆ s ⊸ popnext(s) ⋆ popans(s)
    toppx: top ⋆ s ⊸ s ⋆ topans(s)
endmodule
```

The functions "topans" and "popans" calculate the answer of the stack to user commands and the function "popnext" determines the new stack after a "pop".

This example demonstrates how a process can be specified entirely in the algebraic part – the process part is responsible only for the interaction. Such an approach has the drawback that no information is provided about how the algebraic equations shall be applied. This is discussed in more detail on page 211.

By substituting variable "s" in the above process axioms by terms of the sort stack it can be shown that the processes specified by modules stack_3 and stack_1 are isomorphic. ◀

6.3 The Bag

208 Example THE BAG, INTUITIVELY
Intuitively, a bag is a multiset or a black box with the following behaviour:

(1) There exists an "empty" bag.

(2) We may "in" elements into a bag which increases the number of copies inside of the bag by one.

(3) We may "out" elements from a bag which reduces the number of copies inside the bag by one.

(4) If we "out" an element more often than the number of copies available in the bag, we get an error.

Compare also the bags of Definition 239. ◀

In the following, we assume that in the module of the bag elements a non-equality predicate "neq" is defined with values in the booleans. data_3 and data_4 are examples of such modules:

```
module data_3 is
    sorts data, boolean
    operators
        d1: → data
        d2: → data
        neq: data × data → boolean
    variables
    daxioms
        neq(d1,d1) = true
        neq(d1,d2) = false
        neq(d2,d1) = false
        neq(d2,d2) = true
endmodule
```

```
module data_4 is
    sorts data, boolean
    operators
        0: → data
        S: data → data
        neq: data × data → boolean
    variables data n, m;
    daxioms
        neq(0,0) = true
        neq(0,S(n)) = false
        neq(S(n),0) = false
        neq(S(n),S(m)) = neq(n,m)
endmodule
```

209 Example BAG AS PROCESS, ALGEBRAICALLY (1)
The bag interacts with its user via the following interaction tokens:

Token	Issued by	Meaning
in(d)	user	bag, please "in" data element d
out(d)	user	bag, please "out" data element d
inok	bag	I performed an "in" successfully
outok	bag	I performed an "out" successfully
outerr	bag	I performed an "out", and an error occurred

Here we split the interaction tokens into two sorts, "u_in" for tokens which are generated by the user as input to the bag, and "u_out" for tokens which are output by the bag.

In this specification, the entire behaviour of the bag is determined by algebraic equations. "outan" calculates the answer to an "out" and "remove" determines the corresponding change of the bag.

```
module bag_1 is
    imports data_3
    sorts data, bag, u_in, u_out, boolean
    states bag, u_in, u_out
    init empty
    operators
        empty: → bag
        incon: data × bag → bag
        in: data → u_in
        out: data → u_in
        inok: → u_out
        outok: → u_out
        outerr: → u_out
        outan: data × bag → u_out
        remove: data × bag → bag
    variables data a, b; bag x;
    daxioms
        ass: incon(a,incon(b,x)) = incon(b,incon(a,x))
        outan1: if a neq b then outan(a,incon(b,x)) = outan(a,x)
        outan2: outan(a,empty) = outerr
        outan3: outan(a,incon(a,x)) = outok
        remove1: if a neq b then remove(a,incon(b,x)) = incon(b,remove(a,x))
        remove2: remove(a,empty) = empty
        remove3: remove(a,incon(a,x)) = x
    paxioms
        inp: in(a) ⋆ x ⊸ incon(a,x) ⋆ inok
        outp: out(a) ⋆ x ⊸ outan(a,x) ⋆ remove(a,x)
endmodule
```

◀

210 Example OBSERVATIONAL EQUIVALENCE

An interesting variant of the specification "bag_1" is obtained by dropping the equation

ass: incon(a,incon(b,x)) = incon(b,incon(a,x))

Then, the sort "bag" would distinguish the constructor terms "incon(a,incon(b,x))" and "incon(b,incon(a,x))". With above equation, we implement the bag states by a kind of *commutative monoid* which in Example 239 was identified as a bag. Without this equation, we implement the bag states by an algebra of *lists*. This structure contains more information than necessary, since it not only stores information on the number of copies of data elements in the bag, but also remembers the order in which they were inserted.

These two variants lead to different processes, but from another point of view, they may be considered *equivalent*. Under the requirement that the bags will interact with other processes only via the interaction tokens of the sorts "u_in" and "u_out", no process is able to detect any difference between these two variants of a bag. In algebraic specification a similar phenomenon is known by the name of *observational equivalence*, compare [EGLer], [Hen91].

The typical user of a bag will be a process like the following:

```
module bag_user_1 is
```

```
    imports data_3
    sorts data, u_in, u_out, user
    states u_in, u_out
    init free
    operators
        free: → user
        in: data → u_in
        out: → u_in
        inok: → u_out
        outok: → u_out
        outerr: → u_out
    variables data d;
    paxioms
        u_in: free ⊸ in(d)
        u_out: free ⊸ out(d)
        u_inok: inok ⊸ free
        u_outok: outok ⊸ free
        u_outerr: outerr ⊸ free
endmodule
```

Process "bag_user_1" is not able to detect any difference between bag_1 with and without the considered equation but a process which may analyze the ***internal sort*** "bag" could detect a difference. For example, such a process would generate the state "incon(d1,incon(d2,empty))" and then provide a transition which only may execute in state "incon(d2,incon(d1,empty))". This may be accomplished by adding the following process axioms to "bag_user_1":

```
gen: free ⊸ incon(d1,incon(d2,empty))
test: incon(d2,incon(d1,empty)) ⊸ found_equality
```

However, after the above modification, this process may no longer be considered a user process for the bag.

This example shows that a notion of ***process equivalence*** which is based on testing must ***shield the internal implementational details*** from being tempered with. This may be accomplished by requiring that the testing process does not use states which belong to the internals of the tested process, i.e. only the intended interaction tokens may be used for an interaction. The next example illustrates that this may not be sufficient:

Assume we add the process axiom

```
red: in(d) ⋆ in(d) ⊸ in(d)
```

to the specification of bag_1. Since neither bag_user_1 nor any similar user processes of the bag generate a state in which this transition becomes enabled, they may not distinguish bag_1 with or without transition "red". Nevertheless, there exists a process which interacts with the bag only via the prescribed interface sorts u_in and u_out but which does not respect the interaction protocol: This process generates multiple "in" states, and by testing whether the bag "loses" data elements – which is the case, if axiom "red" is present – this process may distinguish the above mentioned variants of the bag.

Therefore, it is not enough to restrict processes which shall test implementations for equivalency to interactions via the *intended interface*; it may also be necessary that they respect the *intended interaction protocol.* ◀

211 Example BAG AS PROCESS, ALGEBRAICALLY (2)

In this specification of the bag the behaviour is also determined algebraically, but with an additional operator which counts the occurrences of a data element. Assume that the module "nat" contains a greater-or-equal predicate "ge" and an equality predicate "eq" on the naturals.

```
module bag_2 is
   imports data_3, nat
   sorts data, bag, u_in, u_out, nat, boolean
   states bag, u_in, u_out
   init empty
   operators
      empty: → bag
      incon: data × bag → bag
      in: data → u_in
      out: data → u_in
      inok: → u_out
      outok: → u_out
      outerr: → u_out
      outan: data × bag → u_out
      remove: data × bag → bag
      count: data × bag → nat
   variables data a, b; bag x;
   daxioms
      count1: count(a,incon(a,x)) = count(a,x) + 1
      count2: if a neq b then count(a,incon(b,x)) = count(a,x)
      count3: count (a,empty) = 0
      outan4: if count(a,x) ge 1 then outan(a,x) = outok
      outan5: if count(a,x) eq 0 then outan(a,x) = outerr
      remove1: if a neq b then remove(a,incon(b,x)) = incon(b,remove(a,x))
      remove2: remove(a,empty) = empty
      remove3: remove(a,incon(a,x)) = x
   paxioms
      in: in(a) ⋆ x ⊸ incon(a,x) ⋆ inok
      out: out(a) ⋆ x ⊸ outan(a,x) ⋆ remove(a,x)
endmodule
```

The initial algebras which are specified in bag_1 and bag_2 are isomorphic. This can be shown by a rigorous equational proof using structural induction on the constructor terms. Since the process axioms of both specifications are identical, bag_1 and bag_2 describe the same process. ◀

212 Example BAG AS PROCESS, FIRST ORDER

In this variant the entire behaviour is specified in the *process part*. This requires additional states and the calculation of the reaction of the bag must be *programmed explicitly*, step by step. In this sense, state search(x,a,y) denotes that we are trying to determine whether element "a" may be removed from the bag, that we already have looked at all elements accumulated in bag "y", and that we still have to look at the elements in bag "x". State con(x,y) is used for concatenating split bags again, and it supplies an answer to the user. State neqt(a,b) contains the information that "a" and "b" are different data elements. Since this specification shall contain no data axioms at all, even none concerning the data elements, this information must be specified in the process axioms of the data modules. Below, the running examples data_1 and data_2 are presented with the respective adaptions made:

module data_5 **is**
  sorts data, cond
  states cond
  init start
  operators
    d1: $\rightarrow$ data
    d2: $\rightarrow$ data
    start: $\rightarrow$ cond
    neqt: data $\times$ data $\rightarrow$ cond
  variables
  paxioms
    n1: start $\multimap$ neqt(d1,d2) $\star$ start
    n2: start $\multimap$ neqt(d2,d1) $\star$ start
endmodule

module data_6 **is**
  sorts data, cond
  states cond
  init start
  operators
    0: $\rightarrow$ data
    S: data $\rightarrow$ data
    start: $\rightarrow$ cond
    neqt: data $\times$ data $\rightarrow$ cond
  variables data n, m;
  paxioms
    nt1: start $\multimap$ neqt(0,S(n)) $\star$ start
    nt2: start $\multimap$ neqt(S(n),0) $\star$ start
    nt3: neqt(n,m) $\multimap$ neqt(S(n),S(m))
endmodule

module bag_3 **is**
  imports data_5
  sorts data, bag, u_in, u_out, calc, cond
  states bag, u_in, u_out, calc, cond
  init empty
  operators
    empty: $\rightarrow$ bag
    incon: data $\times$ bag $\rightarrow$ bag
    in: data $\rightarrow$ u_in
    out: data $\rightarrow$ u_in
    inok: $\rightarrow$ u_out
    outok: $\rightarrow$ u_out
    outerr: $\rightarrow$ u_out
    search: bag $\times$ data $\times$ bag $\rightarrow$ calc
    con: bag $\times$ bag $\rightarrow$ calc
  variables data a, b; bag x, y;
  paxioms
    ins: in(a) $\star$ x $\multimap$ incon(a,x) $\star$ inok
    outs: out(a) $\star$ x $\multimap$ search(x,a,empty)
    search1: search(incon(a,x),b,y) $\star$ neqt(a,b) $\multimap$ search(x,b,incon(a,y))
    search2: search(incon(a,x),a,y) $\multimap$ con(x,y)
    search3: search(empty,b,y) $\multimap$ y $\star$ outerr
    con1: con(empty,empty) $\multimap$ empty $\star$ outok
    con2: con(incon(a,x),empty) $\multimap$ incon(a,x) $\star$ outok
    con3: con(empty,incon(a,x)) $\multimap$ incon(a,x) $\star$ outok
    con4: con(incon(a,x),incon(b,y)) $\multimap$ con(x,incon(a,incon(b,y)))
endmodule

Proving the equivalency of this process with bag_1 is a *rather sophisticated task*. First, it requires a suitable ***notion of abstraction*** from the internal sorts "calc" and "cond". On this problem, compare Section 5.4. Second, it requires ***fairness assumptions***, since otherwise it is possible that the data module data_6 engages in an endless generation of information on the non-equality of natural numbers, consistently neglecting instances of the transitions "nt2" and "search2" in favour of instances of "nt1" and "nt3". Due to the unbounded non-determinism contained in the axioms "nt1" and "nt2" which allows the arbitrary choice of a natural number, and due to the structure of "nt3", we must assume ***global fairness***[2] in order to ensure that the required non-equality information eventually is generated.

This non-deterministic generation of non-equality information can be modified to obtain the following deterministic process which is the "most executable" version of all specifications of a bag in this section. It closely resembles a specification in an imperative programming language.

In this specification the state temp(a,x,b,y) stores the values of a, x, b, and y, until the value of test(a,b) has been determined; then this state is destroyed by the transitions temp1 or temp2. This examplifies a number of important features of the specification mechanism: All storage cells are ***allocated and deallocated automatically*** and in an ***automatic*** fashion as part of the specification of the transitions. Therefore, the life-time of every variable is explicitly visible and no compile-time analysis is necessary. Furthermore, ***neither explicit memory management, nor heap management, nor garbage collection*** is required, allowing a more efficient and less error-prone programming style.

```
module data_7 is
   sorts data, cond
   states cond
   operators
      0: → data
      S: data → data
      test: data × data → cond
      positive: → cond
      negative: → cond
   variables data n, m;
   paxioms
      t1: test(0,0) ⊸ positive
      t2: test(0,S(n)) ⊸ negative
      t3: test(S(n),0) ⊸ negative
      t4: test(S(n),S(m)) ⊸ test(n,m)
endmodule

module bag_4 is
   imports data_7
   sorts data, bag, u_in, u_out, calc, cond
   states bag, u_in, u_out, calc, cond
   init empty
   operators
      empty: → bag
      incon: data × bag → bag
```

[2]Compare Example 138.

in: data → u_in
out: data → u_in
inok: → u_out
outok: → u_out
outerr: → u_out
search: bag × data × bag → calc
con: bag × bag → calc
temp: data × bag × data × bag → calc
variables data a, b; bag x, y;
paxioms
ins: in(a) ⋆ x ⊸ incon(a,x) ⋆ inok
outs: out(a) ⋆ x ⊸ search(x,a,empty)
searcht: search(incon(a,x),b,y) ⊸ temp(a,x,b,y) ⋆ test(a,b)
templ: temp(a,x,b,y) ⋆ positive ⊸ con(x,y)
temp2: temp(a,x,b,y) ⋆ positive ⊸ search(x,b,incon(a,y))
search3: search(empty, b, y) ⊸ y ⋆ outerr
conl: con(empty,empty) ⊸ empty ⋆ outok
con2: con(incon(a,x),empty) ⊸ incon(a,x) ⋆ outok
con3: con(empty,incon(a,x)) ⊸ incon(a,x) ⋆ outok
con4: con(incon(a,x),incon(b,y)) ⊸ con(x,incon(a,incon(b,y)))
endmodule ◀

213 Remark Declarative and Procedural Specification
The above examples show that a behaviour may be described in three ways:

In *purely procedural specifications* only process axioms are used and there are no equations or data axioms at all. "bag_3" from Example 212 is such a specification. The specified initial algebra is identical with the term algebra of the signature, it is *decidable* and every term represents a normal form. If such a specification is implemented, *no equational reasoning* must be performed. The entire behaviour of the process is programmed in an explicit manner. This often makes it necessary to introduce a large number of *internal states* which code intermediate steps of the computation. Therefore, it is not easy to determine how the process interacts with its environment because one must *abstract* from a large number of internal states and transitions. Depending on the form of the process axioms, such a specification can be highly non-deterministic and may pose problems with respect to fairness.

In *purely declarative specifications* only data axioms are used. Process axioms are needed only to produce states for the interaction with other processes. "bag_1" from Example 209 is such a specification. Since the process axioms describe only the *interaction* with the environment, this part of the process can easily be understood. On the other hand, equational reasoning is necessary to understand the output of the system. There are no a priori normal forms by which the initial algebra may be represented. It is possible that the algebra is not decidable and that the E-unification which is required for an execution[3] of the specification cannot be performed algorithmically. Equivalence proofs for such specifications use equational and inductive techniques which are better understood than the abstraction from internal states which is required in the case of purely procedural specifications.

Hybrid specifications contain process as well as data axioms.

[3]See Section 6.9 in the present chapter.

These facets of our calculus allow *transformations* between the data and the process aspect. For example, we may begin with a purely declarative specification. Then we may develop a decision algorithm for the initial algebra and successively transform equations into process axioms. Thus, the specification language supports *stepwise operational refinements* which lead from a purely declarative specification to a purely procedural and directly executable one.

From a functional programming point of view all "impure" non-functional features which exhibit *side-effects are concentrated* in the process axioms of a purely declarative specification. Thus the specification language elegantly solves the problem of pure functional languages [FH80] by offering a compact method for introducing *state* and *state-change*. ◀

6.4 Interrupts

214 Example TRAFFIC LIGHT WITHOUT INTERRUPT
Imagine a traffic light which is functioning normally until an *interrupt* is generated after which it flashes yellow. To keep the specification simple, the traffic light will be modeled by three states, each corresponding to a situation where exactly one light is turned on, and a fourth state in which no light is illuminated. Situations where more than one light is turned on will not be considered. This leads to the following signature:

module signature **is**
  sorts state
  states state
  init red
  operators
    red: $\rightarrow$ state
    yellow: $\rightarrow$ state
    green: $\rightarrow$ state
    dark: $\rightarrow$ state
endmodule

For the normal operation of the traffic light we will introduce a program counter:

module traffic_light_1 **is**
  imports signature, nat
  sorts prog_counter
  states prog_counter
  init pc(1)
  operators
    pc: nat $\rightarrow$ prog_counter
  paxioms
    p1: pc(1) $\star$ red $\multimap$ pc(2) $\star$ yellow
    p2: pc(2) $\star$ yellow $\multimap$ pc(3) $\star$ green
    p3: pc(3) $\star$ green $\multimap$ pc(4) $\star$ yellow
    p4: pc(4) $\star$ yellow $\multimap$ pc(1) $\star$ red
endmodule

For the interrupt handler a separate program counter is used:

```
module handler_1 is
   imports signature, nat
   sorts handler_prog_counter
   states handler_prog_counter
   operators
      hpc: nat → handler_prog_counter
   variables state s;
   paxioms
      h1: hpc(1) ⋆ s ⊸ hpc(2) ⋆ yellow
      h2: hpc(2) ⋆ yellow ⊸ hpc(3) ⋆ dark
      h3: hpc(3) ⋆ dark ⊸ hpc(2) ⋆ yellow
endmodule
```

◀

In the following, we will discuss various possibilities to *couple* the normal operation and the interrupt handler with an agent which raises the interrupt. The first idea is to have the interrupting agent generate an interrupt token "int":

215 Example INTERRUPT, VARIANT 1
In this situation the interrupting agent may be as follows:

```
module agent_1 is
   sorts agent
   states agent
   operators
      start: → agent
      int: → agent
   paxioms
      i1: start ⊸ int
endmodule
```

Now axioms must be provided which activate the interrupt handler when state "int" is generated. For this purpose the program counter from the normal operation is converted into the program counter of the interrupt handler. In assembler programming this is realized by a jump instruction:

```
variables nat i;
activate: pc(i) ⋆ int ⊸ hpc(1)
```

This solution has the drawback that the interrupt handler not necessarily is activated. After the generation of the state "int", the transition "activate" is permanently enabled but in an unfair maximal behaviour this may consistently be *ignored*. In the present form of the model there is no reasonable justification for a fairness assumption like in Example 43 of the erroneous channel. Furthermore, in this specification the controller of the traffic light is a *non-deterministic* system which it usually is not. Finally, there is no possibility for the controller to *check* whether an interrupt is pending or not, since only the presence of token "int" may be detected, but not its absence. ◀

These problems may be solved by a modification of the interrupting agent:

216 Example TESTING FOR INTERRUPTS BY MODIFIED AGENT
In this version, the agent does not produce a state whose presence signals an interrupt but it toggles an interrupt flag:

```
module agent_2 is
    sorts agent
    states agent
    operators
    init no_pending
        is_pending: → agent
        no_pending: → agent
    paxioms
        i1a: no_pending ⊸ is_pending
endmodule
```

The traffic light and the interrupt handler are coupled quite easily to this agent:

```
module traffic_light_2 is
    imports signature, nat, agent_2, handler_2
    sorts prog_counter
    states prog_counter
    init pc(1)
    operators
        pc: nat → prog_counter
    paxioms
        p1a: no_pending ⋆ pc(1) ⋆ red ⊸ no_pending ⋆ pc(2) ⋆ yellow
        p2a: no_pending ⋆ pc(2) ⋆ yellow ⊸ no_pending ⋆ pc(3) ⋆ green
        p3a: no_pending ⋆ pc(3) ⋆ green ⊸ no_pending ⋆ pc(4) ⋆ yellow
        p4a: no_pending ⋆ pc(4) ⋆ yellow ⊸ no_pending ⋆ pc(1) ⋆ red
endmodule
```

```
module handler_2 is
    imports signature, nat
    sorts handler_prog_counter
    states handler_prog_counter
    init hpc(1)
    operators
        hpc: nat → handler_prog_counter
    variables state s;
    paxioms
        h1: is_pending ⋆ hpc(1) ⋆ s ⊸ is_pending ⋆ hpc(2) ⋆ yellow
        h2: is_pending ⋆ hpc(2) ⋆ yellow ⊸ is_pending ⋆ hpc(3) ⋆ dark
        h3: is_pending ⋆ hpc(3) ⋆ dark ⊸ is_pending ⋆ hpc(2) ⋆ yellow
endmodule
```

The interrupt handler is already initialized with its program counter. As soon as the agent toggles the interrupt state from "no_pending" to "is_pending" the normal execution is suspended and the handler may start its activity. By providing transitions for "no_pending" and "is_pending" it is also possible to check for interrupts.

Also in this example *fairness* and *non-determinism* are an issue since there is a conflict between transaction i1a of the interrupting agent and the transitions of the traffic light. In this situation, a fairness assumption is reasonable:

Assuming that the traffic light controller and the interrupting agent are executing on two different processors and are sharing only their access to the interrupt line[4], the non-determinism is due to ***race conditions***. It is reasonable to require that the relative computing speed of the two processors do not differ too much. This corresponds to a fairness assumption which guarantees that the interrupt is serviced within a certain latency time.

Should the interrupting agent and the traffic light be executing on the same processor, then a scheduler which assigns the CPU alternatingly to these two tasks must be added to the specification. This scheduler will make the system deterministic and it must guarantee that the interrupt will be serviced with a latency time which is less or equal to the time slice of the scheduler. ◀

6.5 Data Replication

Distributed systems often use ***data replication***, [BHG87], [Cap90]: This means that a single logical datum is stored in two or more physical copies which may be accessed concurrently. In geographically distributed systems, time-consuming remote accesses can be replaced by local ones. Therefore, replication can lead to a significant increase in performance.

On the other hand, sophisticated algorithms must guarantee that the value of the several physical copies stays consistent during write operations. Consistency often is defined as ***one-copy-serializability*** [BHG87]: The algorithm which coordinates the access to the physical copies must behave in a way that an application program is not able to detect that a certain datum is stored in several replicated physical copies instead of a single one.

The problems associated with data replication are well understood and there are many algorithms which maintain the mentioned form of consistency [Cap90]. In the following, the problem of data replication is treated in our process theory. A simple ***locking mechanism*** is introduced and the ***primary site algorithm*** will be specified. To keep our presentation simple, a restriction to two copies and two users will be adopted. This may easily be generalized to an arbitrary number of copies and users, both of which may even appear as a parameter in the specification.

Assume that module "datadef" contains the specification of the values which shall be stored. For initialization purposes an operator "default" supplies a default data value. An example for such a module where the default value is "d0" is presented below:

module datadef **is**
    sorts data
    operators
        d0: $\rightarrow$ data
        d1: $\rightarrow$ data
        d2: $\rightarrow$ data
        default: $\rightarrow$ data
    daxioms
        def: default = d0
endmodule

217 Example SINGLE COPY, SINGLE USER STORE

[4]whose state is represented by is_pending and no_pending.

The following specifies a store of one physical copy of a datum which is serving one user.

```
module data_store_1 is
    imports datadef
    sorts data, user, storage
    states user, storage
    init stored(default)
    operators
        stored: data → storage
        write: data → user
        read: → user
        written: → user
        is: data → user
    variables data x, y;
    paxioms
        write(x) ⋆ stored(y) ⊸ stored(x) ⋆ written
        read ⋆ stored(x) ⊸ stored(x) ⋆ is(x)
endmodule
```
◀

In order to serve two users, the commands of the two users and the respective return codes must be distinguished. This may be accomplished as follows:

218 Example SINGLE COPY, TWO USERS STORE
The following specifies a store of one physical copy of a datum serving two users. Note the separate interfaces for the two users. As outlined above, the specification could also be parametrized to serve an arbitrary number of users.

```
module data_store_2 is
    imports datadef
    sorts data, user_1, user_2, storage
    states user_1, user_2, storage
    init stored(default)
    operators
        stored: data → storage

        /* interface for user 1 */
        write_u_1: data → user_1
        read_u_1: → user_1
        written_u_1: → user_1
        is_u_1: data → user_1

        /* interface for user 2 */
        write_u_2: data → user_2
        read_u_2: → user_2
        written_u_2: → user_2
        is_u_2: data → user_2
    variables data x, y;
    paxioms
        /* axioms for user 1 */
        write_u_1(x) ⋆ stored(y) ⊸ stored(x) ⋆ written_u_1
        read_u_1 ⋆ stored(x) ⊸ stored(x) ⋆ is_u_1(x)

        /* axioms for user 2 */
```

```
    write_u_2(x) ⋆ stored(y) ⊸ stored(x) ⋆ written_u_2
    read_u_2 ⋆ stored(x) ⊸ stored(x) ⋆ is_u_2(x)
endmodule
```
◀

Similarly, if two copies are stored these copies must be distinguished:

219 Example TWO COPIES, TWO USERS STORE, ERRONEOUS

The following specifies a store of two physical copies of a datum which *transparently* is serving two users:

There are two copies "a" and "b" of the datum which may be accessed via an interface defined by the sorts "int_a" and "int_b".

Transparency means that the user cannot see which copy he is accessing, nor that he is accessing replicated copies. Therefore, the user interface of this specification must be identical to the one used in data_store_2.

Copy a	Copy b
sort int_a	sort int_b
write_at_a	write_at_b
read_at_a	read_at_b
written_at_a	written_at_b
is_at_a	is_at_b

User 1	User 2
sort user_1	sort user_2
write_u_1	write_u_2
read_u_1	read_u_2
written_u_1	written_u_2
is_u_1	is_u_2

The main task of the algorithm is to map states like write_u_1 or read_u_2 which describe the interaction with the user to states like write_at_a or read_at_b which describe the interaction with the physical copies.

```
module data_store_3 is
    imports datadef
    sorts data, user_1, user_2, storage_at_a, storage_at_b, int_a, int_b
    states user_1, user_2, storage_at_a, storage_at_b, int_a, int_b
    init stored_at_a(default) ⋆ stored_at_b(default)
    operators
        /* interface to copy a */
        stored_at_a: data → storage_at_a
        write_at_a: data → int_a
        read_at_a: → int_a
        written_at_a: → int_a
        is_at_a: data → int_a

        /* interface to copy b */
        stored_at_b: data → storage_at_b
        write_at_b: data → int_b
        read_at_b: → int_b
        written_at_b: → int_b
        is_at_b: data → int_b

        /* interface for user 1 */
        write_u_1: data → user_1
        read_u_1: → user_1
        written_u_1: → user_1
        is_u_1: data → user_1

        /* interface for user 2 */
```

```
        write_u_2: data → user_2
        read_u_2: → user_2
        written_u_2: → user_2
        is_u_2: data → user_2
    variables data x, y;
    paxioms
        /* axioms for accessing the physical copy a */
        write_at_a(x) ⋆ stored_at_a(y) ⊸ stored_at_a(x) ⋆ written_at_a
        read_at_a ⋆ stored_at_a(x) ⊸ stored_at_a(x) ⋆ is_at_a(x)

        /* axioms for accessing the physical copy b */
        write_at_b(x) ⋆ stored_at_b(y) ⊸ stored_at_b(x) ⋆ written_at_b
        read_at_b ⋆ stored_at_b(x) ⊸ stored_at_b(x) ⋆ is_at_b(x)

        /* axioms for serving user requests */
        ··· see text ···
endmodule
```

Now process axioms must be provided which map the (logical) read/write requests of the two users "1" and "2" to the (physical) read/write requests on the two copies "a" and "b".

A first idea is to implement a write request of a user by writing both copies and a read request by reading the nearest copy. Suppose, that copy "a" resides at the site where the user process "1" is running and that similarly "b" and "2" are near to each other. This would lead to the following axioms:

```
write_u_1 (x) ⊸ write_at_a (x) ⋆ write_at_b (x)
write_u_2 (x) ⊸ write_at_a (x) ⋆ write_at_b (x)
read_u_1 ⊸ read_at_a
read_u_2 ⊸ read_at_b
```

The write request shall return when both copies have been written, and the read request when the one copy has been read. This already poses a problem: How does the system know in the state written_at_1 ⋆ written_at_2 whether to acknowledge user 1 or user 2.

For this purpose two states, pending_u_1 and pending_u_2 are introduced which remind the system about pending acknowledgement:

```
write_u_1 (x) ⊸ write_at_a (x) ⋆ write_at_b (x) ⋆ pending_u_1
write_u_2 (x) ⊸ write_at_a (x) ⋆ write_at_b (x) ⋆ pending_u_2
read_u_1 ⊸ read_at_a
read_u_2 ⊸ read_at_b
written_at_a ⋆ written_at_b ⋆ pending_u_1 ⊸ written_u_1
written_at_a ⋆ written_at_b ⋆ pending_u_2 ⊸ written_u_2
is_at_a(x) ⊸ is_u_1(x)
is_at_b(x) ⊸ is_u_2(x)
```

This algorithm for data replication is *not correct*. To show this, a user process is constructed which can distinguish the non-replicating store from the replicating one:

Let user 1 write the value d1 and user 2 write the value d2. Then, when both write requests have returned, the user process reads the logical datum and obtains the values r_1 and r_2. Finally, these two values are compared. One expects that r_1 and r_2 are equal and that the common value is either d1 or d2, depending on which process was served first. For the non-replicating store this is in fact the case. However, for the replicating store this is not necessarily true.

Formally, this distinguishing process may be defined by the following module:

module disting **is**
  imports datadef
  sorts data, user_1, user_2, proc, notify
  states user_1, user_2, proc, notify
  init start_1 ⋆ start_2
  operators
    start_1: → proc
    start_2: → proc
    write_is_done: → proc
    okay: → notify
    error: → notify
  paxioms
    start_1 ⊸ write_u_1(d1)
    start_2 ⊸ write_u_2(d2)
    written_u_1 ⋆ written_u_2 ⊸ write_is_done
    write_is_done ⊸ read_u_1 ⋆ read_u_2
    is_u_1(d1) ⋆ is_u_2(d1) ⊸ okay
    is_u_1(d2) ⋆ is_u_2(d2) ⊸ okay
    is_u_1(d1) ⋆ is_u_2(d2) ⊸ error
    is_u_1(d2) ⋆ is_u_2(d1) ⊸ error
endmodule ◀

The problem behind this replicating implementation is that two user-write-requests may overlap, leaving the two copies in an inconsistent state. This inconsistent state as such is not yet a problem, since sophisticated read mechanisms might still produce a correct behaviour[5]. However, in the above example the required consistency criterion is not satisfied.

One of many solutions for this problem is to guarantee that a write request does not overlap with another write request nor with a read request. This may be accomplished by using *locks*.

220 Example SIMPLE LOCKING MECHANISM
Locks are a technique for *synchronizing* two or more processes. Intuitively, the idea may be described as follows:

(1) Every process may "obtain" a shared lock or an exclusive lock and, once obtained, it may "return" this lock again.

(2) If any process has obtained an exclusive lock, no other process may be "holder" of a shared nor of an exclusive lock at the same time.

(3) Shared locks may "coexist" at the same time.

(4) In order to fulfill conditions (2) and (3), a process must "apply" for a lock. If the

[5]For example, version numbers or time stamps could be used to determine up-to-date copies.

process can be "holder" of the lock without contradicting these conditions, then[6] it is "granted" the lock. Otherwise the process must wait[7] until other processes return locks.

Above, the notion "at the same time" suggests that a lock will be managed centrally by a single, sequential process the time base of which provides a meaning to "at the same time". Similar to data replication, it is not important that locks actually are managed in a centralized way. The correctness criterion for a distributed lock manager is that it cannot be distinguished from a centralized one by a user process which adheres to the agreed interaction protocol.

To keep the lock manager simple, assume that the processes which use the lock manager behave well in the following sense:

(1) A process only tries to obtain a lock, if it presently does not hold a lock.

(2) A process only tries to return a lock, if it actually is holding a lock.

Consequently, the lock manager need not take care of the owners of locks. Its state space only must encode the number of shared locks and the information whether an exclusive lock is present or not. Furthermore, it need not generate error messages like "you are trying to obtain a lock although you are already owning one", or "you are trying to perform an unlock operation although you are not actually holding a lock". It is easy to get rid of these restrictions but this would unnecessarily obscure the ideas which shall be demonstrated in the following. As a further simplification a lock manager for two processes is specified. Again it is an easy task to extend it to handle an arbitrary, dynamically varying number of processes.

Figure 6.1 exhibits a kind of *interaction graph* for this lock manager: The vertices represent the states of the lock manager and the edges illustrate the possible state changes. The edges are labeled with the user-interaction-token which triggered the respective state change. Additionally, the edges could be labeled with the user-interaction-token which was generated in response to this state change.

Such an *interaction graph* can be drawn for every process which exhibits a *quasi-sequential* behaviour. Here, it may be used – instead of a correctness proof – to learn more about the behaviour of the lock manager.

The states in the graph and in the specification have the following interpretation:

[6]A crucial part of the algorithm is that the transition from checking whether a lock can be granted to actually granting this lock is "atomic". This requirement prevents that first it is checked that two processes may obtain an exclusive lock and then both are granted exclusive locks in contradiction to condition (2). In our solution this requirement is guaranteed by a conflict for the state of the lock manager.

[7]Lock managers of real operating systems must provide time-out features: Otherwise the entire system could run into a deadlock if a single process crashes before returning the lock it is holding. All this can be modeled in our theory but is omitted here for simplicity.

State	Interpretation
elock_1	user 1 requests an exclusive lock
slock_1	user 1 requests a shared lock
unlock_1	user 1 requests an unlock
granted_1	user 1 is granted the requested lock
unlockok_1	the unlock operation requested by user 1 was performed
unlock	there exists no lock
slocked	there exists one shared lock
sslocked	there exist two shared locks
elocked	there exists an exclusive lock

```
module lock_1 is
  sorts user_1, user_2, state
  init unlocked
  operators
    /* interface for user 1 */
    elock_1: → user_1
    slock_1: → user_1
    unlock_1: → user_1
    granted_1: → user_1
    unlockok_1: → user_1

    /* interface for user 2 */
    elock_2: → user_2
    slock_2: → user_2
    unlock_2: → user_2
    granted_2: → user_2
    unlockok_2: → user_2

    /* state of the lock */
    unlocked: → state
    elocked: → state
    slocked: → state
    sslocked: → state
  paxioms
    /* axioms for user 1 */
    elock_1 ⋆ unlocked ⊸ elocked ⋆ granted_1
    slock_1 ⋆ unlocked ⊸ slocked ⋆ granted_1
    slock_1 ⋆ slocked ⊸ sslocked ⋆ granted_1
    unlock_1 ⋆ elocked ⊸ unlocked ⋆ unlockok_1
    unlock_1 ⋆ slocked ⊸ unlocked ⋆ unlockok_1
    unlock_1 ⋆ sslocked ⊸ slocked ⋆ unlockok_1

    /* axioms for user 2 */
    elock_2 ⋆ unlocked ⊸ elocked ⋆ granted_2
    slock_2 ⋆ unlocked ⊸ slocked ⋆ granted_2
    slock_2 ⋆ slocked ⊸ sslocked ⋆ granted_2
    unlock_2 ⋆ elocked ⊸ unlocked ⋆ unlockok_2
    unlock_2 ⋆ slocked ⊸ unlocked ⋆ unlockok_2
    unlock_2 ⋆ sslocked ⊸ slocked ⋆ unlockok_2
endmodule
```

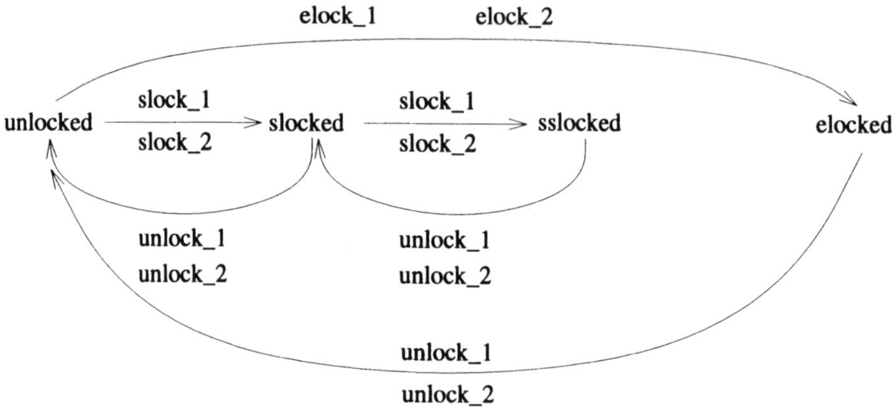

Figure 6.1: Interaction Graph of a Lock Manager.

This lock manager may ***block a process*** from proceeding: Suppose the lock manager is in state "unlocked" and two processes are requesting an exclusive lock. This produces a conflict which is resolved by a non-deterministic choice. Let this conflict be resolved in favour of the first process. Then let the first process return the lock and, even before the second process is granted the lock, let it again request a lock. This behaviour may continue, so that the second process is infinitely waiting for a lock.

This phenomenon is easily resolved: If the two processes are running on two processors then ***fairness assumptions*** on the relative speeds of the two processors guarantee that both processes may obtain a lock after some maximal latency time. If they are running on one processor this must be guaranteed by the scheduler. ◀

With the help of the lock manager, a correct replication algorithm can be designed:

221 Example PRIMARY SITE ALGORITHM
The specification reads as follows:

```
module data_store_4 is
  imports data_store_3, lock_1
  sorts pend_1, pend_2
  init
  operators
    wpending_u_1: → pend_1
    rpending_u_1: → pend_1
    wpending_u_2: → pend_2
    rpending_u_2: → pend_2
  variables data x;
  paxioms
    /* axioms for write of user 1 */
    write_u_1(x) ⊸ elock_1 ⋆ keep_u_1(x)
    keep_u_1(x) ⋆ granted_1 ⊸ write_at_a(x) ⋆ write_at_b(x) ⋆ wpending_u_1
    written_at_a ⋆ written_at_b ⋆ wpending_u_1 ⊸ written_u_1 ⋆ unlock_1

    /* axioms for write of user 2 */
```

```
    write_u_2(x) —o elock_2 ⋆ keep_u_2(x)
    keep_u_2(x) ⋆ granted_2 —o write_at_a(x) ⋆ write_at_b(x) ⋆ wpending_u_2
    written_at_a ⋆ written_at_b ⋆ wpending_u_2 —o written_u_2 ⋆ unlock_2

    /* axioms for read of user 1 */
    read_u_1 —o slock_1 ⋆ rpending_u_1
    granted_1 ⋆ rpending_u_1 —o read_at_a ⋆ rpending_u_1
    is_at_a(x) ⋆ rpending_u_1 —o is_u_1(x) ⋆ unlock_1

    /* axioms for read of user 2 */
    read_u_2 —o slock_2 ⋆ rpending_u_2
    granted_2 ⋆ rpending_u_2 —o read_at_b ⋆ rpending_u_2
    is_at_b(x) ⋆ rpending_u_2 —o is_u_2(x) ⋆ unlock_2
endmodule
```
◄

Renaming operators and *parametrization mechanisms* may be designed to allow a *reuse* of existing specification texts:

222 Example PARAMETRIC SPECIFICATION
For the above application a *parametrized module*

data_store_p (data, user, storage, stored, write, read, written, is)

may be defined by the text defining module data_store_1. Then module data_store_2 can be defined by importing two suitably parametrized versions:

```
module data_store_2 is
  imports data_store_p (data, user_1, storage, stored, write_u_1, read_u_1, written_u_1, is_u_1)
  imports data_store_p (data, user_2, storage, stored, write_u_2, read_u_2, written_u_2, is_u_2)
```

In general, parametrization and composition of modules is a rather difficult matter, see [BHK90]. ◄

6.6 The Alternating Bit Protocol

The *alternating bit protocol* (=ABP) is a communication protocol which has been widely used as an example for formal specification techniques [MV93], [Vaa89], [Vaa90].

A process, called the "sender", wants to transmit data to a process, called the "receiver". For this purpose, two communication channels are available: "data" for transmitting data from the sender to the receiver, and "acknowledge" for returning acknowledgements from the receiver to the sender. Both channels are unreliable in the fail-safe failure mode: They may choose non-deterministically not to transmit a message at all or to transmit a message completely, but if a channel transmits something this transmission is correct. Furthermore, the failures of a channel may be detected. The channel does not modify transmitted data, it does not interchange or duplicate data, nor does it deliver data which never were sent.

These assumptions are reasonable since there exist many established techniques which can guarantee that an arbitrary faulty channel behaves in the above described manner, e.g. checksums, message counters, time out mechanism, etc.

The purpose of the alternating bit protocol is to hide faults of the mentioned type from its user by a suitable retransmission mechanism.

In the following, an intuitive specification, a linear logic specification, and a process algebraic specification of the alternating bit protocol is provided.

223 Example ABP, INTUITIVE SPECIFICATION
Compare Figure 6.2 for the port names.

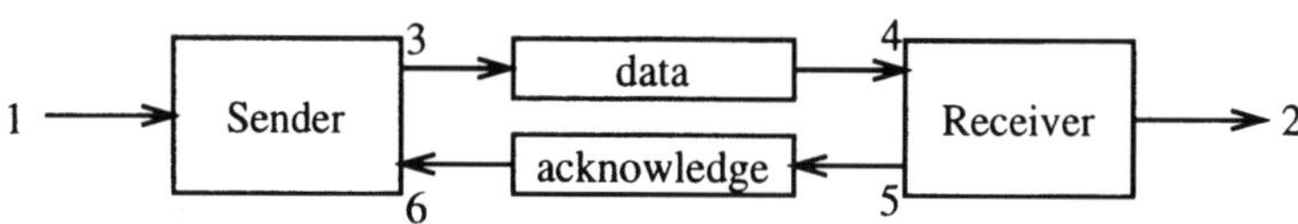

Figure 6.2: Situation of the Alternating Bit Protocol.

The protocol accepts data at port 1 and returns the transmitted data at port 2. Both sender and receiver have a state bit b, which initially is set to 0. The sender starts by accepting a data element d at port 1. It generates a message consisting of this data element d and of its own state bit b and transmits (d, b) along the data channel to the receiver. If the receiver gets the message (d, b), it extracts the state bit b and compares it to its own state bit. If the two bits match, this represents a successful communication: The receiver extracts the data element d, hands it over to port 2, sends the state bit as an acknowledgement to the sender, and flips its state bit. When the sender receives this bit, it flips its own state bit and reads the next datum at port 1.

If either the sender or the receiver receive a bit which is different from their own state bit, or if they receive an error indication from the channel, then the fault tolerant part of the protocol must be activated: If the sender detects a problem, it retransmits the data and the state bit; if the receiver detects a problem, it does not send its present state bit but its negation to the sender, and it does not flip its state bit. ◀

224 Example ABP, LINEAR LOGIC SPECIFICATION
In the linear logic specification, the following sorts will be used:

Sort	Interpretation
user	user interface
s_state	state of the sender
r_state	state of the receiver
dc_state	state of the data channel
ac_state	state of the acknowledgement channel
data	sort which shall be transmitted
bit	sort of the sender and receiver state bits

The states have the following meaning:

State	Interpretation
send(d)	request from user to protocol to transmit data d
rcvd(d)	answer from protocol to user that d was transmitted
s_ready	sender is ready to accept a request from the user
r_ready	receiver is ready to accept data from the channel
s_bit(b)	the state bit of the sender is b
r_bit(b)	the state bit of the receiver is b
s_frame(d,b)	message to be sent from sender to receiver
r_frame(d,b)	message as received from receiver
data_err	transmission error on data channel
keep(d)	data to be kept by sender for possible retransmission
s_ack(b)	acknowledgement bit to be transmitted from receiver to sender
r_ack	acknowledgement bit as received by sender
ack_err	transmission error on acknowledgement channel

The states s_ready and r_ready provide some kind of *flow control* which guarantees that data is delivered to the protocol only at a rate at which it can be handled by the ABP. This is necessary since no additional buffers are specified.

```
module abp is
    imports data, bit
    sorts dc_state, ac_state, s_state, r_state, user
    states dc_state, ac_state, s_state, r_state, user
    init s_ready ⋆ r_ready ⋆ s_bit(0) ⋆ r_bit(0)
    operators
        /* user interface */
        send: data → user
        rcvd: data → user

        /* state of the sender */
        s_ready: → s_state
        s_bit: bit → s_state
        keep: data → s_state

        /* state of the receiver */
        r_ready: → r_state
        r_bit: bit → r_state

        /* state of the data channel */
        s_frame: data × bit → dc_state
        r_frame: data × bit → dc_state
        d_err: → dc_state

        /* state of the acknowledgment channel */
        s_ack: bit → ac_state
        r_ack: bit → ac_state
```

ack_err: $\rightarrow$ ac_state

variables data d; bit b;
paxioms
/* behaviour of sender */
send_it: send(d) $\star$ s_bit(b) $\multimap$ s_frame(d,b) $\star$ s_bit(b) $\star$ keep(d)
no_ack: keep(d) $\star$ s_bit(b) $\star$ ack_err $\multimap$ s_frame(d,b) $\star$ s_bit(b) $\star$ keep(d)
wrong_ack: keep(d) $\star$ s_bit(b) $\star$ r_ack(neg(b)) $\multimap$ s_bit(b) $\star$ keep(d) $\star$ s_frame(d,b)
success: keep(d) $\star$ s_bit(b) $\star$ r_ack(b) $\multimap$ s_bit(neg(b)) $\star$ s_ready

/* behaviour of data channel */
mess_ok: s_frame(d,b) $\multimap$ r_frame(d,b)
mess_err: s_frame(d,b) $\multimap$ d_err

/* behaviour of ack channel */
ack_ok : s_ack(b) $\multimap$ r_ack(b)
ack_err: s_ack(b) $\multimap$ ack_err

/* behaviour of receiver */
rcve_it: r_ready $\star$ r_bit(b) $\star$ r_frame(d,b) $\multimap$ rcvd(d) $\star$ s_ack(b) $\star$ r_bit(neg(b))
phase_err: r_bit(b) $\star$ r_frame(d,neg(b)) $\multimap$ s_ack(neg(b)) $\star$ r_bit(b)
data_err: r_bit(b) $\star$ d_err $\multimap$ s_ack(neg(b)) $\star$ r_bit(b)
endmodule ◀

In order to demonstrate the practical usability of our process theory for the prototyping of communication protocols a subset of the specification language was implemented [KK94]. This language subset and the developed simulator is described in more detail in Section 6.9.

Below, the specification text of the alternating bit protocol in this language is presented. In order to facilitate the scanning, parsing, and compilation phases, the syntax of the subset is different from the syntax of the specification language employed so far. The correct translations between the two syntactical respresentations is straightforward.

225 Example IMPLEMENTATION OF THE LINEAR LOGIC SPECIFICATION
The specification is given by the following text:

```
TOKENDEF

data:      d1, d2;
bit:       off, on, neg(bit);
user:      send(data), rcvd(data);
s_state:   s_ready, s_bit(bit), keep(data);
r_state:   r_ready, r_bit(bit);
dc_state:  s_frame(data,bit), r_frame(data,bit), d_err;
ac_state:  s_ack(bit), r_ack(bit), ack_err;

VARDEF

d:  data;
b:  bit;

RULESDEF

initial:          init -o s_ready # r_ready # s_bit(off) # r_bit(off);
```

```
send_it(d,b):       send(d) # s_bit(b) -o s_frame(d,b) # s_bit(b) # keep(d);
no_ack(d,b):        keep(d) # s_bit(b) # ack_err -o s_frame(d,b) # s_bit(b) # keep(d);
wrong_ack_on(d):    keep(d) # s_bit(on) # r_ack(off) -o
                      s_bit(on) # keep(d) # s_frame(d,on);
wrong_ack_off(d):   keep(d) # s_bit(off) # r_ack(on) -o
                      s_bit(off) # keep(d) # s_frame(d,off);
success_on(d):      keep(d) # s_bit(on) # r_ack(on) -o s_bit(off) # s_ready;
success_off(d):     keep(d) # s_bit(off) # r_ack(off) -o s_bit(on) # s_ready;

mess_ok(d,b):     s_frame(d,b) -o r_frame(d,b);
mess_err(d,b):    s_frame(d,b) -o d_err;

ack_ok(b)  :      s_ack(b) -o r_ack(b);
ack_err(b):       s_ack(b) -o ack_err;

rcve_it_on(d):    r_ready # r_bit(on) # r_frame(d,on) -o
                  rcvd(d) # s_ack(on) # r_bit(off);
rcve_it_off(d):   r_ready # r_bit(off) # r_frame(d,off) -o
                  rcvd(d) # s_ack(off) # r_bit(on);
phase_err_on(d):  r_bit(on) # r_frame(d,off) -o s_ack(off) # r_bit(on);
phase_err_off(d):r_bit(off) # r_frame(d,on) -o s_ack(on) # r_bit(off);
data_err_on:    r_bit(on) # d_err -o s_ack(off) # r_bit(on);
data_err_off:   r_bit(off) # d_err -o s_ack(on) # r_bit(off);

send_submit_d1: s_ready -o send(d1);
send_submit_d2: s_ready -o send(d2);
rcv_consume_d1: rcvd(d1) -o r_ready;
rcv_consume_d2: rcvd(d2) -o r_ready;
```

The simulator is an experimental prototype and presently does not contain a rewriting system. Therefore, the algebraic equations

$$\text{neg(off)} = \text{on} \qquad\qquad \text{neg(on)} = \text{off}$$

had to be incorporated into the process axioms which resulted in two versions for the rules `wrong_ack`, `success`, `rcve_it`, `phase_ok`, and `phase_err`.

On a Sparcstation2 the unoptimized simulator achieves a speed of 4500 transitions per second. In the following, a screen log of a sample run consisting of 12 transitions is presented. It illustrates the following steps in the execution of the alternating bit protocol:

(1) Initialization.

(2) The user of the protocol submits the value d1 for transmission by the protocol.

(3) The sender hands over the data frame consisting of the value d1 and the present state bit whose value is "off" to the data channel.

(4) The data channel correctly transmits these two entities.

(5) The receiver picks up these two entities from the channel.

(6) The user of the protocol consumes the correctly transmitted value d1 from the buffer in the receiver.

(7) The acknowledgment channel makes an error in the transmission of the acknowledgement bit.

(8) The sender receives this error message and prepares a new s_frame for retransmission.

(9) The data channel correctly transmits these two entities.

(10) The receiver identifies these two entities as a retransmission due to the phase shift in the value of the state bit of the sender and prepares for the transmission of an acknowledgment without handing the received data to the user, thus preventing duplication of the data.

(11) This time the acknowledgment is transmitted correctly.

(12) The sender accepts this acknowledgment and flips its state bit.

(13) Abort of the simulation.

In this example, the non-deterministic choice in steps 2, 4, 6, 7, and 11 was performed by the user. A *random mode* is available which performs this choice by consulting a random number generator.

```
male:/cap/simulator% simulate abp2.p

Syntax check ... okay
Compilation  ... okay
Linking      ... okay

Running program: abp2.p
======================

Choose interactive (0) or random (1) mode: 0

1-----------------------------------------------------------------------

Token(s):
  init

Rule(s):
    0: initial

Choose a rule: 0

The chosen rule(s) is (are) executed.

2-----------------------------------------------------------------------
```

```
Token(s):
  s_ready  s_bit(off)  r_ready  r_bit(off)

Rule(s):
    0: send_submit_d1
    1: send_submit_d2

Choose a rule: 0

The chosen rule(s) is (are) executed.

3-------------------------------------------------------------------------

Token(s):
  send(d1)  s_bit(off)  r_ready  r_bit(off)

Rule(s):
    0: send_it(d1, off)

Choose a rule: 0

The chosen rule(s) is (are) executed.

4-------------------------------------------------------------------------

Token(s):
  s_bit(off)  keep(d1)  r_ready  r_bit(off)  s_frame(d1,off)

Rule(s):
    0: mess_ok(d1, off)
    1: mess_err(d1, off)

Choose a rule: 0

The chosen rule(s) is (are) executed.

5-------------------------------------------------------------------------

Token(s):
  s_bit(off)  keep(d1)  r_ready  r_bit(off)  r_frame(d1,off)

Rule(s):
    0: rcve_it_off(d1)

Choose a rule: 0

The chosen rule(s) is (are) executed.

6-------------------------------------------------------------------------

Token(s):
```

```
  rcvd(d1)  s_bit(off)  keep(d1)  r_bit(on)  s_ack(off)

Rule(s):
    0: ack_ok(off)
    1: ack_err(off)
    2: rcv_consume_d1

Choose a rule: 2

The chosen rule(s) is (are) executed.

7-----------------------------------------------------------------------------

Token(s):
  s_bit(off)  keep(d1)  r_ready  r_bit(on)  s_ack(off)

Rule(s):
    0: ack_ok(off)
    1: ack_err(off)

Choose a rule: 1

The chosen rule(s) is (are) executed.

8-----------------------------------------------------------------------------

Token(s):
  s_bit(off)  keep(d1)  r_ready  r_bit(on)  ack_err

Rule(s):
    0: no_ack(d1, off)

Choose a rule: 0

The chosen rule(s) is (are) executed.

9-----------------------------------------------------------------------------

Token(s):
  s_bit(off)  keep(d1)  r_ready  r_bit(on)  s_frame(d1,off)

Rule(s):
    0: mess_ok(d1, off)
    1: mess_err(d1, off)

Choose a rule: 0

The chosen rule(s) is (are) executed.

10----------------------------------------------------------------------------
```

```
Token(s):
  s_bit(off)  keep(d1)  r_ready  r_bit(on)  r_frame(d1,off)

Rule(s):
    0: phase_err_on(d1)

Choose a rule: 0

The chosen rule(s) is (are) executed.

11---------------------------------------------------------------------------

Token(s):
  s_bit(off)  keep(d1)  r_ready  r_bit(on)  s_ack(off)

Rule(s):
    0: ack_ok(off)
    1: ack_err(off)

Choose a rule: 0

The chosen rule(s) is (are) executed.

12---------------------------------------------------------------------------

Token(s):
  s_bit(off)  keep(d1)  r_ready  r_bit(on)  r_ack(off)

Rule(s):
    0: success_off(d1)

Choose a rule: 0

The chosen rule(s) is (are) executed.

13---------------------------------------------------------------------------

Token(s):
  s_ready  s_bit(on)  r_ready  r_bit(on)

Rule(s):
    0: send_submit_d1
    1: send_submit_d2

Choose a rule: q
abp2: User abort!
male:/cap/simulator%
```

The states of all cooperating protocol components are logged in every step and allow the user to follow every detail in the execution. ◄

To compare this approach to existing ones we will present a specification of the ABP in the specification formalism PSF which was built on top of process algebra ACP. The following is taken from [MV93], where details on process algebra and the employed specification formalism may be found. The presented simulations are from the author.

226 Example ALTERNATING BIT PROTOCOL IN PSF
In the PSF specification, the sort BIT contains the state bits, the sort DATA contains the application data and the sort ERROR comprises a token informing about error conditions. The employed parametrized atomic actions are discussed below:

Atoms beginning with letter "s" are defined for the sender, atoms beginning with letter "r" are defined for the receiver. For example, r3(d,b) means "receive datum d and bit b at port 3". If receive and send actions of identical structure and port number occur simultaneously, they can be seen from the outside as a single communication action. The respective atoms begin with the letter "c" for communication.

Atom	Interpretation
s2(d)	send datum d at port 2
s3(d,b)	send datum d and bit b at port 3
s4(d,b)	send datum d and bit b at port 4
s4(error)	send error token at port 4
s5(b)	send bit b at port 5
s6(b)	send bit b at port 6
s6(error)	send error token at port 6

Below, the operator **encaps** denotes the encapsulation operator ∂ from process algebra which guarantees that the receive and send actions must synchronize within the protocol process and must not interfere with any other action. Compare [Vaa90] for the details.

The following specification text from [MV93] describes the ABP:

```
process module ABP
begin
exports
  begin
    atoms
      c3, c4 : BIT # DATA
      c4, c6 : ERROR
      c5, c6 : BIT
    processes
      ABP
    end
  imports
    Sender, Receiver, ChannelK, ChannelL
  sets of atoms
    H = s3(b,d), r3(b,d), s4(b,d), r4(b,d), s4(error), r4(error),
    s5(b), r5(b), s6(b), r6(b), s6(error), r6(error) | b in BIT, d in DATA
  communications
    s3(b,d) | r3(b,d) = c3(b,d) for b in BIT, d in DATA
    s4(b,d) | r4(b,d) = c4(b,d) for b in BIT, d in DATA
    s4(error) | r4(error) = c4(error)
    s5(b) | r5(b) = c5(b) for b in BIT
    s6(b) | r6(b) = c6(b) for b in BIT
    s6(error) | r6(error) = c6(error)
```

```
    definitions
        ABP = encaps(H, S || K || R || L)
end ABP

data module Bits
begin
    exports
        begin
            sorts
            BIT
        functions
            0 : -> BIT
            1 : -> BIT
            inv : BIT -> BIT
        end
    equations
        [1] inv(0) = 1
        [2] inv(1) = 0
end Bits

process module ChannelK
begin
    exports
        begin
            atoms
            r3, s4 : BIT # DATA
            s4 : ERROR
        processes
            K
        end
    imports
        Bits, Data, Errors
    definitions
        K = (r3(0,d0) . (skip . s4(0,d0) + skip . s4(error)) . K +
        r3(1,d0) . (skip . s4(1,d0) + skip . s4(error)) . K) +
        (r3(0,d1) . (skip . s4(0,d1) + skip . s4(error)) . K +
        r3(1,d1) . (skip . s4(1,d1) + skip . s4(error)) . K)
end ChannelK

process module ChannelL
begin
    exports
        begin
            atoms
                r5, s6 : BIT
                s6 : ERROR
            processes
                L
            end
    imports
        Bits, Errors      definitions
        L = r5(0) . (skip . s6(0) + skip . s6(error)) . L +
        r5(1) . (skip . s6(1) + skip . s6(error)) . L
```

```
end ChannelL
data module Data
begin
    exports
        begin
            sorts
                DATA
            functions
                d0 : -> DATA
                d1 : -> DATA
            end
end Data
data module Errors
begin
    exports
        begin
            sorts
                ERROR
            functions
                error : -> ERROR
        end
end Errors
process module Receiver
begin
    exports
        begin
            atoms
                r4 : BIT # DATA
                r4 : ERROR
                s2 : DATA
                s5 : BIT
            processes
                R
        end
    imports
        Bits, Data, Errors
        processes
            R : BIT
    variables
        b : -> BIT
        definitions
            R = R(1) . R(0) . R
            R(b) = ( r4(b,d0) + r4(b,d1) + r4(error)) . s5(b) . R(b) +
            r4(inv(b),d0) . s2(d0) . s5(inv(b)) +
            r4(inv(b),d1) . s2(d1) . s5(inv(b))
end Receiver
process module Sender
begin
    exports
```

```
    begin
      atoms
        r1 : DATA
        s3 : BIT # DATA
        r6 : BIT
        r6 : ERROR
      processes
        S
    end
  imports
    Bits, Data, Errors
    processes
      S : BIT
      S, T : BIT # DATA
    variables
      d : -> DATA
      b : -> BIT
    definitions
      S = S(0) . S(1) . S
      S(b) = r1(d0) . S(b,d0) + r1(d1) . S(b,d1)
      S(b,d) = s3(b,d) . T(b,d)
      T(b,d) = (r6(inv(b)) + r6(error)) . S(b,d)
      + r6(b)
end Sender
```

With the help of the process simulator of the PSF toolkit traces can be generated. Below, the trace of the same transactions as in Example 225 is presented. Since no states are logged, this output is shorter but it contains considerably less information on the states of the components of the protocol.

```
atom r1(d1)
com. c3(0, d1)
     skip<4>
com. c4(0, d1)
atom s2(d1)
com. c5(0)
     skip<9>
com. c6(error)
com. c3(0, d1)
     skip<4>
com. c4(0, d1)
com. c5(0)
     skip<8>
com. c6(0)
```

For a more detailed analysis, the PSF simulator provides the states of the processes and subprocesses as well as the execution history in the following table. For an explanation of the format, the PSF manuals [Mau92] must be consulted.

PID	PPID	STATUS	FLAGS	HEAD
0	---	S 1		<a>
1	0	S 2	T	<e>
2	1	S 2	T	
3	2	S 2	T	
5	3	S 1	T	<a>
18	2	S 1	T	<a>
29	1	S 1	T	<a>
47	18	S 1	T	<a>
69	3	S 2	T	<a>
70	69	S 2	T	<a>
71	70		T E	r3(1, d1)
72	70		T E	r3(0, d1)
73	69	S 2	T	<a>
74	73		T E	r3(1, d0)
75	73		T E	r3(0, d0)
76	47	S 2	T	<a>
77	76		T E	r4(1, d1)
78	76	S 2	T	<a>
79	78		T E	r4(1, d0)
80	78	S 2	T	<a>
81	80		T E	r4(error)
82	80	S 2	T	<a>
83	82		T E	r4(0, d1)
84	82		T E	r4(0, d0)
87	29	S 2	T	<a>
88	87		T E	r5(1)
89	87		T E	r5(0)
90	5	S 2	T	<a>
91	90		T	r1(d1)
92	90		T	r1(d0)

◀

227 Remark FAIRNESS AND THE ABP
In order to prove that the alternating bit protocol is eventually able to transmit a data value correctly, several fairness assumptions are necessary. First, the two channels must not always choose to exhibit erroneous behaviour and thus both must be assumed to be fair. Furthermore, global fairness as described in Example 138 must be assumed since a successful communication requires a situation where first the data channel and then the acknowledgment channel functions correctly. If the channels alternatingly produce errors they both exhibit a fair behaviour but successful communication is not possible. ◀

6.7 Workflow Management

Workflow management provides a framework for controlling and executing tasks within an organization [Rei93]. A software system which helps a company with work flow management offers the possibility to define the various steps within a business process. This includes the specification of sequential and parallel tasks and of activities which may depend on certain conditions. Usually these descriptions contain information on the data which are necessary to fulfill the respective tasks. The software system now delegates the individual tasks, together with the required data, to the available employees. This distribution of jobs takes into account the various different abilities of the employees, their present work load and availability. It sets up and enforces deadlines and informs the administrator in case of problems.

In the following, we will demonstrate how our process theory may be used to specify a business process. In our example we will model the handling of insurance claims. We will develop the specifications only to a point, where the *general idea* becomes visible, our solution is by no means complete.

Informally the *business process* may be described as follows:

(1) Check whether the insurance policy was valid at the time of the event causing the claim.

(2) Check whether the claim is covered by this type of policy.

(3) (1) and (2) may be performed in parallel.

(4) If any of the two checks (1) and (2) results in a negative answer, write a rejection letter and consider the task as performed.

(5) If both of the two checks (1) and (2) result in a positive answer, then have an expert evaluate the sum which the insurance company is going to pay. Then pay this sum and consider the task as performed.

Sort	Interpretation
claimid	identification of the claim
type	type of claim
desc	description of claim
polid	identifier and data of the insurance policy
time	time of the event, which caused the claim
pers	identification of the person making the claim
money	amount of money for payment
action	state governing the actions of the workflow system

module workflow **is**
  sorts claimid, type, desc, polid, time, pers, money, action
  states action
  init
  operators
    claim: claimid × type × desc × polid × time × pers → action
    info: claimid × type × desc × polid × time × pers → action
    check_validity: claim × polid × time → action
    claim_valid: claimid → action
    claim_invalid: claimid → action
    check_covered: claimid × type × desc × polid → action
    claim_covered: claimid → action
    claim_notcovered: claimid → action

```
        proceed: claim → action
        reject: claim × pers → action
        determine_sum: claimid × type × desc → action
        sum_is: money × claimid → action
        pay_sum: money × pers → action
        reject_letter: claim × pers → action
    variables claimid cl; type ty; desc de; polid po; time ti; pers pe; money mo;
    paxioms
        /* process the incoming claim, allow parallelism */
        claim(cl,ty,de,po,ti,pe) ⊸ check_validity(cl,po,ti) ⋆
                check_covered(cl,ty,de,po) ⋆ info(cl,ty,de,po,ti,pe)
        /* deal with results of both checks */
        claim_valid(cl) ⋆ claim_covered(cl) ⊸ proceed(cl)
        claim_invalid(cl) ⋆ claim_covered(cl) ⊸ reject(cl,pe)
        claim_valid(cl) ⋆ claim_notcovered(cl) ⊸ reject(cl,pe)
        claim_invalid(cl) ⋆ claim_notcovered(cl) ⊸ reject(cl,pe)
        /* deal with negative result of one check */
        claim_invalid(cl) ⋆ check_covered(cl,ty,de,po) ⊸ reject(cl,pe)
        claim_notcovered(cl) ⋆ check_validity(po,ti) ⊸ reject(cl,pe)
        /* proceed */
        proceed(cl) ⋆ info(cl,ty,de,po,ti,pe) ⊸ determine_sum(cl,ty,de) ⋆ info(cl,ty,de,po,ti,pe)
        sum_is(mo,cl) ⋆ info(cl,ty,de,po,ti,pe) ⊸ pay_sum(mo,pe)
        reject(cl,pe) ⋆ info(cl,ty,de,po,ti,pe) ⊸ reject_letter(cl,pe)
endmodule
```

The following modules describe the interface of the workflow manager with the employees who perform the required tasks. These modules do not specify how the employees actually perform their tasks, but describe their possible answers in a non-deterministic fashion. Hence, they may not be considered as an *operational description* but as an *interface specification*.

```
module check_validity is
    paxioms
        check_validity(cl,po,ti) ⊸ claim_valid(cl)
        check_validity(cl,po,ti) ⊸ claim_invalid(cl)
endmodule
module check_covered is
    paxioms
        check_covered(cl,ty,de,po) ⊸ claim_covered(cl)
        check_covered(cl,ty,de,po) ⊸ claim_notcovered(cl)
endmodule
module determine_sum is
    paxioms
        determine_sum(cl,ty,de) ⊸ sum_is(mo,cl)
endmodule
```

Note, how the flow of business data is contolled: Every module only gets that information which is required for accomplishing the requested task. The state "info" saves all information which is associated with the ongoing transaction until it is completed.

6.8 FIBONACCI Numbers

The FIBONACCI numbers are provided by the function $f : \mathbb{N}_0 \to \mathbb{N}_0$ which is defined by the following equations:

$$f(0) = 0 \qquad f(1) = 1 \qquad \forall n \in \mathbb{N}_0 : \ f(n+2) = f(n+1) + f(n)$$

These equations may be used in several kinds to calculate the n–th FIBONACCI number. *Naive recursion* resolves the recursive equation $f(n+2) = f(n+1) + f(n)$ by leftmost-innermost reduction steps: For the calculation of $f(4)$ the values of $f(3)$ and $f(2)$ are required. Although the calculation of $f(2)$ is contained in the calculation of $f(3)$, the value $f(2)$ is calculated twice. For large n this leads to an exponential explosion of unnecessary calculations. More *refined recursion* employs memoization [FH80] so that the function f is calculated at most once for every argument. *Iterative computations* start with $f(0)$ and compute $f(n)$ for increasing n.

These three approaches are examples for how computational work with the above equational specification may be coordinated for obtaining a result. In the following, a non-deterministic process is defined which subsumes all possible computations with these numbers. It interacts with its user via the states "in(n)" and "out(m)". The process is presented the state "in(n)", performs some calculations, and finally generates the state "out(m)" with m = f(n). Internally, the process uses states "res(m,q)" to encode partial results with q = f(m).

module fibon **is**
  sorts nat, internal, user
  states internal, user
  init res(0,0) ⋆ res(1,1)
  operators
    in: nat → user
    out: nat → user
    res: nat × nat → internal
  variables nat n, m, p, q;
  paxioms
    answ: in(n) ⋆ res(n,p) ⊸ out(p) ⋆ res(n,p)
    cont: res(m,p) ⋆ res(m+1,q) ⊸ res(m+2,p+q) ⋆ res(m,p) ⋆ res(m+1,q)
    disc: res(m,p) ⊸ **1**
endmodule

This process is highly non-deterministic and describes all possible forms in which FIBONACCI numbers may be calculated with the help of above formula.

For the calculation of $f(4)$, the initial state is res(0,0) ⋆ res(1,1) ⋆ in(4).

The *iterative calculation* corresponds to the following sequence of states:

res(0,0) ⋆ res(1,1) ⋆ in(4)

res(0,0) ⋆ res(1,1) ⋆ in(4) ⋆ res(2,1)

res(0,0) ⋆ res(1,1) ⋆ in(4) ⋆ res(2,1) ⋆ res(3,2)

$$\text{res}(0,0) \star \text{res}(1,1) \star \text{in}(4) \star \text{res}(2,1) \star \text{res}(3,2) \star \text{res}(4,3)$$

$$\text{res}(0,0) \star \text{res}(1,1) \star \text{res}(2,1) \star \text{res}(3,2) \star \text{res}(4,3) \star \text{out}(3)$$

The *recursive calculations* produces after its first calulcation step the state

$$\text{res}(0,0) \star \text{res}(1,1) \star \text{in}(4) \star$$
$$(\forall \text{p} : \forall \text{q} : \text{res}(2,\text{p}) \star \text{res}(3,\text{q}) \multimap \text{res}(4, \text{p}+\text{q}) \star \text{res}(2,\text{p}) \star \text{res}(3,\text{q}))$$

The higher order state ($\forall$ p: $\forall$ q: res(2,p) $\star$ res(3,q) $\multimap$ res(4, p+q) $\star$ res(2,p) $\star$ res(3,q)) indicates that the calculation of the value of $f(4)$ is resolved into the calculation of $f(2)$ and $f(3)$. Since the results of these calculations are not yet known they are represented by bound variables.

In the strict technical sense, this universally quantified formula with bound variable represents neither a specification formula nor can it be interpreted as a state of a higher order transition system. On the other hand, the above interpretation of this state is reasonable. By applying other rules of the specification, this state can be rewritten until the base cases $f(0)$ and $f(1)$ of the recursion are encountered. Therefore, this example provides further ideas for an extension of the theory.

The above specification is highly non-deterministic and there are various ways to coordinate a computation. This suggests the development of mapping or coordination languages which allow to choose a specific computation from the large number of possible computations. [Led91] discusses such an approach in the framework of functional and imperative programming.

6.9 An Implementation

In order to test the linear logic specification formalism, a subset of the specification language, called PETRIL, was implemented in an experimental prototype [KK94]. With the goal of obtaining an operational prototype within a short period of time, a number of restrictions were made: Presently only finitely branching first order specifications without algebraic equations may be used. Work is in progress to enhance the implementation. Also some syntactic adaptions were necessary to allow the quick development of scanners, parsers, and compilers using the UNIX tools `lex` and `yacc`.

The implementation consists of a *compiler* which translates a linear logic specification into a C program which allows the construction of the branching structure of the specified process by an interactive exploration guided by the user.

First, the linear logic specification is scanned, parsed, and the resulting information is stored in a *complicated table structure*. Then this information is linked with a C program which contains *specification invariant functions*, for example a unification algorithm, a token and rule manager, and a simple ASCII user interface. *Extension hooks* are provided to extend the ASCII interface to a graphical user interface, to enhance the unification by an equational theory and a rewrite system, and to add more sophisticated token and rule managers which can deal with higher order specifications and certain fairness assumptions.

The main loop of the simulator inspects the current global state which is a bag of tokens. It tests which transitions are enabled by comparing sub-bags of the global state with the preconditions of the transitions. If the preconditions contain variables, then the unification algorithm is started to determine the most general unification for the variables. As a further restriction to the language, all variables which occur in the postconditions of a transition must also occur in the preconditions. Hence, no non-deterministic choice of an element of the employed algebras is possible. In the case of finite sorts, this restriction may be removed by a transformation of the specification, but infinitely branching non-determinism presently cannot be used. Then the user may choose a single transition or a concurrently enabled bag of transitions, the system determines the new global state and writes it to the screen.

Thus, the user may ***explore the branching structure*** of the specified process. A non-deterministic choice is resolved either by the interactive decisions of the user or with the help of a random number generator. Since the algorithm does not operate on the basis of string comparisons but is driven by a complicated table structure, it is fairly fast.

Present experiences with this prototypical implementation contain several versions of the stack, the bag, and the alternating bit protocol[8]. It is intended to significantly expand this implementation into a ***prototyping tool for distributed and parallel processes***.

[8]More details on this implementation are provided in Section 6.6.

Chapter 7

Categorical Interpretations

The value of rubbish is unaltered by translation, interpretation, or citation.

D. GUNTON,
Gunton's Laws of Librarianship.

During our work on the present process theory, ideas and constructions from category theory often provided important hints for our design decisions. In this chapter, the connections between transition systems, linear logic, and categories are illustrated.

Section 1 contains some general remarks on the *correspondence* of different syntactic frameworks, illustrated by the example of the CURRY-HOWARD correspondence.

Section 2 presents a categorical interpretation of *transition systems*.

Finally, Section 3 outlines connections between *linear logic* and category theory.

7.1 The CURRY-HOWARD Correspondence

The development of higher order transition systems in Chapter 4 was guided by the concept of a closed category. In functional programming, this categorical construction is an elegant framework for dealing with higher order functions. For further technical information on these categories, see Appendix C.

The famous CURRY-HOWARD correspondence [GLT88] is a most fascinating connection of rather different areas of theoretical computer science. It demonstrates that the following formalisms are only different syntactic incarnations of the same concept:

(1) Cartesian closed categories (CCC) [LS86].

(2) Intuitionistic, first order logic [Gol84].

(3) Type theory [AL91].

(4) Typed lambda calculus [LS86].

(5) The categorical abstract machine[1] [Cou90b].

(6) The theory of functional programming [FH80].

The achievements of the CURRY-HOWARD correspondence and of similar *connections between formalisms* is the following: Although the connected formalisms are just different syntactic incarnations of the same idea, there are certain problems which can be more easily solved in the framework of one specific formalism than in the related ones. For example, it is decidable whether two expressions which denote morphisms in the free cartesian closed category are equal. To prove this, the decision problem is translated into typed lambda calculus where a decision algorithm easily can be derived [LS86]. Hence, there also must be a decision algorithm in the framework of category theory, i.e. a canonical rewrite system for the equational theory of morphisms of a cartesian closed category. However, such a rewrite system has not yet been found [LS86], [Hue90].

Higher order transition systems may be equiped with the structure of a symmetric monoidal closed category (SMC), which is a very close relative of the CCC. Apart from technical details, the definition of an SMC[2] is obtained by a minor modification of the definition of a CCC[3]: A condition which requires the *unique* decomposition of certain objects is replaced by a condition which only requires the *existence* of a decomposition and no longer contains the uniqueness requirement. See Remark 372 for a synopsis.

This raises the question which effect this minor modification has on the other formalisms, and whether they may be adapted so that a similar connection of formalisms may be constructed. This question has not yet been settled entirely. However, it seems that the modification in the definition of a CCC may indeed be carried over accordingly to the other concepts so that we obtain the following connection:

(1) Symmetric, monoidal closed categories (SMC) [Mac88], [Tro92].

(2) Positive fragment of linear logic [Gir87a], [Tro92].

(3) A refined theory of linear types for functional programming [Wad90].

(4) Linear lambda calculus, [Abr90a].

(5) The linear abstract machine, [Laf88], [Tro92].

(6) The theory of higher order transition systems.

All the obtained concepts promise most interesting results: Linear types, for example, allow an efficient implementation of functional programming languages without garbage collection mechanisms. The linear abstract machine provides an operational mechanism for such an implementation.

Our application of this interconnection is the development of a process specification language which is based on linear logic. The following observations [Gol84], [LS86] allow us to connect categories, formal systems and transition systems:

[1] A kind of mathematical machine, similar to the TURING machine.
[2] See Remark 372.
[3] See Definition 369.

Category theory deals with two kinds of entities: *Objects*, and *morphisms*. Every morphism is associated with a pair of objects.

Logical systems also deal with two kinds of entities: *Formulae*, and *proofs*. Assume that for a formal system in pure deduction notation[4] $\varphi \vdash \psi$ holds. Usually there exists a set of rules, from which this fact may be derived in a more or less formal manner. This process of deriving $\varphi \vdash \psi$ is called a *proof*. Proofs may be represented in various different formats. In Appendix B, for example, the proofs in the formal system of linear logic are represented by proof trees. More generally speaking, a proof of the fact $\varphi \vdash \psi$ is some formal entity which is associated with the pair (φ, ψ) of formulae.

Also *transition systems* deal with two kinds of entities: *States* and *transitions*. In typed transition systems, a transition is associated with exactly one pair of states.

Thus it is reasonable to propose the following correspondence:

Transition Systems	Logic	Categories
States	Formulae	Objects
Transitions	Proofs	Morphisms

In a more *computational interpretation*, we may think of states (or formulae, or objects) as states of a computational device, and of transitions (or proofs, or morphisms) as executing programs which transform one state of the device into another.

7.2 Transition Systems

In this section it is illustrated[5] how every entity from the theory of transition systems may be interpreted in the language of category theory.

The set of *states* corresponds to the class of objects.

The set $\mathfrak{T}(A, B)$ of *transitions* from a state A to a state B corresponds to the class of morphisms from object A to object B.

The *identity transitions* id_A with $A \xrightarrow{id_A} A$ correspond to the identity morphisms $id_A : A \rightarrow A$.

The *composition* of transitions corresponds to the composition of morphisms.

A *functor* of a category operates on objects and morphisms in a form so that the information which is contained in the composition of morphisms is conserved[6] across the application of the functor. In a transition system, a bifunctor constructs a new state $\mathcal{F}(A, B)$ from two states A and B. For every pair of transitions $t : A \longrightarrow A'$ and $u : B \longrightarrow B'$ from the states A and B to the states A' and B', the functor provides a transition $\mathcal{F}(t, u) : \mathcal{F}(A, B) \longrightarrow \mathcal{F}(A', B')$ between the states which are obtained by an application of the functor.

Intuitively speaking, a *covariant* functor conserves causality and the direction of time,

[4]See Section B.1.

[5]i.e. we will point out the links but we will not redo the entire theory.

[6]For a more technical definition, compare the diagrams in Figures C.5 and C.6 which illustrate the action of a functor on a closed diagram.

whereas a *contravariant* functor reverses causality and the direction of time: If $t : A \longrightarrow B$ is a transition, then A is the cause and B the consequence of the state change. A contravariant functor produces a transition $\mathcal{F}(t) : \mathcal{F}(B) \longrightarrow \mathcal{F}(A)$ in which the transformed cause plays the role of the consequence and vice versa. The action of a contravariant functor on a composition of transitions further supports this interpretation.

The functors which play an important role in process theory are the *product bifunctor* and the *exponential bifunctor*.

The pair $(\cdot, \odot)$ of operations on states and transitions represents the *product bifunctor* which is covariant in both arguments[7]. $\cdot$ is the object part of the functor and $\odot$ is its morphism part. The functor is defined by

$$\mathcal{F} : \hat{S} \times \hat{S} \to \hat{S} \qquad \mathcal{F}(A, B) := A \cdot B$$

$$\mathcal{F} : \mathfrak{T}(A, B) \times \mathfrak{T}(C, D) \to \mathfrak{T}(A \cdot C, B \cdot D) \qquad \mathcal{F}(f, g) := f \odot g$$

The pair $(\twoheadrightarrow, \rightarrowtail)$ of operations on states and transitions represents the *exponential bifunctor* which is contravariant in its first and covariant in its second argument[8]. $\twoheadrightarrow$ is the object part of the functor and $\rightarrowtail$ is its morphism part. The functor is defined by

$$\mathcal{E} : \ddot{S} \times \hat{S} \to \hat{S} \qquad \mathcal{E}(A, B) := A \twoheadrightarrow B$$

$$\mathcal{E} : \mathfrak{T}(A, B) \times \mathfrak{T}(C, D) \to \mathfrak{T}(B \twoheadrightarrow C, A \twoheadrightarrow D) \qquad \mathcal{E}(f, g) := f \rightarrowtail g$$

The peculiar variance of this bifunctor, contravariant in the first and covariant in the second argument, follows technically from its definition[9]. Intuitively, it can be interpreted in the following way: As outlined in Remark 145, this bifunctor corresponds to a gap closing transformation. From the point of view of the gap, the left hand argument lies in the past and the right hand argument lies in the future. Thus, the treatment of time in both arguments is different.

A *natural transformation* connects two functors $\mathcal{F}_1$ and $\mathcal{F}_2$ in a structure preserving way[10]. We obtain the same state whether we first apply the functor $\mathcal{F}_1$ and then the natural transformation or if we apply the natural transformation first and then the functor $\mathcal{F}_2$.

The *abstraction* Λ corresponds to the abstraction in the definition of a closed category[11].

In higher order transition systems the property that the abstraction is a natural transformation is equivalent to the *naturality condition*. It demonstrates that in a transition $t : A \cdot B \longrightarrow C$ and in the corresponding transition $\Lambda(t)_{A,B,C} : A \longrightarrow (B \twoheadrightarrow C)$ not only the same information is represented, but that the mode in which these two transitions are associated stays the same, even if state changes[12] for A, B, and C are considered.

[7] Compare Definition 344.

[8] Compare Definition 345.

[9] Compare Proposition 370, Remark 145, and Proposition 152.

[10] Compare the commuting diagrams in Figures C.7 and C.8.

[11] Compare Definition 365, Definition 369, and Remark 372.

[12] For this interpretation, consult the diagram of the naturality condition in Figure 4.1 and the commuting diagram in Figure C.23.

A *first order transition system* is a symmetric monoidal category[13] in which the canonical natural transformations are all equal to the identity.

The *coherence theorem*[14] for symmetric monoidal categories demonstrates that the construction principles of a first order transition system are not a source[15] of non-determinism.

A *higher order transition system* is a symmetric monoidal closed category[16] in which the canonical natural transformations are all equal to the identity.

The *coherence theorem*[17] for symmetric monoidal closed categories is connected with the conjecture that the construction principles of a higher order transition system are not a source[18] of non-determinism.

Categorical structure theory was applied to all forms of processes and transition systems during the preparation of this work. For example, it is possible to define categories of processes and of first and higher order transition systems. In the extension order[19] of a behaviour the concept of a retraction[20] was helpful.

7.3 Linear Logic

A categorical interpretation of a logical system is constructed along the following ideas:

The *objects* of the category represent the formulae of the logic. They are syntactic constructions or words of a language, without any interpretation. In this sense, the formulae $(\varphi \star \psi)$ and $(\psi \star \varphi)$ of linear logic are different objects, despite the fact that they have the same semantics and that it is possible to prove their equivalence.

The *morphisms* of the category represent the proofs of the deduction system: If there exists a morphism from object φ to object ψ, then there exists a proof for $\varphi \vdash \psi$. In addition to deduction, a categorical interpretation of a logical system also contains a calculus of proofs, since two transitions between the same two states can be equal or not.

As discussed in more detail in Section 5.1, there are the following correspondences between operators in transition systems, in category theory, and in linear logic:

Transition S.	Category	Linear Logic
$\circ$	$\circ$	
$\cdot$	$\cdot$	$\star$
$\odot$	$\odot$	
$\twoheadrightarrow$	$\Rightarrow$	$\multimap$
$\rightarrowtail$	$\Rightarrow$	

[13]Compare Remark 372.
[14]Compare Theorem 373.
[15]See Theorem 165.
[16]Compare Remark 372.
[17]Compare Theorem 374.
[18]Compare the discussion in Remark 166.
[19]Compare Definitions 30 and 75.
[20]In fact the injections of the respective definitions are *retractions* and not *monomorphisms* of behaviours. This is essential for the definition of the extension.

In favour of a better readability of this section we will not distinguish these operators in the same strict sense as in Section 5.1 but rather allow some mild form of overloading.

A category contains for every object φ the *identity morphism* id_φ. Hence in every logic for which there is a categorical interpretation, one can show that $\varphi \vdash \varphi$ holds for every formula φ.

The *composition* of morphisms corresponds to the transitivity of the deduction relation. Assume that $f : A \to B$ and $g : B \to C$ are morphisms. In the logic this corresponds to the deductions $A \vdash B$ and $B \vdash C$. The morphisms f and g may be interpreted as proofs for these deductions. In the category the morphisms f and g can be composed to the morphism $g \circ f : A \to C$. In the logic the deduction rule (*tra1*) from Theorem 256 allows to derive the deduction $A \vdash C$. The composed morphism $g \circ f$ can be interpreted as proof for this deduction. Thus, the composition operator of the category corresponds to the deduction rule (*tra1*) of the logic.

Category	**Logic**
$f : A \to B$	f proof of $A \vdash B$
$g : B \to C$	g proof of $B \vdash C$
$g \circ f : A \to C$	$g \circ f$ proof of $A \vdash C$

Functors of the category represent rules of the deduction system: They take objects and morphisms as arguments and return new objects and morphisms. Thus functors construct at the same time new formulae and new proofs.

For example, the *product bifunctor* $(\cdot, \odot)$ acts on morphisms in the following form $\odot : \mathfrak{M}(A,B) \times \mathfrak{M}(C,D) \to \mathfrak{M}(A \odot C, B \odot D)$. If $f : A \to B$ and $g : C \to D$ are morphisms, then this corresponds to two deductions $A \vdash B$ and $C \vdash D$ of which f and g are the proofs. In linear logic, the deductions $A \vdash B$ and $C \vdash D$ can be used to derive the deduction $A \star C \vdash B \star D$. In a monoidal category, the product bifunctor can be used to construct the morphism $f \odot g : A \cdot C \to B \cdot D$.

Category	**Logic**
$f : A \to B$	f proof of $A \vdash B$
$g : C \to D$	g proof of $C \vdash D$
$f \odot g : A \odot C \to B \odot D$	$f \odot g$ proof of $A \star C \vdash B \star D$

A similar interpretation is possible for the *exponential functor* $(\twoheadrightarrow, \rightarrowtail)$:

Category	**Logic**
$f : B \to A$	f proof of $B \vdash A$
$g : C \to D$	g proof of $C \vdash D$
$f \rightarrowtail g : (A \twoheadrightarrow C) \to (B \twoheadrightarrow D)$	$f \rightarrowtail g$ proof of $(A \multimap C) \vdash (B \multimap D)$

The *natural transformations* of the category provide additional morphisms and thus additional proofs. Due to the naturality condition[21], they also provide means for restructuring proofs.

[21]Compare the diagrams in Figures C.7 and C.8.

For example, the natural isomorphism $\mathbf{A}_{A,B,C}$ supplies an isomorphism $A \odot (B \odot C) \rightarrow (A \odot B) \odot C$. Furthermore, if $f : A \rightarrow A'$, $g : B \rightarrow B'$ and $h : C \rightarrow C'$ are morphisms it produces the commuting diagram which is shown in Figure 7.1.

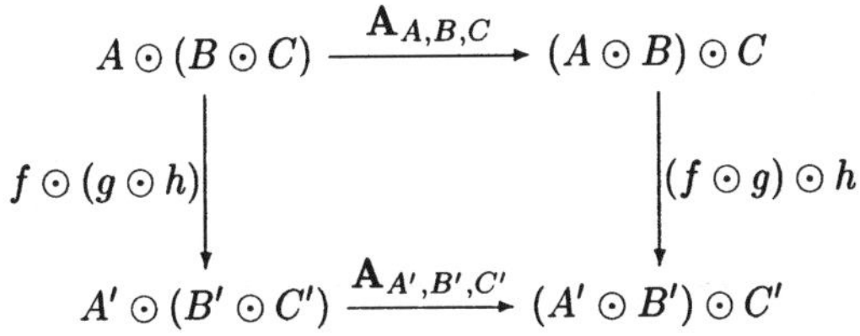

Figure 7.1: Natural Isomorphism for Associativity.

Logically, this can be interpreted as follows: Assume f, g and h are proofs for $A \vdash A'$, $B \vdash B'$, and $C \vdash C'$, respectively. Then there exists a proof $f \odot (g \odot h)$ for $A \star (B \star C) \vdash A' \star (B' \star C')$ and a proof $(f \odot g) \odot h$ for $(A \star B) \star C \vdash (A' \star B') \star C'$. These two proofs can be converted into each other with the help of the natural isomorphisms provided by $\mathbf{A}$.

A categorical interpretation of a logical system provides a *calculus of proofs*. In addition to the information whether a deduction[22] exists, it also contains information on the equality of proofs. Therefore, it is an excellent arena for discussing proof normalization.

Furthermore, a categorical interpretation also allows an interesting observation on the *quality* of proofs: Assume two formulae A and B are equivalent in the sense that $A \vdash B$ and $B \vdash A$. Then there exist morphisms $f : A \rightarrow B$ and $g : B \rightarrow A$. However, these morphisms not necessarily are inverse to each other. From a computational point of view, proofs which are isomorphisms can be "undone" during proof search: Assume that during the construction of a proof, a proof step f is added to the current proof expression, extending the current goal from formula A to formula B. Later it turns out that backtracking is necessary and that we have to go back from formula B to formula A. This always can be done by adding a suitable morphism to the current proof expression. If the original morphism f was an isomorphism, it is possible to add its inverse to the proof expression which is the same as removing the original application of f. This removal produces shorter proofs.

Similar observations connect *non-determinism* in transition systems, *coherence*[23] in categories, and the existence of *more than one proof* for a single deduction.

[22]and thus a proof, i.e. a morphism.

[23]Compare Theorms 373 and 374.

Chapter 8

Conclusions, Challenges, and Opportunities

If a professor thinks what matters most
Is to have gained an academic post
Where he can earn a livelihood, and then
Neglect research, let controversy rest
He's but a pitty tradesman at the best,
Selling retail the work of other men.

KALIDASA, between 200 BC and 400 AD,
In J. BROUGH, *Poems from the Sanskrit.*

This chapter contains a survey of the achieved results as well as a list of open questions, conjectures, and ideas for future work.

Section 1 summarizes the *results* which have been achieved in the present work.

Section 2 addresses open *theoretical questions* and Section 3 discusses *practical problems*.

Finally, Section 4 contains some *speculative ideas*.

8.1 What Has Been Achieved?

In the present work we discussed the well-known *transitional theory* of non-deterministic sequential processes. We *extended* this theory to distributed systems by adapting the *causality mechanisms* from PETRI-net theory and by including ideas of colored PETRI-nets.

We introduced the concept of a *branching structure* which provided additional insight into the behaviour of distributed systems. Especially, we analyzed *confusion* and *fairness*. Next, we added *higher order features* to our concept of transition systems.

In order to obtain a logical framework for reasoning about processes and a language for specifying them we provided a connection between the positive fragment of *linear logic* and higher order transition systems and proved correctness and completeness results.

Finally, we illustrated the practical application of the achieved calculus in a number of *examples*.

According to Section 1.4, our process theory can be *classified* as follows: It is a *partial order theory* which places its *emphasis on states* rather than on state changes or actions, and it is an *observational theory* based on a mechanism of *causality* rather than a compositional theory.

Interaction of Parallelism and Non-Determinism

We presented a calculus which allows an *orthogonal, independent treatment* of parallelism and non-determinism. *Parallelism* is treated in the sense of a partial order theory and thus is well adapted to the physical concept of time as well as to the experiences made with distributed systems. *Non-determinism* arises from conflicts in preconditions. Both concepts have a natural place in the calculus. The *interaction* of parallelism and non-determinism may be described by using the concept of a *branching structure*.

Higher Order Process Theory

We presented a *higher order* process theory which allows the construction and modification of a process *during execution*. This feature is not available in other process theories.

In PETRI-*net jargon*, the higher order features can be explained as follows: PETRI-nets are *static* – as soon as the topology of the net has been defined it remains the same throughout the firing of transitions. Our new higher order concepts are *dynamic* – they may be used to change the topology of the net by the firing of higher order transitions. This new concept goes beyond the modeling capabilities of PETRI-nets.

Higher order constructions can be employed in *practical specifications*, as this was done in Example 206 for the stack and in Section 6.8 for the FIBONACCI numbers. They provide additional possibilities for *structuring the control flow* of a process.

Higher order concepts allow a concept of *partial execution*, i.e. an analogon of *partial evaluation*.

Multiparadigmatic Programming

We introduced a single formalism which allows both a *procedural* and a *declarative* style of specification. Thus, it is possible to program in and transform between these two programming styles in the same formalism and even within a single program. The advantage of this kind of multiparadigmatic programming was discussed in detail in Remark 213. Below, we present additional applications to functional and algebraic programming.

Pure functional languages are free of side-effects and have no notion of a state. Therefore, it is impossible to perform any I/O operations, since the related functions have side-effects;

in fact, the only reason why I/O functions are called is to execute the associated side-effect. There are two solutions to this problem:

The first solution is to define *impure functional languages* which contain only a small number of constructs which are not free of side-effects. As a consequence, all functions which contain these constructs may inherit their side-effects. Thus, the main advantage of functional programming, i.e. not having to worry about states or side-effects, and the possibility of using the principle of referential transparency, is lost. The second solution is to *include everything* which could destroy referential transparency in the argument list of the function. This approach is theoretically the correct approach but for practical applications it is a burden, since now the entire disk, the screen, and even the user who provides the program with input data must be included at least conceptually into the argument list of every function, because the output of the functions may depend on all these elements.

Our specification method provides a more elegant solution: All the function definitions of a functional program which are *free of side-effects* are transformed into equations. The parts which *produce side-effects* are modeled with the help of the process axioms. Thus, we obtain a specification where the side-effect free and the side-effect producing parts of a process are immediately visible. However, when transforming functional definitions into equations, the information in which direction and in which order the equations must be used is lost. In a certain sense this transformation adds non-determinism to the program. This problem may easily be solved:

A very attractive possibility is to employ *coordination constructs* which furnish the information how the equations must be used in order to obtain efficiently the results of the original functional program. Below, this option is discussed in more detail.

The second solution is to transform the functional program into a purely procedural specification. For modeling the execution of every functional definition, a special kind of internal states is used. By distinguishing these states from other states it is possible to recognize where side-effects occur.

In *algebraic programming languages* like OBJ [GWMF93], a specification consists only of equations. A general equational reasoner is used to apply the equations during the calculation. Depending on the structure of the equations, programs in such algebraic languages can be very inefficient when the equational reasoner uses an inefficient proof strategy. Once more our specification formalism can be used for supplying additional *rewrite information*: By transforming equations into process axioms, directional information can be provided. It is even possible to simulate the effect of the KNUTH-BENDIX procedure [Wir90] in these transformations. Again, successive refinements from one programming style to another are supported in the framework of our process calculus.

Specification of Behavioural and Data Aspects

Our process theory incorporates the *behavioural* and the *data* aspects of a process. These aspects may be transformed into each other. The data aspect was part of the calculus from the beginning, i.e. it has not been added at a later stage, and it has a natural place in the calculus.

State Driven Programming

Our process calculus supports a *state driven* style of programming. As argued in the following, this programming paradigm is a most adequate approach for modeling real systems.

In many present specification and programming languages, the *linear structure* of the specification text imposes a *linear semantics*. For example, in imperative languages, the sequence in which the statements occur in the specification text determines the sequence of the execution of these statements. Thus, the programmer is lead to a *linear mode of reasoning* about programs: "First this statement is executed, then procedure so-and-so is activated, etc." Parallel or non-deterministic languages further complicate this.

This kind of linearity poses a *software engineering problem*: In the case of conditionals, loop, or jump statements, a human reader is required to dig in the listing in search of the "else" part of a conditional statement or of the beginning of a loop. If the specification is non-deterministic then all possible choices must be considered. In the case of parallel programs this is still more difficult since now the human reader must perform the same work in an n-fold way, n being the degree of parallelism. Furthermore, all possibilities of relative execution speeds must be considered.

This becomes worse in the specification of *reactive systems*. For example, advanced window systems like X11 have a complicated notification system: Events like moving the mouse or typing on the keyboard must be routed from the operating system to the correct client programs which are responsible for the activation of callback functions, event handlers, or other kinds of functions. These callback functions and event handlers are declared upon startup of the program or during its execution. Their activation may depend not only on the sequential execution of the program but also on the history of mouse and keyboard events, on the speeds by which these requests could be served, on the remaining items in the event queue, etc. The program no longer is executed in a sequential mode and it is often extremely difficult to understand in which way the thread of execution switches to interrupt routines, event handlers, or callback functions, and back to the main program.

The *state driven programming paradigm* is based on the idea that certain actions may take place if certain (pre)conditions are satisfied. Every real system functions according to this paradigm and thus this paradigm provides a most adequate approach to process specification.

From this point of view, linear[1] programming languages require the *introduction of an artificial state*, a so-called line number or program counter, and the execution of statements must be made dependent on the present value of this counter. The decades of training and programming in linear languages can mislead us to believe that this is a natural concept. However, only real world systems whose behaviour is controlled by clocks give rise to such a counter. Most real world systems[2] are asynchronous and not of the above mentioned kind.

In *object oriented programming*, the concept of sequential execution is replaced by the invocation of methods and the exchange of messages between objects. For many purposes this provides a much better model for describing real world systems. In our process theory the

[1] In *linear* logic the term "linear" addresses properties of proofs. This is in no way connected with the linearity of programming languages.

[2] Computers are controlled by a clock and thus linear languages are well suited to describe them. However, the *ultimate* goal of computer science is not to describe computers but to describe real world systems with the help of computers. The technical details of the employed tools, i.e. computers, should be hidden from the user as far as possible. This observation accounts for what at first sight seems a paradox.

execution of a transition depends on the presence of certain local states. This proves to be a most adequate form of modeling real systems. In the case of numerical computations this technique is equivalent to the *data-flow paradigm*: Here the local states whose presence allow the execution of an assignment statement are the values which are required for the calculation of the value which shall be assigned. Many examples in this work can be interpreted in this fashion. In the case of *reactive systems*, our approach directly supports a *state driven* type of programming. The following formula illustrates the basic idea:

$$\texttt{button\_pressed} \star \texttt{mouse\_at}(\texttt{x},\texttt{y}) \star \texttt{is\_in\_button\_rectangle}(\texttt{x},\texttt{y},\texttt{buttonid}) \multimap$$

$$\multimap \texttt{activate\_handler}$$

In order to further verify the above claims, more examples of reactive systems must be specified and prototyped in our process specification calculus. Communication protocols, user interfaces, interrupts, and notification systems provide excellent candidates.

Basic and Derived Concepts

It was shown by a considerable number of examples that the mechanism of *local causality* which is employed in our process calculus is able to specify all those process behaviours which in most other calculi are derived by composition operators. Furthermore, we argued that this state based causality is closer to real world systems than an action based approach using composition operators.

We conclude that *local causality* is a more basic feature in process theory than composition operators, that the composition operators should be understood in terms of local causality, and that causal mechanisms should be essential concepts in future programming languages.

Logic Framework for PETRI-Net-Like Systems

We provided a *logical framework* for PETRI-net-like systems by connecting transition systems with linear logic. Although the axiomatizations for a number of interesting relations have not yet been developed fully, a complete and correct deduction system which may be applied to PETRI-nets has been presented.

Modularity for PETRI-Net-Like Systems

We presented a concept which provides a modular approach to processes and may be applied to PETRI-nets as well. If several processes are combined to a new process the possible interactions may be localized by analyzing the set of states of both processes.

Categorical Interpretation

A process, i.e. a distributed transition system, is the structural kernel of both a first and a higher order transition system. The latter two are special *categories*. Thus, there is a close connection between process theory and category theory which provides a number of fascinating insights when a concept from process theory is interpreted categorically and vice versa. During the development of our process theory, many design decisions were guided by this connection.

This correspondence of process theory with category theory, and subsequently with type theory[3] and labeled linear logic, is a kind of CURRY-HOWARD *isomorphism for processes*.

With this interpretation it is now fairly easy to explain to a category theorist what a process is: It is the structural kernel of a symmetric monoidal closed category.

Comparison with Process Algebraic Approaches

In contrast to process algebra ACP [BK90] and similar specification techniques like CCS [Mil89] or CSP [Hoa85], our process specification technique allows an *orthogonal treatment* of parallelism and non-determinism.

In process algebra, the state of a system is determined by the previous actions and the *algebraic transformations* which were used to access them in the process expression. Although process algebra allows to model the state of a system explicitly by using the state operator [BW90a] or by employing parametrized actions [GP90], [GP91], the algebraic structure of the process expression plays the most important role. This contribution to the state is difficult to interpret. In our process specification technique, the states of all components of the system always are explicitly modeled.

On the other hand, process algebra is an *equational calculus* which may be handled more easily than a logical calculus. In contrast to linear logic techniques, there is a considerable number of worked examples [Vaa90], [BW90a], [MV93], [Vaa89], [BK86], an elegant notion for abstraction [BW90a] and for fairness [BBK87], and an equational axiomatization for these notions. Several properties of algebraic processes are known to be decidable, the respective algorithms have been studied [CPS91], there exists a specification language and a software engineering environment [MV93], [Mau92].

Many of the above mentioned points are due to the *compact and aesthetic structure* of process algebra and the relatively large number of person-years already spent in process algebraic research.

Comparison with PETRI-Net Techniques

Several aspects of our specification technique are *similar* to PETRI-nets.

PETRI-nets are based upon a *graph concept*, and in order to formalize nets, a more convenient description of the net structure is necessary. Our specification technique starts immediately with a *formal description* and a *logical framework*. Therefore, our approach is better suited for reasoning with specifications, as well as for machine-assisted or machine-conducted transformations and analyses of specifications.

PETRI-nets are *static* and it is not possible to describe a modification of the net during its execution. Linear logic specifications can directly make use of the higher order constructions

[3] A category gives rise to a type theory on the objects and morphisms, and a process gives rise to a type theory on the states and transitions.

and hence allow a *dynamic* description of processes.

PETRI-net theory often is introduced by using the concept of a *marking* which corresponds to a *global state*. As discussed in Sections 2.2 and 3.1, global concepts are inadequate for distributed systems. Our specification technique ***avoids global concepts*** and uses them only as an a posteriori explanation mechanisms for confusion and fairness phenomena.

Comparison with Event Structures

Similar to event structures, our approach is a partial order theory which is able to treat parallelism and non-determinism as orthogonal concepts. On the other hand, in our approach parallelism and non-determinism are consequences of a *single basic mechanism*, i.e. local causality, whereas in event structures non-determinism is described by an *additional conflict relation*. Further research and comparisons like those made in [BMC64] and [Pra94a] must analyze the connections between these two approaches.

8.2 Theoretical Problems

The history of process calculi shows that the initial design of the formalism *stimulates further research* and that the clarification of important questions usually continues for several decades. We believe that the results which we have achieved in the present work justify further research for clarifying a considerable number of very interesting questions which arise in the present state of this calculus.

In this section, theoretical questions are discussed and in the subsequent section, practical issues are outlined.

Loose Semantics

In this work, an *initial semantics* approach[4] was used: In Definition 146 of the algebra of higher order states and in Definition 147, initial algebras were used. Similarly, the set of transitions as it is defined in Definitions 153 and 132 is constructed by using initial semantics. This approach has the advantage that the higher order transition system which is specified by a process specification can be constructed by syntactic means, and that it is unique up to isomorphism.

One should develop and analyze other approaches to the semantics of process specification. Especially a closer look at *loose semantics* would be interesting. Such an approach could also help solving the problem in the footnote to Theorem 188. This problem arose since in our approach to higher order transition systems the states which are obtained by interpreting the formulae $c \multimap (e \multimap f)$ and $c{\star}e \multimap d{\star}f$ are necessarily different. In loose semantics they could be identified. On the other hand, in a loose approach the connection between specifications and specified transition systems would be more complex.

[4] Compare [Wir90] for a more detailed discussion of initial and loose semantics.

Geometry of Non-Determinism

Non-determinism and parallelism can interact in very complicated ways. We believe that not yet all possible forms of this interaction are fully understood and thus suggest to develop a *classification* of confusion and of other non-deterministic phenomena. Possible categories for such a classification were suggested in Section 3.7. However, the forms of non-determinism in this section were detected by experimentation and they constitute no systematic classification. Therefore, it is necessary to define a notion when two *non-deterministic situations* shall be considered equal. For example, we could consider two conflict introducing confusions as equal, iff the respective conflict sets have the same number of elements.

Branching Structure

Given a finite branching structure which is only labeled by transitions, *how many states* are required to construct this branching structure? This question arises, if a system shall have a certain parallel and non-deterministic structure and one wants to realize this structure and determine its complexity. A similar question arises for *infinite behaviours*. A further problem which is associated with the branching structure is addressed in the footnote to Definition 72.

Additive and Exponential Operators, Linear Negation

The *additive operators* of linear logic were not used in the present approach. Their interpretation in Section B.7 indicates that they contain non-deterministic aspects. The *exponential operators* of linear logic were used to deal with sets of non-logical axioms. Their interpretation in Section B.7 indicates that they may be used as a kind of storage of states as well as for reproducing states.

It should be studied how the mentioned operators can be included into the framework of higher order transition systems which advantages – or disadvantages – this might cause, and how this would affect the power of the present specification technique. Furthermore, the role of linear negation must be analyzed.

Quantified Formulae

In Section 6.8, *universally quantified expressions containing bound variables* were used to represent intermediate states of a recursive computation. Quantified formulae are not yet part of our specification language nor may they be interpreted as states of a higher order transition system. On the other hand, Section 6.8 motivates that they can play an important and very useful role for understanding recursion.

The notion of transition systems shall be extended in order to handle *states containing bound variables*. The theory shall be extended to allow the analysis of *intermediate states* of *recursive computations*.

Axiomatization of Linear Logic

There is a large number of *different axiomatizations* of the deduction relation of linear logic, see [Lin92], [Tro92], [LS94], [Sch91]. The effect of choosing a specific axiomatization on theoretical and practical work is not yet fully understood.

Many references introduce operators as syntactic abbreviations and not by providing deduction rules. [Tro92] for example defines $\varphi + \psi :=\sim ((\sim \varphi) \star (\sim \psi))$. These abbreviations may be transformed into deduction rules but the resulting syntactic properties are not fully known. The present approach is sufficient for theoretical purposes, for practical reasoning with large linear logic specifications it leads to problems.

There are several examples which show that even slight extensions and modifications of linear logic can have far reaching and astonishing consequences. Recently, H. SCHELLINX demonstrated that the differences between *intuitionistic* and full linear logic are larger than the research community had conjectured for a long time [Lin92]. In [Tro92] it is explained how the important property of cut-elimination can be lost in a *calculus of two implications*. Linear logic has been studied only for a very short period of time and therefore, the main references [Gir87a] and [Tro92] employ different notations and contain non-obvious typographical errors in critical parts, e.g. [Tro94], [Tro].

Another interesting syntactical problem in linear logic is the *axiomatization of the exponentials*. There seems to be no deduction system in which the two exponents ! and ? are axiomatized in separate deduction rules from which the duality of these two operators may be deduced.

Thus, it is necessary to give a *complete overview of the axiomatizations* of linear logic and to *study their syntactic properties*.

Decidability of Linear Logic

For supporting linear logic process specification, *decision algorithms*, *proof tools*, and *proof checkers* are required.

Unfortunately, full linear logic is undecidable [Tro92], [Lin92]. Only recently some fragments could be shown to be decidable and the respective complexity classes could be determined [Kan91b], [Kan91a], [LMSS92], [LMSS90], [LS92], [LW92], [Lin92]. These papers provide deep theoretical insight into the deduction process of linear logic, but since they operate with encodings of decision problems by TURING machines, the decision procedures can hardly be used by the *working programmer* who wants to specify a system with linear logic.

Abstraction and Process Equivalence

During many phases of the software life cycle suitable notions of *process equivalence* are required for abstracting from details which are irrelevant for the specific phase: Especially testing, debugging, and validating a program requires techniques for abstracting from implementation details.

In Section 5.4 we illustrated some approaches to notions of process abstractions. In algebraic process calculi there exists a considerable number of similar notions, compare [Hüt94],

[BW90a], [BR83]. Hand in hand with practical experiences such notions should be developed for the presented process theory. Furthermore, correct and complete axiomatizations for the resulting relations must be found.

Properties of Processes

There are a number of important properties of processes, the most prominent being their equivalence with other processes. It should be studied which properties of processes can be *described* in linear logic. For example, in imperative programming it is known that the property of program termination cannot be described in first order HOARE logic [Apt81]. Furthermore, it is known that first order HOARE logic is a relatively complete deduction system for properties of processes [Apt81]. Similar results should be developed for process theory.

In classical recursion theory, the theorems of RICE and RICE-SHAPIRO discuss which properties of a function can be *decided* or *proven* by an analysis of an arbitrary program for this function, [Odi92], [EL92]. Similarly, it is necessary to determine which properties of a process can be decided or proven by an analysis of an arbitrary specification of this process. Due to the power of our concept we expect limitations which are similar to the theorems of RICE and RICE-SHAPIRO in classical recursion theory.

For the practical description of process properties a logic but also a practical engineering approach should be developed for processes, similar as this was done for computational programs by E. DIJKSTRA in [Dij76] and D. GRIES in [Gri89].

Recursion Theory for Processes

In classical recursion theory, there exist the concepts of a *universal function* and of *partial evaluation*. Based upon these concepts, a number of very interesting results have been achieved, for example the generation of compilers and interpreters, [EL92], [Gom92].

The higher order approaches of classical recursion theory and of the presented process theory are very similar. In fact, the concept of a partial evaluation also exists in our process theory. We conjecture that also a *universal process* can be defined. This is a process $\mathcal{U}$ with the following property: For every process $\mathcal{P}$ there exists a state p such that the process $\mathcal{U}$ exhibits the same behaviour as process $\mathcal{P}$ if $\mathcal{U}$ is initialized in state p. The phrase "the same behaviour" requires some kind of abstraction from implementation details.

We believe that along this line of reasoning similar results as in classical recursion theory can be developed and that these results may similarly be employed for a semi-automatic generation of compilers and interpreters. We believe that this field will become a most interesting research area.

Other Specification Techniques

The present form of higher order transition systems was descibed by using linear logic. It should be studied to what extent transitional structures can be described by using *modal* or *temporal* logic or with the help of *algebraic* means. Ideas for such approaches can be found in [Eme90], [Mer92], [McD82], [Sti92], [Krö87], [vBvES93], [Ber89].

Higher Order Techniques

There are still many open problems concerning higher order techniques. Especially the question whether the higher order formalisms would add non-determinism could not be answered in a satisfactory way. Furthermore, it is interesting to study whether the higher order concepts add power to our specification technique or whether all higher order constructions can be reduced to the first order.

Fairness

In process algebra, KOOMEN*'s fair abstraction rule* provides an elegant axiomatization of fairness [BBK87]. A similar logical approach to fairness is required for our process calculus.

In distributed transition systems, fairness notions were increasingly difficult to define formally. Thus, modal logic should be used to define and reason about fairness [Sti92].

8.3 Practical Problems

G. GOOS mentions the following problem with formal methods [Goo92]:

> Indem man die Begriffe "mit mathematischer Genauigkeit formulierte Spezifikationen" und "mathematisch abgekürzte Verifikation" durch die griffigeren Wörter "formale Spezifikation, formale Verifikation, formale Programmableitung" und was es sonst noch alles an 'Formalem' gibt, ersetzt, verschweigt man dem Publikum, daß man eigentlich die schwierigste Aufgabe noch vor sich hat. Diese besteht nur zu einem kleinen Teil darin, die Methodik weiterzuentwickeln. Vor allem geht es darum, die Methodik in eine solche Form zu bringen, daß sie für Tausende von Entwicklern in der industriellen Praxis, die die benötigten mathematischen Grundlagen während ihrer Ausbildung nicht gelernt haben, anwendbar wird. Formeln sind zwar das A und O einer präzisen Spezifikation, namentlich, wenn sie rechnergestützt weiterverarbeitet werden soll. Der Mensch denkt aber weiterhin in Bildern, nicht in Formeln. Gefragt ist daher die umgangssprachliche oder graphische Aufbereitung der Formeln ohne Verlust an Genauigkeit, wie sie die Mathematik seit Jahrhunderten vorführt.

In English this quotation reads as follows:

> By replacing the notions "specifications which are formulated with mathematical precision" and "mathematically shortened verification" by the fancier words "formal specification, formal verification, formal program derivation" and by whatever other 'formal notion', one does not tell the audience that the most difficult part of the task still lies ahead. The overall task consists only to a small part of

the improvement of the method. The main part is to transform the method into such a form that it can be applied by thousands of developers in the industrial environment who have not learned the required mathematical tools during their studies. Formula are the essence of a precise specification, especially if it shall be transformed with the help of a machine. However, man still thinks in images, not in formulae. Therefore, the intuitive or graphical form of these formulae must be obtained without losing precision; a task which has been successfully performed in mathematics for centuries.

Software Engineering Environments

Before a specification method can be put to work, suitable *languages* and *language processors* as well as *libraries* for routine tasks must be developed. The practical application of a specification technique requires also *support for the various phases of the life-cycle*, i.e. an environment which supports the editing, storing, archiving, and managing of specification texts, which helps debugging and simulating specified processes, which supports performance analysis and prediction, and which offers a variety of other convenient functions. Furthermore, proof checkers and generators are necessary.

Implementation

In Section 6.9 we presented an implementation of a subset of our specification language. Input was provided to the simulator by choosing a set of concurrently enabled rules and the resulting state changes were implemented by printing symbolic representations of the states onto the screen.

It is necessary to develop *more sophisticated forms* of implementations. The states of a process should not be represented by symbolic expressions on an ASCII interface but rather be directly associated with physical states of the computer, for example the disk, the screen, or the CPU. Clicking the mouse button or typing on the keyboard should submit input to the system at a lower level than that of a read-eval-print loop.

The present restrictions on the implementation should be removed in order that higher order specifications, equations, and infinitely branching non-determinism may be used. Features like E-unification [Kni89], coordination constructs for more efficient equational reasoning, and rewrite systems should be available. As outlined in Section 6.8, higher order specifications tend to be highly non-deterministic. Concepts like the *cut* from PROLOG and *backtracking mechanisms* could help solving this problem. Furthermore, it is necessary to find possibilities for integrating *fairness assumptions* into the implementations.

Executable Specifications and Prototyping

As we were able to show with the help of our implementation, our process calculus can be used to construct *executable specifications* and to prototype protocols and more general systems. It is necessary to gain more experience in the practical use of this approach in the context of large software systems. This also requires more support by tools. The definition and implementation of a *virtual machine concept* for our specification formalism could be helpful.

Coordination and Refinement

A *non-deterministic specification* defines more than one maximal behaviour which all are considered correct. In Section 6.8, the example of a numerical computation is presented which offers a considerable number of possibilities for computing the value of a function. Only in very rare cases, the final implementation still is non-deterministic. Nevertheless, during earlier stages of the software life-cycle the non-determinism contained in an under-specification provides an important form of flexibility.

Mapping or *coordination* concepts should be developed which would allow choosing a specific maximal behaviour. Ideally, this refinement of the specification can be made by furnishing additional information, i.e. coordination constructs, and not by modifying the existing specification. With these coordination constructs it should not be possible to change the process in a manner contradicting the original specification.

In [Led91] a pragmatic approach to mapping languages for functional and imperative programming is discussed.

Libraries

In the present form of our calculus, the user is required to specify many low-level details. For example, it is necessary to explicitly define and increment a program counter, if a sequential execution of commands is required. Also the mechanisms which attach an interrupt handler to an interrupting agent and which enable and disable interrupts during the execution of a program always follow the same principles. It is therefore necessary to provide these mechanisms in the form of *macros*, *libraries*, and *language extensions*.

The above constructions can be translated into their low-level linear logic form, or they can be implemented directly in hardware. For example, it is by far more efficient to map a program counter to the corresponding hardware register of the CPU than to interpret it in software. Similarly, an interrupt mechanism should be implemented by using the hardware interrupt controller. It is only possible to benefit from such *highly efficient implementations* if the programmer uses these macros, libraries, and language extensions which have been adapted to the employed hardware. It is difficult to imagine a preprocessor which scans a linear logic specification, detects automatically the states playing the role of a program counter, and maps them to the respective hardware register of the CPU.

Our specification formalism is quite powerful and it may be used to specify many complicated distributed and parallel algorithms for semaphors, mutual exclusion, synchronization, election of a leader, distributed consensus, distributed locking, transaction management, scheduling, and so on. In practical applications these algorithms pose big problems: They require low-level interfacing with the operating system, and sometimes even non-documented features must be used. They are very susceptible to programming errors and most difficult to debug. The common imperative programming languages are not very well adapted to this task. Our process theory provides a solution to this problem: A *library of distributed and parallel control algorithms* can be constructed with the help of our specification formalism.

Programming languages can be coupled with this mechanism, similarly as they are connected to system calls or runtime libraries. Finally, an implementation of the linear logic language or of the mentioned library must be provided.

Transformations and Annotations

Assume, a non-deterministic process has been specified and contains unfair maximal behaviours. Now this process shall be transformed so that it only exhibits fair maximal behaviours. *Transformations* of this kind should be described and studied within our framework. APT and OLDEROG show in [AO91] how such transformations may be carried out in guarded command languages.

For *real-time* applications, annotations must be added to the transitions in order to describe the amount of time which is required for their execution. This may be accomplished in a similar way as in process algebra [BB90b].

In many appliations, also the notion of *real-space* plays an important role. For example, one would like to describe on which computer which distributed service is running, one is interested in migrating processes from one computational agent to another, or one wants to specify which states or data in a system shall be stored on which site, memory, disk, or storage device. It is an interesting task to study how these notions may be added to our calculus. [BB90a] provides examples for such language constructs.

Complexity Issues

In connection with process theory, a number of complexity issues arise: *Structural complexity* as introduced by CHAITIN [Cha88] and KOLMOGOROV [Wat92] studies the amount of text which is required to specify a certain process. *Classical complexity* is associated with the amount of *time, space, and other resources* which is required for performing a certain task.

A specific question of this kind was posed in Example 119, concerning the minimal number of (internal) states which are required to realize a certain process, a certain pattern of non-determinism, or a service specification.

8.4 Speculations

This section contains speculative ideas which could be interesting to study. However, there is less concrete evidence for success than for the questions of the previous sections, and the associated research is less urgent.

A Linear Calculus as Basis of Process Specifications

Lambda calculus plays a prominent role as the fundamental basis of *computational programs* and recursion theory. It should be analyzed in which way a calculus which is developed around the concepts of linear logic is able to play a similar role for *non-computational* programs. In the course of the present work it has been demonstrated that the combination of linear logic and algebraic reasoning is a considerably powerful tool. Hence, it is reasonable to expect that a ***basic calculus for processes*** could emerge from this approach.

Linear Logic and Quantum Physics

To which extent can linear logic be used to model phenomena from quantum physics? There are several interesting observations which indicate that this question is reasonable: [Pra94b] demonstrates connections between quantum logic [Coh89] and linear logic: [Gir87a] provides a semantics for linear logic which closely resembles the semantics of quantum logic. In quantum field the creation, annihilation, and transformation of elementary particles is studied. The associated processes are non-deterministic, satisfy certain discrete conservation laws, and depend on discrete parameters like spin, colour, and strangeness [Ste93]. Linear logic provides a framework in which this may be described much easier than in other discrete calculi.

Processes and Topology

Processes and behaviours are graphs and thus define discrete geometric and topologic properties. It is a well-known problem in relativity theory and differential geometry [MTW73b] that there is no adequate formalism to describe the dynamic change of topological properties of the space-time or general differential manifold as they might be caused by particle creation processes.

The formalism of linear logic was able to provide a mechanism for the description of the dynamic change of a network of states and transitions. Therefore, we expect that it could also provide a framework for dealing with the above mentioned problems.

8.5 Final Remarks

The result of our work consists of a ***formal framework*** for specifying processes together with a collection of representative ***examples*** and an outline on several ***important phenomena*** which may occur in distributed systems.

Apart from the technical results, a number of interesting lessons were learned:

First, distributed systems, when analyzed in detail, prove to be ***much more complicated*** than one might believe. This can be formulated in the following "law":

> LAW OF DISTRIBUTED COMPLEXITY: A distributed system is one order of magnitude more complex than you expect – even if you take into account the Law of Distributed Complexity.

Second, it shows that the rather ***abstract concepts*** of closed categories which often are looked upon with some scepticism and frustration by applied computer scientists, provide an ***elegant framework*** for understanding distributed systems. Increased abstraction may eventually pay back.

Third, it is possible to *bridge the gap between theory and applications* and to provide a calculus which is theoretically sound and can be used for specifying real applications.

Fourth, it is *comparably easy to specify* processes and to derive equivalent specifications by semantic preserving *transformations*, but it is considerably more difficult *to prove the equivalence* of independently generated specification texts.

Finally, during the preparation of this work, the author had to realize that the reflexive Law of Distributed Complexity not only applies to distributed systems themselves but also to books dealing with them.

Appendix A

Mathematical Prerequisites

Et harum scientiarum porta et clavis est Mathematica.

Mathematics is the door and the key to the sciences.

ROGER BACON,
Distinctia Prima, 1267.

In this chapter we will discuss some mathematical notations and prerequisites. Section 1 deals with *order structures* and Section 2 discusses *monoids*.

A.1 Order Structures

In this section, some remarks on order structures are provided. For more details, see [DP90], [Cap93].

228 Definition ORDER RELATIONS
Let I be a set. A binary relation $\Gamma \subseteq I \times I$ on I is called a

non-strict order relation, iff it is reflexive, antisymmetric, and transitive.

strict order relation, iff it is irreflexive, asymmetric, and transitive.

An *order* on a set I is a pair $(\sqsubseteq_I, \sqsubset_I)$ of a non-strict order relation $\sqsubseteq_I$ and a strict order relation $\sqsubset_I$ on I, satisfying the equations:

$$\begin{aligned} \sqsubseteq_I &= \sqsubset_I \cup \Delta_I \\ \sqsubset_I &= \sqsubseteq_I \setminus \Delta_I \end{aligned}$$

◄

Above, $\Delta_I = \{(i,i) \mid i \in I\}$ denotes the *diagonal* of $I \times I$.

It is well known that above equations induce a bijective correspondence between strict order relations $\sqsubset_I$ and non-strict order relations $\sqsubseteq_I$. Therefore, an order, a strict order, and a non-strict order are different aspects of the same concept.

If $(\sqsubseteq_I, \sqsubset_I)$ denotes an order, then the pair of the *dual relations* $(\sqsupseteq_I, \sqsupset_I)$ is an order.

An order is called a *chain* or *linear*, iff its non-strict relation is *linear*: For every pair $(i,j) \in I^2$, (i,j) or (j,i) is an element of the relation.

Two elements i and j are called *comparable* iff (i,j) or (j,i) is an element of the relation.

229 Definition EXTREMAL AND DOMINATING ELEMENTS
Let Γ be a (non-strict) order on a set I. An element $i \in I$ is called

left-extremal	$\forall j \in I\colon (j,i) \in \Gamma \Rightarrow j = i$
right-extremal	$\forall j \in I\colon (i,j) \in \Gamma \Rightarrow j = i$
left-dominating	$\forall j \in I\colon (i,j) \in \Gamma$
right-dominating	$\forall j \in I\colon (j,i) \in \Gamma$

◀

If the sense of larger or smaller shall be connected with an order relation which in this case often is written with the symbols $\sqsubseteq_I, \sqsubset_I, \sqsupseteq_I, \sqsupset_I$, one speaks of *minimal and maximal elements* instead of extremal elements and of *largest and smallest elements* instead of dominating elements. Distinguishing between an order relation and its dual version by using such notions in fact is an additional concept providing an order with some kind of orientation.

230 Definition EXTENSIONS OF ORDER RELATIONS
If Γ_1, Γ_2 are two (non-strict) order relations on a set I, then Γ_2 is called an *extension* of Γ_1, iff

(1) $\Gamma_1 \subseteq \Gamma_2$

(2) $\forall x, y, z \in I:\ (x,z) \in \Gamma_1, (y,z) \in \Gamma_2 \Rightarrow (y,z) \in \Gamma_1$

◀

If an element z already is comprised in an order – appearing on the right side in a pair $(\bullet, z)$ of the order – then an extension may only have additional pairs of the type $(z, \bullet)$ but must not add pairs of the type $(\bullet, z)$.

The concept of a *quasi order* is weaker than the concept of an order and it is obtained by dropping the requirement of antisymmetry. Immediately connected with a quasi order is an equivalence relation and an order on the respective quotient structure:

231 Definition QUASI ORDER
Let I be a set. A binary relation $\Gamma \subseteq I \times I$ on I is called a *quasi order relation*, iff it is reflexive and transitive.

The relation $\sim_\Gamma \ \subseteq I \times I$ defined by $a \sim_\Gamma b \Leftrightarrow (a\Gamma b) \wedge (b\Gamma a)$ is called the *equivalence generated by the quasi order* Γ.

The relation $\hat{\Gamma} \subseteq I\,|_{\sim_\Gamma} \times I\,|_{\sim_\Gamma}$ defined by $[a]\hat{\Gamma}[b] \Leftrightarrow a\Gamma b$ is called the *order generated by the quasi order* Γ. ◀

232 Definition CUT
Let $\sqsubseteq$ be an order relation on a set I. A subset $C \subseteq I$ is called a *cut* of the order relation, iff no two elements of the cut are comparable:

$$\forall x, y \in C : \ (x \sqsubseteq y \vee y \sqsubseteq x) \Rightarrow x = y$$

A cut C is called ***maximal***, iff it is not a true subset[1] of another cut of the order structure. ◀

233 Definition ORDER ON MAXIMAL CUTS
Let $\sqsubseteq$ be an order relation on a set I and $\mathfrak{M}$ the set of all maximal cuts of $\sqsubseteq$.

The *cut order* is the relation $\gtrdot \ \subseteq \mathfrak{M} \times \mathfrak{M}$ on the maximal cuts of $\sqsubseteq$. It is defined for maximal cuts $A, B \subseteq I$ by

$$A \gtrdot B \ :\Leftrightarrow$$

$$(1)\ \forall a \in A,\ b \in B : \ a, b \text{ comparable } \Rightarrow a \sqsubseteq b \quad \text{and}$$

$$(2)\ \exists a \in A,\ b \in B : \ a \sqsubseteq b$$

◀

234 Proposition CUT ORDER IS ORDER
The cut order $\gtrdot$ on the set $\mathfrak{M}$ of all maximal cuts of an order relation $\sqsubseteq$ is itself an order relation.

PROOF:
$\gtrdot$ is *reflexive*:
For a maximal cut A we will show $A \gtrdot A$. Let $a, a' \in A$ be comparable. Then $a = a'$, since A is a cut. For condition (2) we observe that $a \sqsubseteq a$ for all a.

$\gtrdot$ is *transitive*:
Let $A \gtrdot B$ and $B \gtrdot C$. We have to show $A \gtrdot C$. Let $a \in A$, $c \in C$ and a, c comparable. To probe (1), we show $a \sqsubseteq c$. If $a = c$ this is obvious, so we assume that $a \neq c$. For an indirect proof we assume $a \sqsupset c$. For a there exists $b \in B$ such that a and b are comparable (otherwise B would be no ***maximal*** cut). Hence, $a \sqsubseteq b$. For b there exists c' such that b and c' are comparable. Hence, $b \sqsubseteq c'$. Hence, $a \sqsubseteq c'$. With $a \sqsupset c$ we get $c \sqsubseteq c'$. This is in contradiction to the fact that C is a cut. For (2) we have $b \in B$ and $c \in C$ such that $b \sqsubseteq c$. For $b \in B$ there exists $a \in A$ such that a and b are comparable. Hence, $a \sqsubseteq b$ and $a \sqsubseteq c$.

$\gtrdot$ is *antisymmetric*:
We assume $A \gtrdot B$ and $B \gtrdot A$ and prove $A \subseteq B$. Let $a \in A$. There exists $b \in B$ such that a and b are comparable. Hence, $a \sqsubseteq b$ and $b \sqsubseteq a$. Hence $a = b$ and so $a \in B$. ◀

Note, that a cut order may also be defined for non-maximal cuts, but then the cut order not necessarily is transitive and hence not necessarily an order relation.

A.2 Monoids

In the present work the notion of a monoid is frequently employed. This section provides the required background on this algebraic object.

[1]Therefore it is a maximal element in the set of all cuts, order by inclusion.

235 Definition MONOID
A *monoid* is a pair $(M, \circ)$ consisting of a non-empty set M and a binary operation $\circ : M \times M \to M$, satisfying the following two axioms:

(1) The operation of the monoid is *associative*:

$$\forall a, b, c \in M : \ (a \circ b) \circ c = a \circ (b \circ c)$$

(2) There exists an element $n \in M$ so that the following equation holds:

$$\forall m \in M : \ n \circ m = m \circ n = m$$

A monoid is called *commutative*, iff $\forall a, b \in M : a \circ b = b \circ a$. ◀

236 Proposition UNIQUENESS OF NEUTRAL ELEMENT
In a monoid $(M, \circ)$ there is *exactly one* element $n \in M$ which has the property of the *neutral element*, $\forall m \in M : \ n \circ m = m \circ n = m$. ◀

As a consequence of Proposition 236 it is possible to speak of *the* neutral element of a monoid. In order to provide a denotation for the neutral element already as part of the definition of the monoid, the notation $(M, \circ, \varepsilon)$ may be used.

The finite and infinite sequences of elements over a set provide examples for monoids:

237 Example FINITE SEQUENCES
Let S be a set. Then the set $S^\star$ of *finite sequences* of elements of S is defined as the union of all n-tuples of elements of S:

$$S^\star := \bigcup_{n \in \mathbb{N}_0} S^n$$

S^0 by convention is defined as the singleton set $S^0 := \{\varepsilon\}$ containing a special element ε which is not element of S. Let $\circ : S^\star \times S^\star \to S^\star$ be defined by:

$$\forall x \in S^\star : \varepsilon \circ x = x \circ \varepsilon = x$$

$$\forall n, m \in \mathbb{N}, s_i, t_j \in S : (s_0, s_1, \ldots, s_n) \circ (t_0, t_1, \ldots, t_m) = (s_0, s_1, \ldots, s_n, t_0, t_1, \ldots, t_m)$$

Then $(S, \circ)$ is a monoid. Its neutral element is the empty sequence $\varepsilon := ()$. ◀

238 Example FINITE AND INFINITE SEQUENCES
Let S be a set. Then the set S^+ of *finite and infinite* sequences of elements of S is defined as the union of $S^\star$ with the set $Me(\mathbb{N}_0, S)$ of all functions from $\mathbb{N}_0$ to S. The former will be called finite and the latter infinite sequences. Let $\circ : S^+ \times S^+ \to S^+$ be defined on finite sequences as in 237. If ξ and η are two infinite sequences, then

$$\xi \circ \eta := \xi$$

If ξ is an infinite sequence and $(x_0, x_1, \ldots, x_n)$ a finite sequence, then $\xi \circ x := \xi$. The opposite composition $x \circ \xi$ also is an infinite sequence, hence a function from $\mathbb{N}_0$ to S. This function is given by

$$k \mapsto \begin{cases} x_k & \Leftrightarrow \quad 0 \le k \le n \\ \xi(k-n-1) & \Leftrightarrow \quad n < k \end{cases}$$

$(S^+, \circ)$ is a monoid. Its neutal element is the empty sequence $\varepsilon := ()$. ◀

239 Example MULTISETS OR BAGS
Let S be a set. The set of *bags* or *multisets* over S is the set S° of all functions $\rho : S \to \mathbb{N}_0$, whose value $\rho(s)$ is zero for all but finitely many $s \in S$.

Bags are sets, generalized so that they also keep track of a (finite) number of occurrences of their elements. Suitable abbreviations may be used to write down the elements of S°: The empty bag ε represents the function which is zero on all its arguments. For $s_1, s_2, \ldots, s_n \in S$ which not necessarily are pairwise different, the expression $s_1 \cdot s_2 \cdot \ldots \cdot s_n$ represents the bag containing $s_1, s_2, \ldots, s_n$. More formally, this is the function $\rho : S \to \mathbb{N}_0$ where for $s \in S$, $\rho(s)$ is the number of times the element s can be found in the list $(s_1, s_2, \ldots, s_n)$.

Let $\circ : S^\circ \times S^\circ \to S^\circ$ be defined by $(\rho \cdot \sigma)(s) := \rho(s) + \sigma(s)$. Then $(S^\circ, \circ)$ is a commutative monoid. Its neutral element ε is the empty bag, which corresponds to the function on S, which is the zero constant. ◀

240 Example NATURAL NUMBERS
The natural numbers $\mathbb{N}_0$ together with the addition form a commutative monoid. The natural numbers $\mathbb{N}$ without zero together with the multiplication form a commutative monoid. ◀

241 Definition MONOID MORPHISM
Let $(M, \circ, \mu)$ and $(N, \star, \nu)$ be two monoids. A *monoid morphism* is a function $f : M \to N$ such that the following equations hold:

$$\forall x, y \in M : \ f(x \circ y) = f(x) \star f(y)$$

$$f(\mu) = \nu$$

A *monoid comorphism* is a function $g : M \to N$ such that the following equations hold:

$$\forall x, y \in M : \ g(x \circ y) = g(y) \star g(x)$$

$$g(\mu) = \nu$$

A function $f : M \to N$ is a *morphism of commutative monoids*, iff it is a morphism of the monoids.

A function $f : M \to N$ is called an *isomorphism*, iff it is a bijection and its invers f^{-1} also is a monoid morphism. ◀

See Appendix C for further remarks on isomorphisms.

242 Example MONOID MORPHISMS
The *length of a finite list* is the function len : $S^\star \to \mathbb{N}_0$, defined by $\text{len}(\varepsilon) := 0$ and $\text{len}(x_0, x_1, \ldots, x_n) := n + 1$. It is a monoid morphism from $(S^\star, \circ)$ to $(\mathbb{N}_0)$.

The *reversal of a finite list* is the function rev : $S^\star \to S^\star$, defined by $\text{rev}(\varepsilon) := \varepsilon$ and $\text{rev}(x_0, x_1, \ldots, x_n) := (x_n, \ldots, x_1, x_0)$. It is a monoid comorphism from $(S^\star, \circ)$ to $(S^\star, \circ)$. ◀

243 Definition FREE GENERATION OF MONOIDS
A monoid $(M, \circ)$ is said to be *freely generated* by a subset $G \subseteq M$, iff for every monoid $(N, \cdot)$ and function $f : G \to N$, there is a unique morphism $f^\star : M \to N$ which extends f to M, i.e. the following equation holds:

$$\forall g \in G : \; f^\star(g) = f(g)$$

◀

This definition is analogous for commutative monoids.

244 Definition FREE GENERATION OF COMMUTATIVE MONOIDS
A commutative monoid $(M, \circ)$ is said to be *freely generated* by a subset $G \subseteq M$, iff for every commutative monoid $(N, \cdot)$ and function $f : G \to N$, there is a unique morphism $f^\circ : M \to N$ which extends f to M, i.e. the following equation holds:

$$\forall g \in G : \; f^\circ(g) = f(g)$$

◀

See [Wir90] and [EGLer] for a number of equivalent characterizations of *freely generated algebraic structures*. Essentially, a monoid is *freely generated* by a subset G, iff every element of the monoid may be generated by a monoid term constructed from elements in G, and iff the only equations between generator terms are those, which may be deduced by inductive reasoning from the equations of associativity and neutrality.

245 Example FREELY GENERATED MONOIDS
The monoid $(S^\star, \circ)$ of finite lists is freely generated by the lists of length 1. This set of generators may be identified with S.

The monoid $(S^+, \circ)$ is free over the lists of length 1, but it is not generated by them, since the infinite lists cannot be generated by finite ones.

The monoid $(\mathbb{N}_0, +)$ is freely generated by the single element 1.

Let $X := \{id_{\mathbb{R}^2}, p\}$ consist of the identity function of the plane $\mathbb{R}^2$ and of the projection function $p(x, y) := (x, 0)$ onto the x axis. Then X together with the composition of functions forms a monoid. It is generated by the element p, but not freely. The equation $p \circ p = p$ holds, but this is not necessarily true in every monoid. Hence X is not free from additional equations. The free monoid generated by a single element is $(\mathbb{N}_0, +)$.

The commutative monoid S° of *bags* is freely generated by the set of bags with one element. This set can be identified with S. ◀

The reading guide in Section 1.5 lists a number of *references* where more information on algebraic structures in general and on the property of being a free, initial, or terminal model of a structure can be found.

Appendix B

Linear Logic

Gelegentlich ergreifen wir die Feder
Und schreiben Zeichen auf ein weisses Blatt,
Die sagen dies und das, es kennt sie jeder,
Es ist ein Spiel, das seine Regeln hat.

We now and then take pen in hand
And make some marks on empty paper.
Just what they say, all understand.
It is a game with rules that matter.

H. HESSE, *Alphabet*,
Translation by R. ELLIS.

Linear logic was discovered by GIRARD in [Gir87a]. It has gained wide attention in many branches of computer science [Gir87b], [Abr90a], [Roo91], [MOM89]. Several observations have motivated the development of this interesting theory: Linear logic can be seen as a *fine granular version of classical logic* [Tro92], as a logic for reasoning about *states and resources* [GdQ90], as a formal system having more *control on the use of assumptions* than classical logic [OK85], and it also emerges from *proof normalization* [GLT88]. Other applications are *vector addition systems* [Kan94], *logic programming* [AP91], *functional programming* [Abr90a], and *type theory* [Wad90].

In the present work, linear logic is used as a *specification framework for processes*. The required logical background is presented in this Appendix. It is organized as follows:

In Section 1, the *universe of discourse*, i.e. the meta-logical setting, is presented.

Section 2 provides several *intuitive motivations* for linear logic.

In Section 3, the *syntax* and a *natural deduction system* is rigorously defined. This deduction system is the best for an understanding of linear logic, but it has some technical disadvantages.

In Section 4, an *alternate natural deduction axiomatization* is derived. The proofs given in this and the subsequent parts of this chapter will also allow the reader to become more familiar with linear logic and to develop a more intuitive understanding of the underlying concepts.

In Section 5, the *sequent calculus* for linear logic is presented. This axiomatization allows to deduce important conclusions on the decidability of the theory.

In Section 6, the reader finds a collection of *syntactic properties* of linear logic which will be used in this work.

Finally, Section 7 explains how the operators of linear logic may be *interpreted*.

Several of the subsequent results will be proved in the present appendix, for several other statements references are provided.

B.1 Universe of Discourse

Before concentrating on linear logic, the universe of discourse is defined in which logical systems will be discussed. The goal of every logic is the *description* of real world facts or of more abstract mathematical properties by using a language. This requires a set $\mathcal{M}$ containing worlds or *models* which we want to describe, a language $\mathcal{L}$ for expressing facts, and a relation $\models \ \subseteq \mathcal{M} \times \mathcal{L}$. If the relation $M \models l$ holds for a model $M \in \mathcal{M}$ and a formula $l \in \mathcal{L}$, then this formula l describes a fact which is *valid* in the model M.

Semantic reasoning in a model is quite complicated and cannot be automatized by a computer, since it requires qualitative reasoning with objects of human imagination. We therefore would like to replace semantic reasoning by syntactic transformations on the formulae which describe our model. This may be done with a *deduction relation* $\vdash \ \subseteq \mathcal{L} \times \mathcal{L}$. Here $\varphi \vdash \psi$ denotes that it is possible to *deduce* formula ψ from the assumption φ.

How could such a deduction relation help with the task of reasoning about models? Assume, we want to study a model $M \in \mathcal{M}$. For this purpose, we choose a formula φ which is valid in M, i.e. $M \models \varphi$ holds. Now we look for an additional formula ψ for which $\varphi \vdash \psi$ holds. If we have chosen our deduction relation $\vdash$ carefully, the new formula ψ is also valid in M, so that $M \models \psi$ holds. With the help of $\vdash$ we thus have found another valid formula.

The above property which connects the deduction relation $\vdash$ with the model relation $\models$ is called *correctness*. It states that formulae obtained by deduction from valid formulae are also valid. A related property is called *completeness*. It guarantees that every valid formula may be obtained with the help of deduction. Below, this will be made more precise.

If the language $\mathcal{L}$ is decidable and if the relation $\vdash$ is decidable or at least recursively enumerable, then this deduction process may be carried out by a computer program: If $\vdash$ is decidable, then we may write a computer program which accepts two formulae φ and ψ, and which tells us after some calculations, whether $\varphi \vdash \psi$ holds or not. If $\vdash$ is only recursively enumerable, then we can at least expect the computer to print out all formulae φ and ψ for which $\varphi \vdash \psi$ does hold, but it might not be able to determine all those formulae for which this deduction does not hold. Consult [Odi92] and [Coh87] on questions of recursion and decidability.

In case we have a model relation and a deduction relation for which the properties of correctness and completeness hold, and if the deduction relation has one of the mentioned computability properties, then we may *effectively replace human reasoning on models by a computer program* fulfilling the same task.

In the following, we will present a more abstract framework for this concept:

246 Definition LOGICAL SYSTEM
A *logical system* is a quadruple $(\mathcal{L}, \vdash, \mathcal{M}, \models)$ consisting of

(1) a decidable *language* $\mathcal{L}$,

(2) a recursively enumerable relation $\vdash \ \subseteq \mathcal{L} \times \mathcal{L}$, called the *deduction relation*,

(3) a set $\mathcal{M}$ of *models*, and

(4) a relation $\models \ \subseteq \mathcal{M} \times \mathcal{L}$, called the *model relation*.

◀

The deduction relation $\vdash$ and the model relation $\models$ may be connected by the two properties of *correctness* and *completeness*: As outlined in Remark 172, the precise formulation of the completeness property may vary in different logic systems. Below, we present the common formulation for classical logic. The versions for linear logic with transitional semantics is discussed in Theorems 170, 173, and 174.

A logical system is called *correct*, iff every deduced formula is valid. It is called *complete*, iff every valid formula is deducible:

247 Definition CORRECTNESS AND COMPLETENESS
A logical system is called *correct*, iff for all models $M \in \mathcal{M}$ and formulae $\varphi, \psi \in \mathcal{L}$, if $M \models \varphi$ and $\varphi \vdash \psi$, then also $M \models \psi$ holds:

$$\forall \varphi, \psi \in \mathcal{L},\ M \in \mathcal{M} :\ [(M \models \varphi) \wedge (\varphi \vdash \psi)] \Rightarrow (M \models \psi)$$

A logical system is called *complete*, iff the following property holds: If ψ is valid in every model $M \in \mathcal{M}$ in which φ is valid, then ψ may be deduced from φ:

$$\forall \varphi, \psi \in \mathcal{L} :\ [(\forall M \in \mathcal{M} :\ (M \models \varphi) \Rightarrow (M \models \psi)) \Rightarrow (\varphi \vdash \psi)]$$

◀

Since models usually are complicated objects, often a formula $\varphi \in \mathcal{L}$ is used to represent all those models which are characterized by this formula. In this sense, a formula represents the class of models in which it is valid. If a formula ψ is valid in *every* model in which φ is valid, we shall write $\varphi \Vdash \psi$:

248 Definition CONSEQUENCE
The *consequence relation* of a logical system is the relation $\Vdash \ \subseteq \mathcal{L} \times \mathcal{L}$, defined by

$$\varphi \Vdash \psi \ :\Leftrightarrow\ \forall M \in \mathcal{M} :\ (M \models \varphi) \Rightarrow (M \models \psi)$$

◀

By using the consequence relation, the properties of correctness and completeness can be formulated in a more compact and intuitive form:

Correctness: whenever $\varphi \vdash \psi$ then $\varphi \Vdash \psi$

Completeness: whenever $\varphi \Vdash \psi$ then $\varphi \vdash \psi$

For ease of implementation, for technical, historical and other reasons, a variety of different calculi have been developed which serve a similar purpose [Sun83]. Let $\mathcal{L}^*$ be the set of finite sequences of words from $\mathcal{L}$ and let $\mathcal{P}(\mathcal{L})$ be the power set of $\mathcal{L}$. In ***sequent calculi*** a deduction relation of the form $\vdash_s \ \subseteq \mathcal{L}^* \times \mathcal{L}^*$ and in ***natural deduction calculi*** a relation of the form $\vdash_n \ \subseteq \mathcal{L}^* \times \mathcal{L}$ is employed. HILBERT-FREGE ***calculi*** use relations of the form $\vdash_h \ \subseteq \mathcal{P}(\mathcal{L}) \times \mathcal{L}$ or $\vdash_H \ \subseteq \mathcal{P}(\mathcal{L}) \times \mathcal{P}(\mathcal{L})$, and in ***one-sided calculi*** we encounter relations of the form $\vdash^1 \ \subseteq \mathcal{L}$ or $\vdash_1 \ \subseteq \mathcal{L}^*$. For many logical systems, these approaches are equivalent. In Table B.1 this is illustrated for classical propositional logic, by giving the respective versions of the formula "from φ and ψ we may deduce φ or ψ":

Calculus	Deduction	Example
Pure deduction	$\vdash \ \subseteq \mathcal{L} \times \mathcal{L}$	$(\varphi_1 \wedge \varphi_2) \vdash (\psi_1 \vee \psi_2)$
Sequent calculus	$\vdash_s \ \subseteq \mathcal{L}^* \times \mathcal{L}^*$	$\varphi_1, \varphi_2 \vdash_s \psi_1, \psi_2$
Natural deduction	$\vdash_n \ \subseteq \mathcal{L}^* \times \mathcal{L}$	$\varphi_1, \varphi_2 \vdash_n (\psi_1 \vee \psi_2)$
HILBER-FREGE	$\vdash_h \ \subseteq \mathcal{P}(\mathcal{L}) \times \mathcal{L}$	$\{\varphi_1, \varphi_2\} \vdash_h (\psi_1 \vee \psi_2)$
Extended HILBERT-FREGE	$\vdash_H \ \subseteq \mathcal{P}(\mathcal{L}) \times \mathcal{P}(\mathcal{L})$	$\{\varphi_1, \varphi_2\} \vdash_H \{\psi_1, \psi_2\}$
One-sided	$\vdash^1 \ \subseteq \mathcal{L}$	$\vdash^1 ((\varphi_1 \wedge \varphi_2) \Rightarrow (\psi_1 \vee \psi_2))$
One-sided	$\vdash_1 \ \subseteq \mathcal{L}^*$	$\vdash_1 \neg\varphi_1, \neg\varphi_2, \psi_1, \psi 2$

Table B.1: Forms of Deduction Relations.

Not every approach is suitable as universe of discourse for every logical concept. In the natural deduction approach, for example, it is possible to describe systems for which the number of times a formula occurs in the assumptions plays a significant role for the deduction process. Hence, the following two deductions may be distinguished:

$$\varphi, \varphi \vdash \psi \qquad \varphi \vdash \psi$$

In the HILBERT-FREGE approach this is not possible: φ either is element of the set of assumptions or it is not:

$$\{\varphi\} \vdash \psi$$

The pure deduction approach $\vdash$ is the most powerful one and all the other calculi may be represented by it. However, the other approaches are also important since they have interesting or practical structural and computational properties [Llo87], [Gal87].

Deduction relations often are defined by structural induction: First, a number of deductions is given: For example, in propositional logic for every formula φ the deduction $\varphi \vdash \varphi$ holds. Next, several formation rules are presented which allow to derive new deductions. In propositional logic we have the following rule: If we can deduce formula ϑ and formula ψ from a formula φ, i.e. $\varphi \vdash \vartheta$ and $\varphi \vdash \psi$ hold, then we may also deduce the new formula $(\vartheta \wedge \psi)$ from φ, i.e. $\varphi \vdash (\vartheta \wedge \psi)$ holds.

For the definition of deduction relations the following notation is common:

$$\frac{\text{assumptions}}{\text{conclusions}}\ \text{name of the rule}$$

Above formation rule would be represented as

$$\frac{\varphi \vdash \vartheta \qquad \varphi \vdash \psi}{\varphi \vdash (\vartheta \wedge \psi)}\ (\wedge\,\text{introduction})$$

Deductions which hold without any assumption, for example the base of the structural induction rules which define the deduction relation, may also be represented in this formalism. In this case there is no formula on the upper half of the deduction line:

$$\frac{}{\varphi \vdash \varphi}\ (\text{Identity Axiom})$$

Since natural deduction is the most commonly used and intuitively best understood formalism, we will use it in most parts of this work. Hence, the distinguishing index n in the relation $\vdash_n$ will often be dropped.

B.2 Motivations for Linear Logic

249 Remark FINE GRANULAR CLASSICAL LOGIC
In classical logic, many operators can be defined by using different rules of deduction. The conjunction, for example, may be described by the following two introduction rules of the natural deduction calculus: Let Γ and Δ denote sequences of formulae, and let ψ and φ denote single formulae.

$$\frac{\Gamma \vdash \varphi \qquad \Delta \vdash \psi}{\Gamma, \Delta \vdash \varphi \wedge \psi}\ (\wedge I)$$

$$\frac{\Gamma \vdash \varphi \qquad \Gamma \vdash \psi}{\Gamma \vdash \varphi \wedge \psi}\ (\wedge I')$$

In classical logic, both rules are equivalent: Suppose rule $(\wedge I)$ holds and $\Gamma \vdash \varphi$ and $\Gamma \vdash \psi$ are valid. Substituting $\Delta := \Gamma$ in $(\wedge I)$ yields $\Gamma, \Gamma \vdash \varphi \wedge \psi$. With the help of the contraction rule (C) and the permutation rule (P) of the natural deduction calculus

$$\frac{\Gamma, \alpha, \alpha \vdash \beta}{\Gamma, \alpha \vdash \beta} (C)$$

$$\frac{\Gamma, \alpha, \beta, \Delta \vdash \gamma}{\Gamma, \beta, \alpha, \Delta \vdash \gamma} (P)$$

finally $\Gamma \vdash \varphi \wedge \psi$ is obtained. Now suppose rule $(\wedge I')$ holds and $\Gamma \vdash \varphi$ and $\Delta \vdash \psi$ are valid. Using weakening $\Gamma, \Delta \vdash \varphi$ and $\Gamma, \Delta \vdash \psi$ is obtained. With $(\wedge I')$ one finally gets $\Gamma, \Delta \vdash \varphi \wedge \psi$.

Is it possible to design a logic with two different conjunction operators, one of which defined by $(\wedge I)$ and the other by $(\wedge I')$? Above the contraction, weakening, and permutation rules were used to show the equivalence of $(\wedge I)$ and $(\wedge I')$. In a logic with two conjunction operators it is therefore possible to prove the equivalence of these two operators. Only if the contraction and weakening rules are dropped, two non-equivalent conjunction operators are obtained. The permutation rule is retained, since otherwise the order in which assumptions are presented may play a role in the deduction process. This could lead to a too complicated and undesired form of logic [Yet90], [Lin92]. ◄

See [Sun83], [Bor77], [Coh87], [Gal87] for more details on the deduction calculi of classical logic, and [Tro92], [Gir87a], [Hes90] for the development of linear logic and its variants.

250 Remark A LOGIC OF RESOURCE CONSUMPTION
Classical propositional logic cannot be used for reasoning about resources and resource consumption. For example, the following two formulae are tautologies in classical logic. They correspond to the rules of the previous paragraph:

$$[(A \Rightarrow P) \wedge (B \Rightarrow Q)] \Rightarrow [(A \wedge B) \Rightarrow (P \wedge Q)] \quad (\wedge)$$

$$[(A \Rightarrow P) \wedge (A \Rightarrow Q)] \Rightarrow [A \Rightarrow (P \wedge Q)] \quad (\wedge')$$

Now let us try the following interpretation:

A "I own two dollars."
B "I own fifty cents."
P "I can buy a cup of tea for two dollars."
Q "I can buy a piece of cake for fifty cents."

With this interpretation, the first tautology $(\wedge)$ tells us that "If I own two dollars and fifty cents, then I can buy a cup of tea for two dollars and a piece of cake for fifty cents." The second tautology $(\wedge')$ claims that "If I own two dollars I can buy a cup of tea for two dollars and a piece of cake for fifty cents".

The second tautology therefore shows that the conjunction operation of classical logic cannot be used for describing the accumulation of resources. Systems in which resources are generated, used, and consumed cannot be adequately described by classical logic. For this purpose, we need a *"resource-conscious" conjunction operator*, for which only the first formula is deriveable.

With the implication $\Rightarrow$ we face a similar situation. In classical logic an assumption may be used for an arbitrary number of times. This is also reflected by the following formulae which are valid in every model of classical logic.

$$(A \Rightarrow A)$$

$$[(A \wedge A) \Rightarrow A]$$

$$[A \Rightarrow (A \wedge A)]$$

Again, an interpretation of A by "I own three dollars" shows that this conjunction operation must not be interpreted as accumulation of resources and that this implication *does not keep track of the balance of resources.*

Linear logic splits the classical operators into two different kinds: The *multiplicatives* which keep track of resource consumption, and the *additives* which do not. A third kind of operators, the *exponentials* or *modalities*, allow a controlled form of duplication or discarding of resources. The *negation* corresponds to debts of resources.

We will write $\star$ for the *multiplicative conjunction* which corresponds to the accumulation of resources, and $\multimap$ for the *linear implication*. In this sense, $A \star B$ consists of two resources, namely one copy of A and one of B. $A \multimap B$ encodes a situation where we may *consume* one copy of resource A and *produce* one copy of resource B. In linear logic, this will be formalized as follows:

$$[A \star (A \multimap B)] \vdash_{\text{lin}} B$$

$$[A \star A \star (A \multimap B)] \vdash_{\text{lin}} (A \star B)$$

Also *higher order constructions* are possible:

$$(A \multimap B) \star [(A \multimap B) \multimap (C \multimap D)] \vdash_{\text{lin}} (C \multimap D)$$

Resources cannot be discarded at free will and therefore the following formula *does not hold* in linear logic:

$$A \star A \star (A \multimap B) \vdash_{\text{lin}} B \qquad \textbf{(does not hold!)}$$

The expression $(A \multimap B)$ has a conditional nature and may be interpreted as "If I only had A, then I could deduce B". Intuitively the two formulae

$$(A \star B) \multimap C$$

$$A \multimap (B \multimap C)$$

have the meaning:

"If I had A and B, then I could consume and transform them into C"

"If I had A then I could conclude: 'If I only had B, then I could deduce C' "

Both sentences seem to contain the same information. In linear logic this is reflected by the fact that from any of the above formulae one may be deduced from the other:

$$(A \star B) \multimap C \vdash_{\text{lin}} A \multimap (B \multimap C)$$

$$A \multimap (B \multimap C) \vdash_{\text{lin}} (A \star B) \multimap C$$

◀

251 Remark USE OF ASSUMPTIONS IN PROOFS
The deduction calculus of classical logic is based on the following ideas [Cap93]:

- Write down some *axioms* and some instances of *axiom schemes*.
- Choose some of the formulae which already were written down, and apply a ***deduction rule***.
- ***Repeat*** this process. The formulae obtained by this process are the deducible formulae.

If this process is formalized as a sequence or natural deduction calculus, the formulae which were "written down" correspond to a finite sequence of fomulae, compare [Gal87]. Deduction rules like the following ***modus ponens*** do not allow an arbitrary choice of formulae from these sequences but operate with fixed contexts Γ and Δ:

$$\frac{\Gamma \vdash \varphi \Rightarrow \psi \qquad \Delta \vdash \varphi}{\Gamma, \Delta \vdash \psi}$$

Therefore, structural rules must be added to allow arbitrary choice, rearrangement, duplication, and ignorance of rules in a context. They are the rules of ***weakening*** (W), ***contraction*** (C), and ***permutation*** (P):

$$\frac{\Gamma \vdash \vartheta}{\Gamma, \varphi \vdash \vartheta}\,(W)$$

$$\frac{\Gamma, \varphi, \varphi \vdash \vartheta}{\Gamma, \varphi \vdash \vartheta}\,(C)$$

$$\frac{\Gamma, \varphi, \psi, \Delta \vdash \vartheta}{\Gamma, \psi, \varphi, \Delta \vdash \vartheta}\,(P)$$

These rules complicate the search for proofs, since unnecessary rearrangements, weakenings or contractions may lead to complexity explosion. We therefore are interested in a logic where assumptions cannot be duplicated and dropped arbitrarily, but where the use of assumptions is under the control of the logic itself. ◀

252 Remark PROOF NORMALIZATION
In functional programming [FH80], a function

$$f : A \times B \to C$$

of two variables may be associated with a higher order or CURRYied function

$$\Lambda(f) : A \to [B \to C] \qquad \Lambda(f)(a) := \lambda b.f(a,b)$$

essentially containing the "same information". Here $[B \to C]$ denotes a suitable set of functions from B to C.

Interpreting the type constructors $\times$ and $\to$ logically as $\wedge$ and $\Rightarrow$, two equivalent formulae of classical logic are obtained:

$$(A \wedge B) \Rightarrow C$$

$$A \Rightarrow (B \Rightarrow C)$$

This correspondence which translates "equivalent" types into equivalent logic formulae is known as the CURRY-HOWARD isomorphism.

This isomorphism also gives rise to a correspondence between proofs of formulae and functions, or rather lambda expressions of a suitable type. A technically rigorous analysis [LS86] shows a connection between proof normalization in intuitionistic logic and normal forms of typed lambda calculi. [GLT88] argues that this connection cannot be carried over to full classical logic, and it illustrates how an analysis of this situation directly leads to linear logic. ◀

253 Remark NOTATION
In linear logic, two papers have been most influential: The survey [Gir87a] by GIRARD and the lecture notes [Tro92] by TROELSTRA. Both texts have different motivations for the notations they employ. Table B.2 provides an overview of both kinds of symbols. In this text we will make use of the TROELSTRA notation, since we feel that it better reflects the duality of operations.

Operation	Multiplicative		Additive	
	TROELSTRA	GIRARD	TROELSTRA	GIRARD
Conjunction	$\star$	$\otimes$	$\sqcap$	&
Neutral of Conjunction	$\mathbf{1}$	$\mathbf{1}$	$\top$	$\top$
Disjunction	$+$	$\wp$	$\sqcup$	$\oplus$
Neutral of Disjunction	$\mathbf{0}$	$\perp$	$\perp$	$\mathbf{0}$

Table B.2: GIRARD and TROELSTRA Notations.

Operation	Notation
Implication	$\multimap$
Storage	!
Consumption	?

Table B.3: Notations in Linear Logic.

For the negation GIRARD uses $\varphi^{\perp}$, whereas TROELSTRA and the present text will use $\sim \varphi$. The operator notations in Table B.3 coincide for GIRARD and TROELSTRA. ◀

B.3 Syntax and Natural Deduction

In this section we will formally define the *syntax* and the *natural deduction* relation of linear logic.

254 Definition LINEAR LOGIC
$A := \{\star, +, \multimap, \sqcap, \sqcup, !, ?, \sim, \mathbf{1}, \mathbf{0}, \top, \bot, (,)\}$ is the set of *elementary symbols* of linear logic. Let C be another set whose elements are called *theory constants*.

The language $\mathcal{L}[C]$ of *well formed formulae of linear logic over the set* C is generated by the following context free grammar over the set $\{S, B, U\}$ of non-terminals. S is the start symbol and we assume that the non-terminals are not element of C.

$$S \to \mathbf{1} \mid \mathbf{0} \mid \top \mid \bot \mid (SBS) \mid (US)$$

$$B \to \star \mid + \mid \multimap \mid \sqcap \mid \sqcup$$

$$U \to \sim \mid ! \mid ?$$

Furthermore, for every constant $c \in C$, add a rule of the form

$$S \to c$$

◀

We assume that C is well defined by other means. More precisely, let C be a decidable language. If C is infinite, then the traditional framework of context free grammars is left due to the infinitely many rules of the form $S \to c$. This problem easily is corrected with a rule $S \to S'$, where S' is the start symbol of the grammar of C.

In the following, the small Greek letters φ, ψ, ϑ denote single formulae from $\mathcal{L}[C]$, and the capital Greek letters Γ, Δ, Θ represent sequences of formulae. $\varepsilon \in \mathcal{L}[C]^*$ denotes the empty sequence.

The negation and the exponentials shall *bind more closely* than the conjunction and the disjunction, and implication shall be the most weakly binding operator. Therefore the expression $\sim \varphi \star !\psi \multimap \vartheta$ equals $(((\sim \varphi) \star (!\psi)) \multimap \vartheta)$.

255 Definition NATURAL DEDUCTION SYSTEM
The *natural deduction system* of linear logic is given by the relation $\vdash \subseteq \mathcal{L}[C]^* \times \mathcal{L}[C]$ which is defined by structural induction over the context free syntax of $\mathcal{L}[C]$ with the help of the rules in Tables B.4 and B.5. ◀

In the following theorem it is shown that the famous *cut rule*, known from classical logic, holds:

256 Theorem CUT RULE
From $(\multimap I)$ and $(\multimap E)$ it is possible to deduce

$$\frac{\Gamma, \varphi \vdash \psi \qquad \Delta \vdash \varphi}{\Gamma, \Delta \vdash \psi}\,(cut)$$

Especially one has

$$\frac{\Delta \vdash \varphi \qquad \varphi \vdash \psi}{\Delta \vdash \psi}\,(tra)$$

and in an even more special form one obtains

$$\frac{\vartheta \vdash \varphi \qquad \varphi \vdash \psi}{\vartheta \vdash \psi}\,(tra1)$$

PROOF:
Proofs are represented as tree structures which is the accepted technique for natural deduction systems [Gal87].

$$\frac{\dfrac{\Gamma, \varphi \vdash \psi}{\Gamma \vdash \varphi \multimap \psi}\,(\multimap I) \qquad \Delta \vdash \varphi}{\Gamma, \Delta \vdash \psi}\,(\multimap E)$$

The *root* of the tree is depicted at the bottom and shows the formula $\Gamma, \Delta \vdash \psi$ which finally is deduced by the proof. The *leaves* constitute the assumptions $\Gamma, \varphi \vdash \psi$ and $\Delta \vdash \varphi$. ◀

257 Remark PROOF NORMALIZATION AND NATURAL DEDUCTION
Natural deduction systems usually consist of *introduction* and *elimination* rules which describe how a certain operator is introduced or eliminated at the right hand side of $\vdash$. This right hand side of $\vdash$ consists only of a single formula, whereas the left hand side of $\vdash$ consists of a sequence of formulae, with the comma to be interpreted by $\star$. This will be shown rigorously in Proposition 263.

This concept of introduction and elimination rules allows the straight forward definition of *proof normalization*. Intuitively it is clear that a proof which introduces an operator and immediately eliminates it again, may be transformed into a less complex proof which does not utilize this operator at all. For example, in the following segment of a proof tree, $\multimap$ is introduced and eliminated again.

$$\frac{\dfrac{\Gamma, \varphi \vdash \psi}{\Gamma \vdash \varphi \multimap \psi}\,(\multimap I) \qquad \dfrac{}{\varphi \multimap \varphi}\,(Ax)}{\Gamma, \varphi \vdash \psi}\,(\multimap E)$$

The entire proof tree may be collapsed to the single line $\Gamma, \varphi \vdash \psi$. See [GLT88] for more details. ◀

Natural deduction systems have a syntactical structure allowing a direct computational interpretation:

258 Remark COMPUTATIONAL INTERPRETATION OF NATURAL DEDUCTION
The formulae of a deduction can be interpreted as *types* and they can be represented by data structures: Let $a, b \in C$ be two constants which are interpreted by some elementary types α and β and which are represented by the sets A and B. Then the formula $a \star b$ corresponds to the product type $\alpha \times \beta$ and may be represented by the cartesian product $A \times B$ of the

sets. Similarly, the formula $a \multimap b$ corresponds to the functional or *higher order* type $\alpha \to \beta$ which may be represented by a suitable[1] set of functions from A to B.

A deduction $\gamma_1, \gamma_2, \ldots, \gamma_k \vdash \varphi$ corresponds to a suitably typed function $f : \gamma_1 \times \gamma_2 \times \ldots \times \gamma_k \to \varphi$. A deduction rule may be interpreted as a functional which transforms functions into new ones:

For example, the rule

$$\frac{\gamma, \varphi \vdash \vartheta}{\gamma \vdash \varphi \multimap \vartheta} (\multimap I)$$

transforms a function $f : \gamma \times \varphi \to \vartheta$ which corresponds to the deduction $\gamma, \varphi \vdash \vartheta$ into a function $h : \gamma \to [\varphi \to \vartheta]$. This function h may also be defined by applying the abstraction to the function f: $h : \gamma \to [\varphi \to \vartheta]$, $h(x) := \lambda y.f(x, y)$.

There is a similar situation for the rule

$$\frac{\gamma \vdash \varphi \multimap \vartheta \qquad \delta \vdash \varphi}{\gamma, \delta \vdash \vartheta} (\multimap E)$$

If we have a function $f : \gamma \to [\varphi \to \vartheta]$ and a function $g : \delta \to \varphi$, then we can construct a function $h : \gamma \times \delta \to \vartheta$. This function is given by $h(x, y) := [f(x)](g(y))$. $f(x)$ is a function of type $\varphi \to \vartheta$ and $g(y)$ is of type φ. ◀

In Remark 250 it was mentioned that the ***multiplicative operators*** are resource-conscious, whereas the ***additive operators*** and the exponentials are not. This will be shown rigorously in Section B.7 below. If a logic is needed for dealing with resource accumulation only a multiplicative conjunction operator is required. For resource transformations an implication operator must be added. Thus, various ***fragments*** of linear logic will be considered.

259 Definition FRAGMENTS
In linear logic, there exist the ***subsets*** of the language $\mathcal{L}[C]$ of formulae of linear logic as outlined in Tables B.6 and B.7. The ***deduction relations*** for these respective fragments are defined by the rules from Table 255 according to Table B.7. In case the respective fragment is not obvious from the context, the index is used to specify it. ◀

In the tensor fragment, we only deal with the accumulation of resources. For example, there are deductions like $(a \star b) \vdash_t (b \star a)$ and $((a \star b) \star c) \vdash_t (a \star (b \star c))$, but there are no deductions like $a \vdash_t (a \star a)$ or $(a \star a) \vdash_t a$. The positive fragment is an extension of the tensor fragment. Is there a deduction which can be formulated in the tensor fragment and does not hold in the tensor fragment, but holds in the positive fragment? For example, does the deduction $a \vdash_t (a \star a)$ which may be formulated in the tensor fragment but does not hold in the tensor fragment, possibly hold in the positive fragment: $a \vdash_p (a \star a)$? In principle, this is possible, since the passage from the tensor fragment to the positive fragment adds additional rules. In the following, we will analyze such situations.

260 Definition INTERRELATION OF FRAGMENTS
Let $\mathcal{L}_1[C]$ with $\vdash_1$ and $\mathcal{L}_2[D]$ with $\vdash_2$ be two fragments of linear logic. $(\mathcal{L}_2, \vdash_2)$ is called an *extension* of $(\mathcal{L}_1, \vdash_1)$, iff $\mathcal{L}_1[C] \subseteq \mathcal{L}_2[D]$ and $\vdash_1 \subseteq \vdash_2$.

[1] By working out the details of this correspondence it becomes evident that it is necessary to restrict $a \to b$ to all ***continuous*** functions with respect to some topology. For the case of intuitionistic logic the details have been worked out by [LS86] and [Gol84].

Basic Axioms

$$\frac{}{\varphi \vdash \varphi}\,(Ax) \qquad \frac{\Gamma, \varphi, \psi, \Delta \vdash \vartheta}{\Gamma, \psi, \varphi, \Delta \vdash \vartheta}\,(P)$$

Tensor Axioms

$$\frac{}{\varepsilon \vdash \mathbf{1}}\,(\mathbf{1}I) \qquad \frac{\Gamma \vdash \mathbf{1} \qquad \Delta \vdash \varphi}{\Gamma, \Delta \vdash \varphi}\,(\mathbf{1}E)$$

$$\frac{\Gamma \vdash \varphi \qquad \Delta \vdash \vartheta}{\Gamma, \Delta \vdash \varphi \star \vartheta}\,(\star I) \qquad \frac{\Gamma \vdash \varphi \star \vartheta \qquad \Delta, \varphi, \vartheta \vdash \psi}{\Gamma, \Delta \vdash \psi}\,(\star E)$$

Positive Axioms

$$\frac{\Gamma, \varphi \vdash \vartheta}{\Gamma \vdash \varphi \multimap \vartheta}\,(\multimap I) \qquad \frac{\Gamma \vdash \varphi \multimap \vartheta \qquad \Delta \vdash \varphi}{\Gamma, \Delta \vdash \vartheta}\,(\multimap E)$$

Multiplicative Axioms

$$\frac{\Gamma, \varphi \vdash \mathbf{0}}{\Gamma \vdash \sim \varphi}\,(\sim I) \qquad \frac{\Gamma \vdash \sim \varphi}{\Gamma, \varphi \vdash \mathbf{0}}\,(\sim E)$$

$$\frac{\Gamma, \sim \varphi \vdash \psi}{\Gamma \vdash \varphi + \psi}\,(+I1) \qquad \frac{\Gamma \vdash \varphi + \psi \qquad \Delta, \varphi \vdash \mathbf{0}}{\Gamma, \Delta \vdash \psi}\,(+E)$$

$$\frac{\Gamma, \sim \psi \vdash \varphi}{\Gamma \vdash \varphi + \psi}\,(+I2)$$

$$\frac{\Gamma, \sim \varphi \vdash \mathbf{0}}{\Gamma \vdash \varphi}\,(\mathbf{0}E)$$

Table B.4: Natural Deduction Rules, Part 1.

Positive Exponential Axioms

$$\frac{\Gamma \vdash !\varphi \qquad \Delta \vdash \psi}{\Gamma, \Delta \vdash \psi}\ (\text{none}) \qquad \frac{\Gamma \vdash !\varphi \qquad \Delta, \varphi \vdash \psi}{\Gamma, \Delta \vdash \psi}\ (\text{one})$$

$$\frac{\Gamma \vdash !\varphi \qquad \Delta, !\varphi, !\varphi \vdash \psi}{\Gamma, \Delta \vdash \psi}\ (\text{many}) \qquad \frac{\Delta_1 \vdash !\varphi_1 \ \dots\ \Delta_n \vdash !\varphi_n \quad !\varphi_1, \dots, !\varphi_n \vdash \psi}{\Delta_1, \dots, \Delta_n \vdash !\psi}\ (!I)$$

Exponential Axioms

$$\frac{\Gamma \vdash \sim (!(\sim \varphi))}{\Gamma \vdash ?\varphi}\ (?I) \qquad \frac{\Gamma \vdash ?\varphi}{\Gamma \vdash \sim (!(\sim \varphi))}\ (?E)$$

Additive Axioms

$$\frac{}{\Gamma \vdash \top}\ (\top I) \qquad \frac{}{\Gamma, \bot \vdash \varphi}\ (\bot E)$$

$$\frac{\Gamma \vdash \varphi \qquad \Gamma \vdash \vartheta}{\Gamma \vdash \varphi \sqcap \vartheta}\ (\sqcap I) \qquad \frac{\Gamma \vdash \varphi \sqcap \vartheta}{\Gamma \vdash \varphi}\ (\sqcap E1)$$

$$\frac{\Gamma \vdash \varphi \sqcap \vartheta}{\Gamma \vdash \vartheta}\ (\sqcap E2)$$

$$\frac{\Gamma \vdash \varphi}{\Gamma \vdash \varphi \sqcup \vartheta}\ (\sqcup I1) \qquad \frac{\Delta \vdash \varphi \sqcup \vartheta \qquad \Gamma, \varphi \vdash \psi \qquad \Gamma, \vartheta \vdash \psi}{\Gamma, \Delta \vdash \psi}\ (\sqcup E)$$

$$\frac{\Gamma \vdash \vartheta}{\Gamma \vdash \varphi \sqcup \vartheta}\ (\sqcup I2)$$

Table B.5: Natural Deduction Rules, Part 2.

Name	Set	Index	Syntactic Elements
Tensor (t)	$\mathcal{L}_t[C]$	t	$\mathbf{1}, \star$
Positive (p)	$\mathcal{L}_p[C]$	p	$\mathbf{1}, \star, \multimap$
Exponential (e)	$\mathcal{L}_e[C]$	e	$\mathbf{1}, \star, \multimap, !$
Multiplicative (m)	$\mathcal{L}_m[C]$	m	$\mathbf{0}, \mathbf{1}, \star, \multimap, +, \sim$
Multiplicative-exponential (me)	$\mathcal{L}_{me}[C]$	me	$\mathbf{0}, \mathbf{1}, \star, \multimap, +, \sim, !, ?$
Multiplicative-additive (ma)	$\mathcal{L}_{ma}[C]$	ma	$\mathbf{0}, \mathbf{1}, \star, \multimap, +, \sim, \sqcap, \sqcup, \top, \bot$
Full	$\mathcal{L}[C]$		$\mathbf{0}, \mathbf{1}, \star, \multimap, +, \sim, \sqcap, \sqcup, \top, \bot, !, ?$

Table B.6: Fragments of Linear Logic, Language.

Fragment	Rules
Tensor	Basic + Tensor
Positive	Basic + Tensor + Positive
Exponential	Basic + Tensor + Positive + Positive Exponential
Multiplicative	Basic + Tensor + Positive + Multiplicative
Multiplicative-exponential	All except rules containing additives
Multiplicative-additive	All except rules containing exponentials
Full	All rules

Table B.7: Fragments of Linear Logic, Deduction.

An extension $(\mathcal{L}_2, \vdash_2)$ of $(\mathcal{L}_1, \vdash_1)$ is called *conservative*, iff for $\Gamma \in \mathcal{L}_1[C]^*$ and $\varphi \in \mathcal{L}_1[C]$ we have $\Gamma \vdash_1 \varphi \Leftrightarrow \Gamma \vdash_2 \varphi$. ◀

A conservative extension *does not produce new deductions* in the larger fragment for formulae which can be constructed in the smaller fragment. A conservative extension only "affects deduction" for the "new" formulae.

261 Theorem INTERRELATION OF POSTITIVE AND TENSOR FRAGMENT
The *positive* fragment of linear logic is a conservative extension of the *tensor* fragment and the *exponential* fragment is a conservative extension of the *positive* fragment.

PROOF:
This result follows from a careful syntactical analysis of the respective deduction rules. Compare also [GG89]. ◀

B.4 Alternate Natural Deduction

Despite the interesting interpretations mentioned above, our axiomatization also has a number of *disadvantages*. Assume, we want to prove a deduction $\Phi \vdash \psi$ by using the elimination rule $(\star E)$ for backward resolution. In this situation we have to partition the sequence Φ into two sequences Γ and Δ and we must *guess two arbitrary formulae* φ and ϑ. We obtain $\Gamma \vdash \varphi \star \vartheta$ and $\Delta, \varphi, \vartheta \vdash \psi$ as new subgoals of our proof. It is very difficult to guess the two formulae φ and ϑ, since a priori they need not have anything in common with the original goal $\Phi \vdash \psi$.

In this section, a number of alternate axiomatizations are introduced to avoid these problems. Other important results are obtained from the alternate axiomatizations, e.g. a decision algorithm for certain fragments of linear logic.

Generally one should keep in mind that linear logic has been studied only for several years, so there is not yet a didactically optimal way to present it to the reader.

262 Proposition PERMUTATIONS
Let $\Gamma \in \mathcal{L}[C]^\star$ be a sequence of formulae and Δ a permutation thereof. Then $\Gamma \vdash \varphi$ if and only if $\Delta \vdash \varphi$.

PROOF:
Multiple applications of rule (P) and induction on the length of sequences. Applications of this proposition will be denoted by the rule name $(P^\star)$. ◀

The default operator for the left side of $\vdash$ is the multiplicative conjunction $\star$: We may join list elements using $\star$ and we may separate $\star$ expressions into two members of the list:

263 Proposition MULTIPLICATIVE CONJUNCTION
The following two rules hold:

$$\frac{\Gamma,\varphi,\psi,\Delta\vdash\vartheta}{\Gamma,\varphi\star\psi,\Delta\vdash\vartheta}\,(\star L) \qquad \frac{\Gamma,\varphi\star\psi,\Delta\vdash\vartheta}{\Gamma,\varphi,\psi,\Delta\vdash\vartheta}\,(\star L\,\mathrm{inv})$$

PROOF:
First, both rules are shown for the special case, where Δ is the empty list:

$$\dfrac{\dfrac{\dfrac{}{\varphi\star\psi\vdash\varphi\star\psi}\,(Ax) \qquad \Gamma,\varphi,\psi\vdash\vartheta}{\varphi\star\psi,\Gamma\vdash\vartheta}\,(\star E)}{\Gamma,\varphi\star\psi\vdash\vartheta}\,(P)$$

In this proof the left leaves are empty, since $\varphi\star\psi\vdash\varphi\star\psi$ holds under no assumption.

$$\dfrac{\Gamma,\varphi\star\psi\vdash\vartheta \qquad \dfrac{\dfrac{}{\varphi\vdash\varphi}\,(Ax) \qquad \dfrac{}{\psi\vdash\psi}\,(Ax)}{\varphi,\psi\vdash\varphi\star\psi}\,(\star I)}{\Gamma,\varphi,\psi\vdash\vartheta}\,(cut)$$

If Δ is different from the empty list, we first move its formulae into Γ by a repeated use of the permutation rule, then we employ the rules which we just have shown, and finally we use the permutation rule to reconstruct the list Δ. This proof uses inductive reasoning rather than purely equational reasoning, due to the repeated use of the permutation rule. ◀

In the following, a number of alternate rules are derived:

264 Proposition ALTERNATE $(\star E)$ RULE
The rule $(\star L)$ is equivalent to $(\star E)$.

$$\frac{\Gamma,\varphi,\vartheta\vdash\psi}{\Gamma,\varphi\star\vartheta\vdash\psi}\,(\star L)$$

PROOF:
The proof requires (Ax), (P) and (cut).

The $(\star E)\Rightarrow(\star L)$ direction already is shown in Proposition 263. For the reverse direction observe

$$\dfrac{\dfrac{\Delta,\varphi,\vartheta\vdash\psi}{\Delta,\varphi\star\vartheta\vdash\psi}\,(\star L) \qquad \Gamma\vdash\varphi\star\vartheta}{\Gamma,\Delta\vdash\psi}\,(cut)$$

◀

265 Proposition ALTERNATE $(\mathbf{1}E)$ RULE
The following rule $(\mathbf{1}L)$ is equivalent to $(\mathbf{1}E)$:

$$\dfrac{\Gamma \vdash \varphi}{\Gamma, \mathbf{1} \vdash \varphi}\,(\mathbf{1}L)$$

PROOF:
The proof requires (Ax), (P), $(\star R)$ and (tra) (respectively (cut)).
For $(\mathbf{1}L) \Rightarrow (\mathbf{1}E)$ observe

$$\dfrac{\dfrac{\Gamma \vdash \mathbf{1} \qquad \Delta \vdash \varphi}{\Gamma, \Delta \vdash \mathbf{1} \star \varphi}\,(\star I) \qquad \dfrac{\dfrac{\dfrac{\dfrac{}{\varphi \vdash \varphi}\,(Ax)}{\varphi, \mathbf{1} \vdash \varphi}\,(\mathbf{1}L)}{\mathbf{1}, \varphi \vdash \varphi}\,(P)}{\mathbf{1} \star \varphi \vdash \varphi}\,(\star L)}{\Gamma, \Delta \vdash \varphi}\,(tra)$$

and for the other direction one has

$$\dfrac{\dfrac{\dfrac{}{\mathbf{1} \vdash \mathbf{1}}\,(Ax) \qquad \Gamma \vdash \varphi}{\mathbf{1}, \Gamma \vdash \varphi}\,(\mathbf{1}E)}{\Gamma, \mathbf{1} \vdash \varphi}\,(P^\star)$$

266 Proposition ALTERNATE $(\multimap E)$ RULE
The following rule $(\multimap L)$ is equivalent to $(\multimap E)$:

$$\dfrac{\Gamma \vdash \varphi \qquad \Delta, \psi \vdash \vartheta}{\Gamma, \varphi \vdash \psi, \Delta \vdash \vartheta}\,(\multimap L)$$

PROOF:
For $(\multimap E) \Rightarrow (\multimap L)$:

$$\dfrac{\dfrac{\Delta, \psi \vdash \vartheta \qquad \dfrac{\dfrac{}{\varphi \multimap \psi \vdash \varphi \multimap \psi}\,(Ax) \qquad \Gamma \vdash \varphi}{\varphi \multimap \psi, \Gamma \vdash \psi}\,(\multimap E)}{\Delta, \varphi \multimap \psi, \Gamma \vdash \vartheta}\,(cut)}{\Gamma, \varphi \multimap \psi, \Delta \vdash \vartheta}\,(P^\star)$$

For $(\multimap L) \Rightarrow (\multimap E)$:

$$\dfrac{\dfrac{\Gamma \vdash \varphi \multimap \vartheta \qquad \Delta \vdash \varphi}{\Gamma, \Delta \vdash (\varphi \multimap \vartheta) \star \varphi}\,(\star I) \qquad \dfrac{\dfrac{\dfrac{\dfrac{}{\varphi \vdash \varphi}\,(Ax) \qquad \dfrac{}{\vartheta \vdash \vartheta}\,(Ax)}{\varphi, (\varphi \multimap \vartheta) \vdash \vartheta}\,(\multimap L)}{(\varphi \multimap \vartheta), \varphi \vdash \vartheta}\,(P)}{(\varphi \multimap \vartheta) \star \varphi \vdash \vartheta}\,(\star L)}{\Gamma, \Delta \vdash \vartheta}\,(tra)$$

The following theorem will prove very useful:

267 Proposition $(\multimap I)$ IS INVERTIBLE
The rule $(\multimap I\,\text{inv})$ which is the inverse of rule $(\multimap I)$ is equivalent to $(\multimap E)$:

$$\frac{\Gamma \vdash \varphi \multimap \psi}{\Gamma, \varphi \vdash \psi}\,(\multimap I\,\text{inv})$$

PROOF:
The proof requires (Ax) and (cut).

For $(\multimap E) \Rightarrow (\multimap I\,\text{inv})$ observe

$$\frac{\Gamma \vdash \varphi \multimap \psi \qquad \dfrac{}{\varphi \vdash \varphi}\,(Ax)}{\Gamma, \varphi \vdash \psi}\,(\multimap E)$$

For the opposite direction one has

$$\frac{\dfrac{\Gamma \vdash \varphi \multimap \psi}{\Gamma, \varphi \vdash \psi}\,(\multimap I\,\text{inv}) \qquad \Delta \vdash \varphi}{\Gamma, \Delta \vdash \psi}\,(cut)$$

◀

In order to be able to abbreviate a number of proofs and theorems, we introduce the notion of equivalence:

268 Definition EQUIVALENCE
Two formulae φ and ψ are called ***equivalent***, iff $\varphi \vdash \psi$ and $\psi \vdash \varphi$. This will be denoted by $\varphi \simeq \psi$. ◀

It is easy to show, that the equivalence of formulae is an equivalence relation.

269 Proposition CHARACTERIZATION OF EQUIVALENCE
For two formulae φ and ψ, the following is equivalent:

(1) $\varphi \simeq \psi$ (φ and ψ are equivalent, i.e. $\varphi \vdash \psi$ and $\psi \vdash \varphi$).

(2) $\varepsilon \vdash \varphi \multimap \psi$ and $\varepsilon \vdash \psi \multimap \varphi$.

(3) For all sequences Γ one has $\Gamma \vdash \varphi$ if and only if $\Gamma \vdash \psi$.

PROOF:
The proof requires $(\multimap I)$ and $(\multimap I\,\text{inv})$.

(1) $\Rightarrow$ (2) follows directly from $(\multimap I)$. (2) $\Rightarrow$ (1) follows directly from $(\multimap I\text{inv})$. (1) $\Rightarrow$ (3) follows directly from (tra) and (3) $\Rightarrow$ (1) is true a fortiori. ◀

The following theorem shows that the negation $\sim \varphi$ could also be defined as a syntactic abbreviation for $(\varphi \multimap \mathbf{0})$.

270 Proposition DEFINITION OF THE NEGATION
$\sim \varphi \simeq (\varphi \multimap \mathbf{0})$ is equivalent to the rules $(\sim E)$ and $(\sim I)$.

PROOF:
From $\sim\varphi \simeq \varphi \multimap \mathbf{0}$ we may obtain the rules since

$$\dfrac{\dfrac{\Gamma, \varphi \vdash \mathbf{0}}{\Gamma \vdash \varphi \multimap \mathbf{0}}(\multimap I)}{\Gamma \vdash \sim\varphi}(\sim \text{ definition and equivalence properties})$$

and

$$\dfrac{\dfrac{\Gamma \vdash \sim\varphi}{\Gamma \vdash \varphi \multimap \mathbf{0}}(\sim \text{ definition and equivalence properties})}{\Gamma, \varphi \vdash \mathbf{0}}(\multimap I \text{ inv})$$

From the rules one obtains $\sim\varphi \simeq \varphi \multimap \mathbf{0}$ since $\varphi \multimap \mathbf{0} \vdash \sim\varphi$ due to

$$\dfrac{\dfrac{\dfrac{}{\varphi \multimap \mathbf{0} \vdash \varphi \multimap \mathbf{0}}(Ax) \qquad \dfrac{}{\varphi \vdash \varphi}(Ax)}{\varphi \multimap \mathbf{0}, \varphi \vdash \mathbf{0}}(\multimap E)}{\varphi \multimap \mathbf{0} \vdash \sim\varphi}(\sim I)$$

and $\sim\varphi \vdash \varphi \multimap \mathbf{0}$ due to

$$\dfrac{\dfrac{\dfrac{}{\sim\varphi \vdash \sim\varphi}(Ax)}{\sim\varphi, \varphi \vdash \mathbf{0}}(\sim E)}{\sim\varphi \vdash \varphi \multimap \mathbf{0}}(\multimap I)$$

◀

271 Proposition PROPERTIES OF THE NEGATION
The negation is *involutive*: $\sim\sim\varphi \simeq \varphi$.

PROOF:
One observes

$$\dfrac{\dfrac{\dfrac{\dfrac{}{\sim\varphi \vdash \sim\varphi}(Ax)}{\sim\varphi, \varphi \vdash \mathbf{0}}(\sim E)}{\varphi, \sim\varphi \vdash \mathbf{0}}(P)}{\varphi \vdash \sim\sim\varphi}(\sim I)$$

and

$$\dfrac{\dfrac{\dfrac{}{\sim\sim\varphi \vdash \sim\sim\varphi}(Ax)}{\sim\sim\varphi, \sim\varphi \vdash \mathbf{0}}(\sim E)}{\sim\sim\varphi \vdash \varphi}(\mathbf{0}E)$$

◀

Similarly, the next theorem shows that the multiplicative disjunction $\varphi + \psi$ may be defined alternatively as a syntactic abbreviation for $\sim [(\sim \varphi) \star (\sim \psi)]$.

272 Proposition DEFINITION OF THE MULTIPLICATIVE DISJUNCTION
$\varphi + \psi = \sim [(\sim \varphi) \star (\sim \psi)]$ is equivalent to the three rules $(+I1)$, $(+I2)$ and $(E+)$.

PROOF:
$(+I1)$ may be deduced from the definition:

$$\cfrac{\cfrac{\cfrac{\cfrac{\cfrac{\Gamma, \sim \varphi \vdash \psi}{\Gamma, \sim \varphi \vdash \sim\sim \psi}\,\text{(involutive)}}{\Gamma, \sim \varphi, \sim \psi \vdash \mathbf{0}}\,(\sim E)}{\Gamma, \sim \varphi \star \sim \psi \vdash \mathbf{0}}\,(\star L)}{\Gamma \vdash \sim (\sim \varphi \star \sim \psi)}\,(\sim I)}{\Gamma \vdash \varphi + \psi}\,\text{(abbreviation)}$$

For $(+I2)$ this is similar.

For $(+E)$ we have

$$\cfrac{\cfrac{\cfrac{\cfrac{\cfrac{\Gamma \vdash \sim (\sim \varphi \star \sim \psi)}{\Gamma, \sim \varphi \star \sim \psi \vdash \mathbf{0}}\,(\sim E)}{\Gamma, \sim \varphi, \sim \psi \vdash \mathbf{0}}\,(\star L\ \text{inv})}{\Gamma, \sim \varphi \vdash \sim\sim \psi}\,(\sim I)}{\Gamma, \sim \varphi \vdash \psi}\,\text{(involutive)} \qquad \cfrac{\Delta, \varphi \vdash \mathbf{0}}{\Delta \vdash \sim \varphi}\,(\sim I)}{\Gamma, \Delta \vdash \psi}\,(\mathit{cut})$$

For the other direction we show two deductions:

$$\cfrac{\cfrac{\cfrac{\cfrac{\cfrac{\cfrac{}{\varphi + \psi \vdash \varphi + \psi}\,(Ax) \qquad \cfrac{\cfrac{}{\sim \varphi \vdash \sim \varphi}\,(Ax)}{\sim \varphi, \varphi \vdash \mathbf{0}}\,(\sim E)}{\varphi + \psi, \sim \varphi \vdash \psi}\,(+E)}{\varphi + \psi, \sim \varphi \vdash \sim\sim \psi}\,\text{(involutive)}}{\varphi + \psi, \sim \varphi, \sim \psi \vdash \mathbf{0}}\,(\sim E)}{\varphi + \psi, \sim \varphi \star \sim \psi \vdash \mathbf{0}}\,(\star L)}{\varphi + \psi \vdash \sim (\sim \varphi \star \sim \psi)}\,(\sim I)$$

and

$$\dfrac{\dfrac{\dfrac{\dfrac{\dfrac{\dfrac{\dfrac{}{\sim\varphi\vdash\sim\varphi}\,(Ax)\qquad\dfrac{}{\sim\psi\vdash\sim\psi}\,(Ax)}{\sim\varphi,\sim\psi\vdash\sim\varphi\star\sim\psi}\,(\star I)}{\sim\varphi,\sim\psi\vdash\sim\sim(\sim\varphi\star\sim\psi)}\,(\text{involutive})}{\sim\varphi,\sim\psi,\sim(\sim\varphi\star\sim\psi)\vdash\mathbf{0}}\,(\sim E)}{\sim(\sim\varphi\star\sim\psi),\sim\varphi\vdash\sim\sim\psi}\,(P,\sim I)}{\sim(\sim\varphi\star\sim\psi),\sim\varphi\vdash\psi}\,(\text{involutive})}{\sim(\sim\varphi\star\sim\psi)\vdash\varphi+\psi}\,(+I1)$$

◀

273 Proposition ALTERNATE EXPONENTIAL RULES
The following four rules for the storage operator ! are equivalent to their counterparts (none), (one), (many) and $(!I)$:

$$\dfrac{\Gamma\vdash\varphi}{\Gamma,!\psi\vdash\varphi}\,(\text{none}')\qquad\dfrac{\Gamma,\psi\vdash\varphi}{\Gamma,!\psi\vdash\varphi}\,(\text{one}')$$

$$\dfrac{\Gamma,!\psi,!\psi\vdash\varphi}{\Gamma,!\psi\vdash\varphi}\,(\text{many}')\qquad\dfrac{!\Gamma\vdash\varphi}{!\Gamma\vdash!\varphi}\,(!I')$$

PROOF:
A proof for (none′) is presented, the remaining rules are treated similarly.

For (none) ⇒ (none′) one has

$$\dfrac{\dfrac{\dfrac{}{!\psi\vdash!\psi}\,(Ax)\qquad\Gamma\vdash\varphi}{!\psi,\Gamma\vdash\varphi}\,(\text{none})}{\Gamma,!\psi\vdash\varphi}\,(P)$$

For (none′) ⇒ (none) one gets

$$\dfrac{\dfrac{\Gamma\vdash!\varphi\qquad\Delta\vdash\psi}{\Gamma,\Delta\vdash!\varphi\star\psi}\,(\text{none})\qquad\dfrac{\dfrac{\dfrac{\dfrac{}{\psi\vdash\psi}\,(Ax)}{\psi,!\varphi\vdash\psi}\,(\text{none}')}{!\varphi,\psi\vdash\psi}\,(P)}{!\varphi\star\psi\vdash\psi}\,(\star L)}{\Gamma,\Delta\vdash\psi}\,(\mathit{tra})$$

◀

274 Proposition ALTERNATE $(\sqcup E)$ RULE
The following rule $(\sqcup L)$ is equivalent to $(\sqcup E)$:

$$\dfrac{\Gamma,\varphi\vdash\vartheta\qquad\Gamma,\psi\vdash\vartheta}{\Gamma,\varphi\sqcup\psi\vdash\vartheta}\,(\sqcup L)$$

PROOF:
The proofs use (*cut*).

For ($\sqcup E$) $\Rightarrow$ ($\sqcup L$):

$$\frac{\dfrac{}{\varphi \sqcup \vartheta \vdash \varphi \sqcup \vartheta}(Ax) \qquad \Gamma, \varphi \vdash \vartheta \qquad \Gamma, \psi \vdash \vartheta}{\Gamma, \varphi \sqcup \vartheta \vdash \vartheta}(\sqcup E)$$

For ($\sqcup L$) $\Rightarrow$ ($\sqcup E$):

$$\frac{\dfrac{\Gamma, \varphi \vdash \psi \qquad \Gamma, \vartheta \vdash \psi}{\Gamma, \varphi \sqcup \vartheta \vdash \psi}(\sqcup L) \qquad \Delta \vdash \varphi \sqcup \vartheta}{\Gamma, \Delta \vdash \psi}(cut)$$

◀

275 Proposition ALTERNATE ($\sqcap E1$) AND ($\sqcap E2$) RULES
The following two rules ($\sqcap L1$) and ($\sqcap L2$) are equivalent to the two rules ($\sqcap E1$) and ($\sqcap E2$):

$$\frac{\Gamma, \varphi \vdash \vartheta}{\Gamma, \varphi \sqcap \psi \vdash \vartheta}(\sqcap L1) \qquad \frac{\Gamma, \psi \vdash \vartheta}{\Gamma, \varphi \sqcap \psi \vdash \vartheta}(\sqcap L2)$$

PROOF:
The proofs use (*cut*):

For ($\sqcap E1$) $\Rightarrow$ ($\sqcap L1$):

$$\frac{\Gamma, \varphi \vdash \vartheta \qquad \dfrac{\dfrac{}{\varphi \sqcap \vartheta \vdash \varphi \sqcap \vartheta}(Ax)}{\varphi \sqcap \vartheta \vdash \varphi}(\sqcap E1)}{\Gamma, \varphi \sqcap \vartheta \vdash \vartheta}(cut)$$

For ($\sqcap L1$) $\Rightarrow$ ($\sqcap E1$):

$$\frac{\Gamma \vdash \varphi \sqcap \vartheta \qquad \dfrac{\dfrac{}{\varphi \vdash \varphi}(Ax)}{\varphi \sqcap \vartheta \vdash \varphi}(\sqcap L1)}{\Gamma \vdash \varphi}(tra)$$

◀

By using the above theorems it is now possible to jointly replace a number of deduction rules:

276 Theorem ALTERNATE RULE SET
The deduction relation defined by the axiomatization in Tables B.4 and B.5 and the one defined by Tables B.8 and B.9 are equal.

PROOF:
First add the (*cut*) rule to the deduction system. Then use the above results to exchange the old rules against the new, one by one. It is necessary to add the (*cut*) rule to the new deduction system, since rules are removed which are required to prove (*cut*) – and (*cut*) itself is used in most equivalence proofs. ◀

Basic Axioms

$$\frac{}{\varphi \vdash \varphi}(Ax) \qquad \frac{\Gamma, \varphi, \psi, \Delta \vdash \vartheta}{\Gamma, \psi, \varphi, \Delta \vdash \vartheta}(P)$$

Cut

$$\frac{\Gamma, \varphi \vdash \psi \qquad \Delta \vdash \varphi}{\Gamma, \Delta \vdash \psi}(cut)$$

Tensor Axioms

$$\varepsilon \vdash \mathbf{1} \qquad (\mathbf{1}I) \qquad \frac{\Gamma \vdash \varphi}{\Gamma, \mathbf{1} \vdash \varphi}(\mathbf{1}L)$$

$$\frac{\Gamma \vdash \varphi \qquad \Delta \vdash \vartheta}{\Gamma, \Delta \vdash \varphi \star \vartheta}(\star I) \qquad \frac{\Gamma, \varphi, \vartheta \vdash \psi}{\Gamma, \varphi \star \vartheta \vdash \psi}(\star L)$$

Positive Axioms

$$\frac{\Gamma, \varphi \vdash \vartheta}{\Gamma \vdash \varphi \multimap \vartheta}(\multimap I) \qquad \frac{\Gamma \vdash \varphi \qquad \Delta, \psi \vdash \vartheta}{\Gamma, \varphi \multimap \psi, \Delta \vdash \vartheta}(\multimap L)$$

Multiplicative Axioms

$$\sim \varphi :\simeq (\varphi \multimap \mathbf{0})$$

$$\varphi + \psi :\simeq \sim((\sim \varphi) \star (\sim \psi))$$

$$\frac{\Gamma, \varphi \multimap \mathbf{0} \vdash \mathbf{0}}{\Gamma \vdash \varphi}(\mathbf{0}E)$$

Table B.8: Alternate Natural Deduction Rules, Part 1

Positive Exponential Axioms

$$\frac{\Gamma \vdash \varphi}{\Gamma, !\psi \vdash \varphi}\ (\text{none}') \qquad \frac{\Gamma, \psi \vdash \varphi}{\Gamma, !\psi \vdash \varphi}\ (\text{one}')$$

$$\frac{\Gamma, !\psi, !\psi \vdash \varphi}{\Gamma, !\psi \vdash \varphi}\ (\text{many}') \qquad \frac{!\Gamma \vdash \varphi}{!\Gamma \vdash !\varphi}\ (!I')$$

Exponential Axioms

$$?\varphi :\simeq\ \sim (!(\sim \varphi))$$

Additive Axioms

$$\frac{}{\Gamma \vdash \top}\ (\top I)$$

$$\frac{}{\Gamma, \bot \vdash \varphi}\ (\bot E)$$

$$\frac{\Gamma \vdash \varphi \qquad \Gamma \vdash \vartheta}{\Gamma \vdash \varphi \sqcap \vartheta}\ (\sqcap I) \qquad \frac{\Gamma, \varphi \vdash \vartheta}{\Gamma, \varphi \sqcap \psi \vdash \vartheta}\ (\sqcap L1)$$

$$\frac{\Gamma, \psi \vdash \vartheta}{\Gamma, \varphi \sqcap \psi \vdash \vartheta}\ (\sqcap L2)$$

$$\frac{\Gamma \vdash \varphi}{\Gamma \vdash \varphi \sqcup \vartheta}\ (\sqcup I1) \qquad \frac{\Gamma, \varphi \vdash \vartheta \qquad \Gamma, \psi \vdash \vartheta}{\Gamma, \varphi \sqcup \psi \vdash \vartheta}\ (\sqcup L)$$

$$\frac{\Gamma \vdash \vartheta}{\Gamma \vdash \varphi \sqcup \vartheta}\ (\sqcup I2)$$

Note: The term $!\Gamma$ denotes that every element of the sequence Γ has a ! in front of it.

Table B.9: Alternate Natural Deduction Rules, Part 2

B.5 Sequent Calculus

Another approach to linear logic is provided by the *sequent calculus*:

277 Definition SEQUENT CALCULUS
The *sequent calculus* of linear logic is given by the relation $\vdash \ \subseteq \mathcal{L}[C]^* \times \mathcal{L}[C]^*$ which is defined by structural induction over the context free syntax of $\mathcal{L}[C]$ with the help of the rules in Tables B.10 and B.11. ◀

How do the natural deduction relation and the sequent calculus relate? Let $\vdash_s$ denote the deduction relation of the sequent calculus as defined by Tables B.10 and B.11, and let $\vdash_n$ denote the deduction relation of the natural deduction as defined by the axiomatization in Tables B.4 and B.5 or, equivalently, the alternate version given in tables B.8 and B.9.

278 Theorem SEQUENT CALCULUS AND NATURAL DEDUCTION CALCULUS
Let $\varphi_1, \varphi_2, \ldots, \varphi_k$ and $\psi_1, \psi_2, \ldots, \psi_m$ be formulae of linear logic. Then the following three deductions are equivalent:

$$\varphi_1, \varphi_2, \ldots, \varphi_k \vdash_s \psi_1, \psi_2, \ldots, \psi_m$$

$$\varphi_1, \varphi_2, \ldots, \varphi_k \vdash_n \psi_1 + \psi_2 + \ldots + \psi_m$$

$$\varphi_1 \star \varphi_2 \star \ldots \star \varphi_k \vdash_n \psi_1 + \psi_2 + \ldots + \psi_m$$

Furthermore, for empty sequences

$$\varepsilon \vdash_s \psi_1, \psi_2, \ldots, \psi_m \text{ if and only if } \varepsilon \vdash_n \psi_1 + \psi_2 + \ldots + \psi_m$$

In the $+$ and $\star$ expressions above, an arbitrary bracketing may be used due to associativity, compare Proposition 288. ◀

279 Definition PURE DEDUCTION FOR LINEAR LOGIC
The *pure deduction relation* of linear logic is given by the relation $\vdash \ \subseteq \mathcal{L}[C] \times \mathcal{L}[C]$ which is defined by structural induction over the context free syntax of $\mathcal{L}[C]$ with the help of the rules which are obtained from the axiomatizations of the natural deduction or the sequent calculus by an application of Theorem 278. ◀

In the following we *will outline what we gained* by the introduction of the three axiomatizations.

The sequent calculus defined allows *cut elimination*:

280 Theorem CUT ELIMINATION FOR SEQUENT CALCULUS
The deduction relation defined by the rules of the sequent calculus in Tables B.10 and B.11 is equivalent to the deduction relation defined by the same set of rules, but without the cut rule for sequences, (*cuts*). Every proof tree which employs the use of (*cuts*) may be translated into a proof tree which does not employ (*cuts*).

PROOF:
1992 a proof of this theorem was given in the Thesis [Lin92] of LINCOLN. The proof fills 20 pages and uses the same ideas as the cut elimination technique for classical logic studied by PRAWITZ and GENTZEN, [Gal87]: By successive transformations of the proof tree, the use of

Basic Axioms

$$\frac{}{\varphi \vdash \varphi}\,(Ax) \qquad \frac{\Gamma, \varphi, \psi, \Delta \vdash \Theta}{\Gamma, \psi, \varphi, \Delta \vdash \Theta}\,(Ps)$$

Cut

$$\frac{\Gamma \vdash \varphi, \Delta \qquad \Sigma, \varphi \vdash \Theta}{\Gamma, \Sigma \vdash \Delta, \Theta}\,(cuts)$$

Tensor Axioms

$$\frac{}{\varepsilon \vdash \mathbf{1}}\,(\mathbf{1}Rs) \qquad \frac{\Gamma \vdash \Delta}{\Gamma, \mathbf{1} \vdash \Delta}\,(\mathbf{1}Ls)$$

$$\frac{\Gamma, \varphi, \psi \vdash \Delta}{\Gamma, \varphi \star \psi \vdash \Delta}\,(\star Ls) \qquad \frac{\Gamma \vdash \varphi, \Delta \qquad \Sigma \vdash \psi, \Theta}{\Gamma, \Sigma \vdash \varphi \star \psi, \Delta, \Theta}\,(\star Rs)$$

Positive Axioms

$$\frac{\Gamma \vdash \varphi, \Delta \qquad \Sigma, \psi \vdash \Theta}{\Gamma, \varphi \multimap \psi, \Sigma \vdash \Delta, \Theta}\,(\multimap Ls) \qquad \frac{\Gamma, \varphi \vdash \psi, \Delta}{\Gamma \vdash \varphi \multimap \psi, \Delta}\,(\multimap Rs)$$

Multiplicative Axioms

$$\frac{\Gamma \vdash \varphi, \Delta}{\Gamma, \sim\varphi \vdash \Delta}\,(\sim Ls) \qquad \frac{\Gamma, \varphi \vdash \Delta}{\Gamma \vdash \sim\varphi, \Delta}\,(\sim Rs)$$

$$\frac{\Gamma, \varphi \vdash \Delta \qquad \Sigma, \psi \vdash \Theta}{\Gamma, \Sigma, \varphi + \psi \vdash \Delta, \Theta}\,(+Ls) \qquad \frac{\Gamma \vdash \varphi, \psi, \Delta}{\Gamma \vdash \varphi + \psi, \Delta}\,(+Rs)$$

$$\frac{}{\mathbf{0} \vdash \varepsilon}\,(\mathbf{0}Ls) \qquad \frac{\Gamma \vdash \Delta}{\Gamma \vdash \Delta, \mathbf{0}}\,(\mathbf{0}Rs)$$

Table B.10: Sequent Calculus for Linear Logic, Part 1.

Exponential Axioms

$$\frac{\Gamma \vdash \Delta}{\Gamma, !\varphi \vdash \Delta}(W!) \qquad \frac{\Gamma, \varphi \vdash \Delta}{\Gamma, !\varphi \vdash \Delta}(L!)$$

$$\frac{!\Gamma \vdash \varphi, ?\Delta}{!\Gamma \vdash !\varphi, ?\Delta}(R!) \qquad \frac{\Gamma, !\varphi, !\varphi \vdash \Delta}{\Gamma, !\varphi \vdash \Delta}(C!)$$

$$\frac{\Gamma \vdash \Delta}{\Gamma \vdash ?\varphi, \Delta}(W?) \qquad \frac{!\Gamma, \varphi \vdash ?\Delta}{!\Gamma, ?\varphi \vdash ?\Delta}(L?)$$

$$\frac{\Gamma \vdash \varphi, \Delta}{\Gamma \vdash ?\varphi, \Delta}(R?) \qquad \frac{\Gamma \vdash ?\varphi, ?\varphi, \Delta}{\Gamma \vdash ?\varphi, \Delta}(C?)$$

Additive Axioms

$$\frac{}{\Gamma, \bot \vdash \Delta}(\bot Ls)$$

$$\frac{}{\Gamma \vdash \top, \Delta}(\top Rs)$$

$$\frac{\Gamma, \varphi \vdash \Delta}{\Gamma, \varphi \sqcap \psi \vdash \Delta}(\sqcap L1s) \qquad \frac{\Gamma \vdash \varphi, \Delta \qquad \Gamma \vdash \psi, \Delta}{\Gamma \vdash \varphi \sqcap \psi, \Delta}(\sqcap Rs)$$

$$\frac{\Gamma, \psi \vdash \Delta}{\Gamma, \varphi \sqcap \psi \vdash \Delta}(\sqcap L2s)$$

$$\frac{\Gamma, \varphi \vdash \Delta \qquad \Gamma, \psi \vdash \Delta}{\Gamma, \varphi \sqcup \psi \vdash \Delta}(\sqcup Ls) \qquad \frac{\Gamma \vdash \varphi, \Delta}{\Gamma \vdash \varphi \sqcup \psi, \Delta}(\sqcup R1s)$$

$$\frac{\Gamma \vdash \psi, \Delta}{\Gamma \vdash \varphi \sqcup \psi, \Delta}(\sqcup R2s)$$

The terms $!\Gamma$ and $?\Gamma$ denote that every element of the sequence Γ has a ! in front of it, or a ?, respectively.

Table B.11: Sequent Calculus for Linear Logic, Part 2.

the cut rule is propagated upwards towards the leaves, until the cut rule vanishes from the proof tree. ◀

281 Theorem SUBFORMULA PROPERTY
If $\Gamma \vdash \Delta$ is a deduction in the sequent calculus which has a cut free proof which does not use the rules $(C!)$ and $(C?)$, then all formulae in this proof tree are subformulae of the formulae in Γ and Δ.

PROOF:
This is an immediate consequence of the syntactic form of the rules. In the assumptions of all formulae, except for $(cuts)$, $(C!)$ and $(C?)$, the assumptions are subformulae of permutations of the formulae of the consequence. ◀

The subformula property immediately gives us a decision procedure for the multiplicative-additive fragment of linear logic. Due to the subformula property, we may restrict the search for proofs to proof trees which contain formulae and subformulae of the deduction which we want to prove.

282 Theorem MULTIPLICATIVE-ADDITIVE FRAGMENT IS DECIDABLE
Let Γ and Δ be sequences of formulae from the *multiplicative-additive fragment* $\mathcal{L}_{ma}[C]$ of linear logic. Then it is *decidable*, whether $\Gamma \vdash_{ma} \Delta$ holds or not.

PROOF:
Immediately, by above remark. ◀

A decision algorithm for the multiplicative-additive fragment written in PROLOG is described in [LS94] and available by anonymous ftp from `ftp.csl.sri.com`. It was used to check several of the formulae in this work.

As a consequence we obtain:

283 Theorem DECIDABLE FRAGMENTS OF LINEAR LOGIC
The *tensor*, the *positive*, and the *multiplicative* fragments of linear logic are decidable.

PROOF:
A decision procedure is the generation of the finite number of possible proof trees within the multiplicative-additive fragment, followed by a test whether a proof uses exclusively the rules of the respective fragment. ◀

Unfortunately, the problems associated with the exponentials cannot be resolved:

284 Theorem FULL LINEAR LOGIC
Full linear logic is *not* decidable.

PROOF:
The proof is given in [Lin92] and [Tro92] and uses the encoding of an undecidable problem of counter machines in linear logic. ◀

B.6 Syntactic Results

In this section syntactic results are presented which will be used in the subsequent text and which may provide a better intuitive understanding of linear logic. Much of the theory presented here is known and readily available from the literature [Tro92], [Lin92], [Gir87a].

The following theorem allows equational reasoning in linear logic: If we take a formula φ and exchange a subformula by an equivalent one, the resulting formula φ' is equivalent to φ.

285 Proposition EQUIVALENCE IS A CONGRUENCE
The equivalence $\simeq \subseteq \mathcal{L}[C] \times \mathcal{L}[C]$ of formulae is a congruence relation with respect to all operators of linear logic. If $\varphi \simeq \psi$, then

$$\begin{array}{lclclcl} \vartheta \star \varphi & \simeq & \vartheta \star \psi & \quad & \varphi \star \vartheta & \simeq & \psi \star \vartheta \\ \vartheta \multimap \varphi & \simeq & \vartheta \multimap \psi & & \varphi \multimap \vartheta & \simeq & \psi \multimap \vartheta \\ \sim \varphi & \simeq & \sim \psi & & & & \\ \vartheta + \varphi & \simeq & \vartheta + \psi & & \varphi + \vartheta & \simeq & \psi + \vartheta \\ \vartheta \sqcap \varphi & \simeq & \vartheta \sqcap \psi & & \varphi \sqcap \vartheta & \simeq & \psi \sqcap \vartheta \\ \vartheta \sqcup \varphi & \simeq & \vartheta \sqcup \psi & & \varphi \sqcup \vartheta & \simeq & \psi \sqcup \vartheta \\ !\varphi & \simeq & !\psi & & ?\varphi & \simeq & ?\psi \end{array}$$

PROOF:
Without loss of generality we only show one part of the proof, e.g. if $\varphi \vdash \psi$ then $\vartheta \star \varphi \vdash \vartheta \star \psi$. The rest follows by symmetry by reversing the roles of φ and ψ or by proof rearrangement.

$$\dfrac{\dfrac{\dfrac{}{\vartheta \vdash \vartheta}(Ax) \qquad \varphi \vdash \psi}{\vartheta, \varphi \vdash \vartheta \star \psi}(\star I)}{\vartheta \star \varphi \vdash \vartheta \star \psi}(\star L)$$

$$\dfrac{\dfrac{\dfrac{\dfrac{}{\vartheta \multimap \varphi \vdash \vartheta \multimap \varphi}(Ax)}{\vartheta \multimap \varphi, \vartheta \vdash \varphi}(\multimap I \text{ inv}) \qquad \varphi \vdash \psi}{\vartheta \multimap \varphi, \vartheta \vdash \psi}(tra)}{\vartheta \multimap \varphi \vdash \vartheta \multimap \psi}(\multimap I)$$

$$\dfrac{\dfrac{\dfrac{\dfrac{}{\varphi \vdash \varphi}(Ax)}{!\varphi \vdash \varphi}(\text{one}') \qquad \varphi \vdash \psi}{!\varphi \vdash \psi}(tra1)}{!\varphi \vdash !\psi}(!I')$$

For $\sim$, $+$ and $?$ this follows from their definition by $\multimap$, $\star$ and $!$.

$$\dfrac{\dfrac{\dfrac{}{\vartheta \sqcap \varphi \vdash \vartheta \sqcap \varphi}\,(Ax)}{\vartheta \sqcap \varphi \vdash \vartheta} \qquad \dfrac{\dfrac{\dfrac{}{\vartheta \sqcap \varphi \vdash \vartheta \sqcap \varphi}\,(Ax)}{\vartheta \sqcap \varphi \vdash \varphi}\,(\sqcap E2) \qquad \varphi \vdash \psi}{\vartheta \sqcap \varphi \vdash \psi}\,(tra)}{\vartheta \sqcap \varphi \vdash \vartheta \sqcap \psi}\,(\sqcap I)$$

$$\dfrac{\dfrac{\dfrac{}{\vartheta \vdash \vartheta}\,(Ax)}{\vartheta \vdash \vartheta \sqcup \varphi}\,(\sqcup I1) \qquad \dfrac{\psi \vdash \varphi \qquad \dfrac{\dfrac{}{\varphi \vdash \varphi}\,(Ax)}{\varphi \vdash \vartheta \sqcup \varphi}\,(\sqcup I2)}{\psi \vdash \vartheta \sqcup \varphi}\,(tra1)}{\vartheta \sqcup \psi \vdash \vartheta \sqcup \psi}\,(\sqcup L)$$

◀

The following rule allows to shift formulae from the right of $\vdash$ to its left:

286 Proposition SHIFT RULE
The following rule holds:

$$\dfrac{\Gamma \vdash \varphi}{\Gamma, \varphi \multimap \psi \vdash \psi}\,(s-l)$$

A special instance of this rule is the inverse of $(\mathbf{0}E)$:

$$\dfrac{\Gamma \vdash \varphi}{\Gamma, \varphi \multimap \mathbf{0} \vdash \mathbf{0}}\,(\mathbf{0}\text{inv}) \qquad \dfrac{\Gamma \vdash \varphi}{\Gamma, \sim\varphi \vdash \mathbf{0}}\,(\mathbf{0}\text{inv})$$

PROOF:

$$\dfrac{\dfrac{\dfrac{}{\varphi \multimap \psi \vdash \varphi \multimap \psi}\,(Ax) \qquad \Gamma \vdash \varphi}{\varphi \vdash \psi, \Gamma \vdash \psi}\,(\multimap E)}{\Gamma, \varphi \multimap \psi}\,(P)$$

◀

The following theorem shows that formulae of the form $\varphi \multimap \psi$ may be interpreted as state transitions. Note the similarity with the classical rule of *modus ponens* and the different interpretation it obtains in linear logic!

287 Theorem STATE TRANSITION THEOREM
The following deduction rules hold:

$$\dfrac{}{\varphi, \varphi \multimap \psi \vdash \psi}\,(mp)$$

$$\dfrac{}{\varphi \star \vartheta, \varphi \multimap \psi \vdash \psi \star \vartheta}\,(mpL) \qquad \dfrac{}{\vartheta \star \varphi, \varphi \multimap \psi \vdash \vartheta \star \psi}\,(mpR)$$

PROOF:
For the first rule we have

$$\dfrac{\dfrac{}{\varphi \multimap \psi \vdash \varphi \multimap \psi}(Ax) \qquad \dfrac{}{\varphi \vdash \varphi}(Ax)}{\varphi \multimap \psi, \varphi \vdash \psi}(\multimap E)$$

followed by a permutation (P).

The second and third rules are similar in proof and require $(\star I)$. ◀

288 Proposition ASSOCIATIVITY AND COMMUTATIVITY
The operators $\star$, $+$, $\sqcap$ and $\sqcup$ are ***associative*** and ***commutative***:

$$\begin{array}{lcllcl}
(\varphi \star \psi) & \simeq & (\psi \star \varphi) & (\varphi \star (\psi \star \vartheta)) & \simeq & ((\varphi \star \psi) \star \vartheta) \\
(\varphi + \psi) & \simeq & (\psi + \varphi) & (\varphi + (\psi + \vartheta)) & \simeq & ((\varphi + \psi) + \vartheta) \\
(\varphi \sqcap \psi) & \simeq & (\psi \sqcap \varphi) & (\varphi \sqcap (\psi \sqcap \vartheta)) & \simeq & ((\varphi \sqcap \psi) \sqcap \vartheta) \\
(\varphi \sqcup \psi) & \simeq & (\psi \sqcup \varphi) & (\varphi \sqcup (\psi \sqcup \vartheta)) & \simeq & ((\varphi \sqcup \psi) \sqcup \vartheta)
\end{array}$$

◀

Similar as in propositional logic, also in linear logic the negation acts as a ***dualizing operation***. The constant $\mathbf{0}$ may be regarded as ***dualizing object***, since the negation can be defined by $\sim \varphi := (\varphi \multimap \mathbf{0})$. Since negation is involutive, the laws expressing duality may be transformed into laws defining an operator by its dual.

Rearranging results from the literature one obtains the following proposition:

289 Proposition DUALITY IN LINEAR LOGIC

(1) *Multiplicative De Morgan:* The multiplicative conjunction $\star$ and the multiplicative disjunction $+$ are dual to each other:

$$\begin{array}{lcllcl}
\sim(\varphi + \psi) & \simeq & (\sim \varphi) \star (\sim \psi) & \varphi + \psi & \simeq & \sim [(\sim \varphi) \star (\sim \psi)] \\
\sim(\varphi \star \psi) & \simeq & (\sim \varphi) + (\sim \psi) & \varphi \star \psi & \simeq & \sim [(\sim \varphi) + (\sim \psi)]
\end{array}$$

(2) *Additive De Morgan:* The additive conjunction $\sqcap$ and the additive disjunction $\sqcup$ are dual to each other:

$$\begin{array}{lcllcl}
\sim(\varphi \sqcap \psi) & \simeq & (\sim \varphi) \sqcup (\sim \psi) & \varphi \sqcap \psi & \simeq & \sim [(\sim \varphi) \sqcup (\sim \psi)] \\
\sim(\varphi \sqcup \psi) & \simeq & (\sim \varphi) \sqcap (\sim \psi) & \varphi \sqcup \psi & \simeq & \sim [(\sim \varphi) \sqcap (\sim \psi)]
\end{array}$$

(3) *Exponentials:* The two exponentials ! and ? are dual to each other:

$$\begin{array}{lcllcl}
\sim(!\varphi) & \simeq & ?(\sim \varphi) & !\varphi & \simeq & \sim ?(\sim \varphi) \\
\sim(?\varphi) & \simeq & !(\sim \varphi) & ?\varphi & \simeq & \sim !(\sim \varphi)
\end{array}$$

(4) *Implication:* The operator of the implication $\multimap$ is dual to its transposed version $\circ\!-$ which may be defined as syntactic abbreviation by $(\psi \circ\!- \varphi) :\simeq (\varphi \multimap \psi)$:

$$\sim(\varphi \multimap \psi) \simeq (\sim\varphi \circ\!- \sim\psi) \simeq (\sim\psi \multimap \sim\varphi)$$

$$\sim(\varphi \circ\!- \psi) \simeq (\sim\varphi \multimap \sim\psi) \simeq (\sim\psi \circ\!- \sim\varphi)$$

$$(\varphi \multimap \psi) \simeq \sim(\sim\varphi \circ\!- \sim\psi)$$

$$(\varphi \circ\!- \psi) \simeq \sim(\sim\varphi \multimap \sim\psi)$$

(5) *Multiplicative Constants:* The multiplicative constants $\mathbf{1}$ and $\mathbf{0}$ are dual to each other:

$$\sim\mathbf{1} \simeq \mathbf{0} \qquad \sim\mathbf{0} \simeq \mathbf{1}$$

(6) *Additive Constants:* The additive constants $\top$ and $\bot$ are dual to each other:

$$\sim\top \simeq \bot \qquad \sim\bot \simeq \top$$

◀

Negation and deduction are connected by the following theorem:

290 Theorem NEGATION AND DEDUCTION
Negation may be used to exchange the sides of the deduction relation:

$$\frac{\Gamma, \varphi \vdash \psi}{\Gamma, \sim\psi \vdash \sim\varphi}\,(\text{excl}) \qquad \frac{\Gamma, \sim\varphi \vdash \sim\psi}{\Gamma, \psi \vdash \varphi}\,(\text{exc2})$$

◀

291 Theorem PROPERTIES OF THE IMPLICATION
Similar to $\vdash$, also $\multimap$ is "transitive":

$$\frac{}{(\varphi \multimap \psi) \star (\psi \multimap \vartheta) \vdash \varphi \multimap \vartheta}\,(\multimap\, tra)$$

The CURRYING construction from functional programming and intuitionistic type theory can be employed:

$$((\varphi \star \psi) \multimap \vartheta) \simeq (\varphi \multimap (\psi \multimap \vartheta))$$

For the conjunctive multiplicative unit $\mathbf{1}$ the following equivalence holds:

$$\varphi \simeq \mathbf{1} \multimap \varphi$$

The following expressions are pairwise equivalent:

$$\varphi \multimap (\psi \multimap \vartheta) \simeq \psi \multimap (\varphi \multimap \vartheta) \simeq (\varphi \star \psi) \multimap \vartheta \simeq (\psi \star \varphi) \multimap \vartheta$$

The following expressions are pairwise equivalent:

$$\varphi \multimap \psi \simeq (\sim \psi) \multimap (\sim \varphi) \simeq \sim (\varphi \star (\sim \psi)) \simeq (\sim \varphi) + \psi$$

◀

292 Proposition PROPERTIES OF THE MULTIPLICATIVE DISJUNCTION
The following deductions hold:

$$(\varphi + \psi) \star \sim \varphi \vdash \psi \qquad (\varphi + \psi) \star \sim \psi \vdash \varphi$$

◀

293 Proposition PROPERTIES OF THE CONSTANTS

(1) The constants **0**, **1**, $\top$, and $\bot$ serve as *neutral elements*:

Constant	Neutral for	Law				
$\mathbf{0}$	$+$	$\mathbf{0} + \varphi$	$\simeq$	$\varphi + \mathbf{0}$	$\simeq$	φ
$\mathbf{1}$	$\star$	$\mathbf{1} \star \varphi$	$\simeq$	$\varphi \star \mathbf{1}$	$\simeq$	φ
$\top$	$\sqcap$	$\top \sqcap \varphi$	$\simeq$	$\varphi \sqcap \top$	$\simeq$	φ
$\bot$	$\sqcup$	$\bot \sqcup \varphi$	$\simeq$	$\varphi \sqcup \bot$	$\simeq$	φ

(2) The additive constants may be *recovered* from the multiplicative constants with the help of the exponentials:

$$\mathbf{1} \simeq !\top \qquad \mathbf{0} \simeq ?\bot$$

◀

The additive operations allow strengthening and weakening, similar to their classical counterparts:

294 Proposition PROPERTIES OF THE ADDITIVE OPERATIONS
The following rules hold:

$$\varphi \sqcap \psi \vdash \varphi \qquad \varphi \sqcap \psi \vdash \psi \qquad \frac{\Gamma \vdash \varphi \qquad \Gamma \vdash \psi}{\Gamma \vdash \varphi \sqcap \psi}$$

and

$$\varphi \vdash \varphi \sqcup \psi \qquad \psi \vdash \varphi \sqcup \psi \qquad \frac{\varphi \vdash \Gamma \qquad \psi \vdash \Gamma}{\varphi \sqcup \psi \vdash \Gamma}$$

◀

295 Proposition PROPERTIES OF THE EXPONENTIALS
The following rules hold:

$$!\varphi \vdash \mathbf{1} \qquad !\varphi \vdash \varphi \qquad !\varphi \vdash !\varphi \star !\varphi$$

This may also be written in the following form:

$$!\varphi \multimap \mathbf{1} \sqcap \varphi \sqcap (!\varphi \star !\varphi)$$

Dualizing yields the following rules:

$$\mathbf{0} \vdash ?\varphi \qquad \varphi \vdash ?\varphi \qquad ?\varphi \,+\, ?\varphi \vdash ?\varphi$$

This may also be written in the following form:

$$\mathbf{0} \sqcup \varphi \sqcup (?\varphi \,+\, ?\varphi) \vdash ?\varphi$$

◀

The exponentials form a link between additive and multiplicative operations which is similar to duality by negation:

296 Proposition EXPONENTIALS, ADDITIVES, AND MULTIPLICATIVES
The following equivalences hold:

$$!(\varphi \sqcap \psi) \simeq (!\varphi) \star (!\psi) \qquad ?(\varphi \sqcup \psi) \simeq (?\varphi) + (?\psi)$$

◀

297 Proposition EXPONENTIALS AND ADDITIVES
The following equivalences hold:

$$!(\varphi \sqcap \psi) \vdash (!\varphi) \sqcap (!\psi) \qquad !(!\varphi \sqcap !\psi) \simeq !(\varphi \sqcap \psi)$$

$$(?\varphi) \sqcup (?\psi) \vdash ?(\varphi \sqcup \psi) \qquad ?(?\varphi \sqcup ?\psi) \simeq ?(\varphi \sqcup \psi)$$

◀

An important property of linear logic is the fact that we may describe deduction by an operator of the logic itself.

298 Theorem INTERNALIZATION OF DEDUCTION
The deduction $\varphi \vdash \psi$ holds if and only if the deduction $\varepsilon \vdash (\varphi \multimap \psi)$ holds.

PROOF:
We have

$$\frac{\varphi \vdash \psi}{\varepsilon \vdash \varphi \multimap \psi}\,(\multimap I) \qquad \text{and} \qquad \frac{\varepsilon \vdash \varphi \multimap \psi \qquad \dfrac{}{\varphi \vdash \varphi}\,(Ax)}{\varphi \vdash \psi}\,(\multimap E)$$

◀

For the description of special situations, one often adds so-called non-logical axioms to the deduction system which single out these special situations. In linear logic, deduction with non-logical axioms may also be internalized:

299 Definition NON-LOGICAL AXIOMS
Let $A \subseteq \mathcal{L}[C]$ be a set of formulae and let $\vdash_A \;\subseteq \mathcal{L}[C]^* \times \mathcal{L}[C]$ be the deduction relation which is obtained by adding for every $\varphi \in A$ the following rule to the set of rules which define $\vdash$.

$$\frac{}{\varepsilon \vdash \varphi}\,(A)$$

◀

300 Theorem INTERNALIZATION OF NON-LOGICAL AXIOMS
Let $A = \{a_1, a_2, \ldots, a_n\} \subseteq \mathcal{L}[C]$ be a finite set of formulae, let $\Gamma \in \mathcal{L}[C]^*$ be a sequence of formulae and $\varphi \in \mathcal{L}[C]$ a single formula.

$$\Gamma \vdash_A \varphi \text{ if and only if } \Gamma, !a_1, !a_2, \ldots, !a_n \vdash \varphi$$

$$\Gamma \vdash_A \varphi \;\Rightarrow\; \exists k \in \mathbb{N}: \; a_1^k, a_2^k, \ldots, a_n^k \vdash \varphi$$

PROOF:
Assume we have a proof tree for $\Gamma \vdash_a \varphi$. There is a finite number of applications of non-logical axioms in the form of

$$\frac{}{\varepsilon \vdash a_i}\,(A)$$

Replacing them all by

$$!a_i \vdash a_i$$

yields a proof tree for $\Gamma, !a_1, \ldots, !a_n \vdash \varphi$, where several of the $!a_i$ may occur more than once. Reduce the multiple occurrences to single ones by using weakening for !. We thus have obtained a proof tree for $\Gamma, !a_1, !a_2, \ldots, !a_n \vdash \varphi$. This observation also proves the second statement.

Conversely, assume we have a proof tree for $\Gamma, !a_1, !a_2, \ldots, !a_n \vdash \varphi$. Since $\vdash_A \;\supseteq\; \vdash$ we also have $\Gamma, !a_1, !a_2, \ldots, !a_n \vdash_A \varphi$. We have $\varepsilon \vdash_A a_i$ and so $!\varepsilon \vdash_A a_i$ and by $(!I)$ even $!\varepsilon \vdash_A !a_i$. Using (cut) several times we obtain:

$$\frac{\Gamma, !a_1, !a_2, \ldots, !a_n \vdash_A \varphi \qquad !\varepsilon \vdash_A !a_1 \quad !\varepsilon \vdash !a_2 \ldots !\varepsilon \vdash_A !a_n}{\Gamma \vdash_A \varphi}\,(cut^*)$$

◀

For individual deductions $\Gamma \vdash_A \varphi$ an infinite set of non-logical axioms may be reduced to a finite one:

301 Proposition COMPACTNESS FOR NON-LOGICAL AXIOMS
Let $A \subseteq \mathcal{L}[C]$ be a set of formulae, let $\Gamma \in \mathcal{L}[C]^*$ be a sequence of formulae and $\varphi \in \mathcal{L}[C]$ a single formula. Then the following holds:

Whenever $\Gamma \vdash_A \varphi$ holds, then there exists a finite subset $F \subseteq A$ of A, so that $\Gamma \vdash_F \varphi$.

PROOF:
This follows directly from the finiteness of every proof tree. ◀

For $k \in \mathbb{N}_0$ and $\psi \in \mathcal{L}[C]$ let ψ^k denote the k-fold product[2] $\psi \star \psi \star \ldots \star \psi$. Furthermore, let ψ^0 be equal to $\mathbf{1}$ by convention. Then the following may be shown by a detailed analysis of the deduction rules, compare [Sch91]:

302 Proposition REPLACEMENT OF EXPONENTIAL
If for two formulae φ and ψ from the positive fragment the deduction $!\psi \vdash \varphi$ holds, then there exists $n \in \mathbb{N}_0$ so that $\psi^n \vdash \varphi$ holds. ◀

B.7 Interpretation of Linear Formulae

In the second section of this chapter a motivation for *multiplicative* linear logic as a formalism for *accumulating and transforming* resources was given. In the following, a more formal

[2]Due to the associativity of $\star$, the position of the brackets is irrelevant.

approach to this concept is presented.

303 Definition NUMBER OF OCCURRENCES
Let $c \in C$ be a constant. The *number of occurrences of* c is the function $\#_c : \mathcal{L}_m[C] \to \mathbb{Z}$, defined uniquely by the following equations over the constructor terms of multiplicative linear logic $\mathcal{L}_m[C]$:

$$\#_c(\mathbf{0}) = 0$$

$$\#_c(\mathbf{1}) = 0$$

$$\#_c(d) = \begin{cases} 1 & \Leftrightarrow \quad d \in C, d = c \\ 0 & \Leftrightarrow \quad d \in C, d \neq c \end{cases}$$

$$\#_c(\varphi \star \psi) = \#_c(\varphi) + \#_c(\psi)$$

$$\#_c(\varphi + \psi) = \#_c(\varphi) + \#_c(\psi)$$

$$\#_c(\sim \varphi) = -\#_c(\varphi)$$

$$\#_c(\varphi \multimap \psi) = \#_c(\psi) - \#_c(\varphi)$$

The number of occurrences may also be extended to sequences of formulae by

$$\#_c(\varepsilon) = 0$$

$$\#_c(\varphi_1, \varphi_2, \ldots, \varphi_k) = \#_c(\varphi_1) + \#_c(\varphi_1) + \ldots + \#_c(\varphi_k)$$

◀

The number of occurrences is uniquely defined by the above formulae, since the grammar defining $\mathcal{L}_m[C]$ is unique and for each constructor term a suitable equation is given.

The following theorem states that the multiplicative fragment of linear logic is resource-conscious:

304 Theorem INVARIANCE THEOREM
Let all formulae, sequences, and deductions be in the multiplicative fragment of linear logic or in a smaller fragment thereof.

If $\Gamma \vdash \varphi$ holds in the natural deduction calculus, then for every constant $c \in C$, the number of occurrences of c in Γ and in φ is equal. If $\Gamma \vdash \Delta$ holds in the sequent calculus, then for every constant $c \in C$, the number of occurrences of c in Γ and in Δ is equal.

If for every constant $c \in C$ the number of occurrences on the left and on the right side of a deduction $\vdash$ is equal, the deduction is called *balanced*. Hence, another way of expressing this theorem is:

Natural and sequent calculus deductions in the multiplicative fragment are balanced.

PROOF:
The proof is easy and applies structural induction on the rules. In the following, we give a sketch for the natural deduction system.

For the basis of the induction we show that the deductions $\varepsilon \vdash \mathbf{1}$ and $\varphi \vdash \varphi$ are balanced.

For the induction step we assume that the assumptions of a rule are balanced and we deduce that then also its conclusion is balanced. For the rule $(\multimap E)$ given by

$$\frac{\Gamma \vdash \varphi \multimap \psi \qquad \Delta \vdash \varphi}{\Gamma, \Delta \vdash \psi} (\multimap E)$$

this is done as follows: If $\#_c(\Gamma) = \#_c(\psi) - \#_c(\varphi)$ and $\#_c(\Delta = \#_c(\varphi)$, then $\#_c(\Gamma, \Delta) = \#_c(\Gamma) + \#_c(\Delta) = \#_c(\psi) - \#_c(\varphi) + \#_c(\varphi) = \#_c(\varphi)$. For the other rules, this is similar. ◀

The formal properties outlined above suggest the following interpretation of the various fragments of linear logic:

305 Remark INTERPRETATION OF EXPONENTIAL FRAGMENT
The elements of C are resources or "information tokens" which may be combined by operators into more complex tokens.

If φ and ψ are tokens, then $\varphi \star \psi$ is a resource containing one copy of token φ and one copy of token ψ. Associativity, commutativity and neutrality of $\star$ suggest that the *multiplicative conjunction* $\star$ serves as constructor for bags and provides for the accumulation of resources.

$\mathbf{1}$ is the *unit for resource accumulation*. It stands for "no resource" or "empty bag".

The *deduction* operator may be interpreted as indicating a development of formulae in time. For example, a formula $\mathbf{1} \star \varphi$ may develop into formula φ. This is denoted by $\mathbf{1} \star \varphi \vdash \varphi$. However, also $\varphi \vdash \mathbf{1} \star \varphi$ holds. This shows that within a deduction a unit element may arbitrarily be generated or removed via the multiplicative conjunction. We used to denote this as $\mathbf{1} \star \varphi \simeq \varphi$.

$\varphi \multimap$ *psi* is a token which may interact with other tokens in a bag: If a bag contains token φ and token $\varphi \multimap$ *psi*, then we may remove one copy of each and add one copy of ψ, instead. This is denoted by $\varphi \star (\varphi \multimap \psi) \vdash \psi$. *Linear implication* $\multimap$ may be interpreted as resource or token transformation in $\star$ combined collections of tokens.

The storage token $!\varphi$ is a supply of either *no, or one or an arbitrary finite number* of tokens. It is possible that token $!\varphi$ just vanishes from a bag, without leaving any trace: $!\varphi \vdash \mathbf{1}$. We may replace $!\varphi$ by a certain number of tokens φ: $!\varphi \vdash \varphi \star \varphi \star \ldots \star \varphi$. We may also use $!\varphi$ to generate a certain number of tokens φ and still keep $!\varphi$ as a supply for φ for some later time: $\varphi \vdash \varphi \star \varphi \star \ldots \star \varphi \star !\varphi$. ◀

306 Remark INTERPRETATION OF MULTIPLICATIVE-EXPONENTIAL FRAG.
This fragment has *two operators for accumulating* resources, $\star$ and $+$. $+$ is an accumulation operator in the sense that it keeps track of the number of resources:

$$\#_c(\varphi + \psi) = \#_c(\varphi) + \#_c(\psi)$$

However, $\star$ and $+$ are not equal, and we are not able to prove $\varphi \star \psi \simeq \varphi + \psi$.

The *multiplicative conjunction* $\star$ is an accumulation operator for tokens which will be used in transformations. Tokens combined by $\star$ may be regarded as a chemical solution. The tokens φ are swimming around until they find a matching $\varphi \multimap \psi$ token. Then a chemical reaction takes place and both tokens interact. As a result, they both disappear and instead a new token ψ is generated in the solution.

With this intuition in mind, we will now try to understand the chemistry of a token $\gamma + \vartheta$ composed by ***multiplicative disjunction***. Our chemical point of view now becomes asymmetric, because it treats $\star$ as primary and $+$ as secondary operator: If we discuss below the chemistry of a token $\gamma + \vartheta$ in a solution, this means that $\gamma + \vartheta$ is subformula of an expression whose top-level operator is $\star$. If the formula is $\gamma + \vartheta$, then we will regard it as single token chemical solution, or as $\mathbf{1} \star (\gamma + \vartheta)$. Since $\star$ and $+$ are dual to each other, there is no logical reason to give one of the operators more importance than the other. However, since the positive fragment will be the basis for our process logic, we will interprete all remaining operators of linear logic in terms of the positive fragment.

$\varphi \star \psi$ is a supply of one copy of φ and one copy of ψ for use in resource transformations. It is the main constructor for composing our chemical solution of tokens.

$\sim \varphi$ is a *debt* of one copy of token φ.

$\mathbf{1}$ is, as above, the neutral element for $\star$ and $\mathbf{0}$ is the unit for $+$.

$\varphi + \psi$ is a supply of one copy of φ and one copy of ψ *to be used for cancelling debts*. If in our chemical solution we find a token $\varphi + \psi$, then this token may interact with a debt token $\sim \varphi$ to produce a token ψ as result of the following reaction:

$$(\varphi + \psi) \star \sim \varphi \vdash \psi \qquad\qquad (\varphi + \psi) \star \sim \psi \vdash \varphi$$

Obviously, token $\varphi + \psi$ is capable of reacting with other tokens in our chemical solution. We therefore expect that $\varphi + \psi$ may be expressed by $\multimap$. Indeed, in Proposition 272 we deduced:

$$\varphi + \psi \simeq \sim \varphi \multimap \psi \simeq \sim \psi \multimap \varphi$$

Cancelling of debts inside of $+$ formulae is not possible. We cannot show $\alpha + \sim \alpha = \mathbf{0}$ or $\alpha + \sim \alpha = \mathbf{1}$. However, we may show, as one might expect, the following:

$$\alpha = \sim \alpha \simeq \alpha \multimap \alpha = \sim \alpha \multimap \sim \alpha$$

Cancelling of debts inside the chemical solution is only possible, if a token meets a suitable $+$ token. How can we understand the deduction $\varphi \star \sim \varphi \vdash \mathbf{0}$ in the light of this statement? Since $\mathbf{0}$ is the unit element for $+$, the second token $\sim \varphi$ in fact is a debt cancellation token for φ, since it may be written as $\sim \varphi \simeq \sim \varphi + \mathbf{0}$. Consequently, we have $\varphi \star (\sim \varphi + \mathbf{0}) \vdash \varphi$. Due to this deduction we expect $\sim \varphi \simeq (\varphi \multimap \mathbf{0})$ which was shown on a rigorous basis in Proposition 270.

Token transformation with the help of $\multimap$ can also be understood as cancelling debts: $\varphi \multimap \psi$ is a supply of one copy of token ψ and a debt of one copy of φ. However, we only can access these tokens by the mechanism of cancelled debts. Therefore we may cancel the debt in φ to obtain ψ, as described by

$$\varphi \star (\varphi \multimap \psi) \vdash \psi$$

but we also may supply a debt of ψ and have it cancelled by the ψ in $\sim \varphi + \psi$. As result we obtain $\sim \varphi$:

$$\sim \psi \star (\varphi \multimap \psi) \vdash \sim \varphi$$

Deduction is not always symmetric as it is in the case of $\varphi \simeq \varphi \star \mathbf{1}$. The unit $\mathbf{1}$ may be produced and absorbed by the solution at free will, but this is not possible with arbitrary tokens. We may observe the reaction $\alpha \star (\alpha \multimap \beta) \vdash \beta$ but there is no reaction $\beta \vdash \alpha \star (\alpha \multimap \beta)$.

Note that the generation of the unit $\mathbf{1}$ may lead to more complicated reactions: $\mathbf{1} \vdash \alpha \multimap \alpha$. This gives $\alpha \vdash \alpha \star \mathbf{1} \vdash \alpha \star (\alpha \multimap \alpha) \vdash \alpha$.

? is the dual of !. By $\mathbf{0} \sqcup \varphi \sqcup (?\varphi + ?\varphi) \vdash ?\varphi$ it may be interpreted as a wrapper, containing $\mathbf{0}$ or φ or $(?\varphi + ?\varphi)$. This will become more clear in the following paragraph, but still, an intuitive interpretation of ? is rather difficult to give. ◀

307 Remark INTERPRETATION OF THE ADDITIVES

$\varphi \sqcap \psi$ is a token which can be ***used*** as a token φ or as a token ψ, the choice is at the discretion of the user. However, token $\varphi \sqcap \psi$ allows only one choice and supplies only one copy of the chosen token. Token $\varphi \sqcap \psi$ may be ***produced*** from a token γ, if there is the possibility to produce token φ from token γ and the possibility to produce token ψ from token γ. This interpretation is derived from the following properties of the deduction relation:

$$\varphi \sqcap \psi \vdash \varphi \qquad \varphi \sqcap \psi \vdash \psi \qquad \frac{\gamma \vdash \varphi \quad \gamma \vdash \psi}{\gamma \vdash \varphi \sqcap \psi}$$

Token $\varphi \sqcup \psi$ is a token which was ***produced*** from token φ or from token ψ, by placing this token in a wrapper which no longer allows an observer to recognize whether inside of the wrapper he will be able to find token φ or token ψ. Token $\varphi \sqcup \psi$ may be ***used*** only to produce such tokens γ which may be produced from φ alone as well as from ψ alone: During the production process the wrapper will be opened, and the wrapped token, either φ or ψ, will be found inside. Which one this will be cannot be predicted from the outside. This interpretation is derived from the following properties of the deduction relation:

$$\varphi \vdash \varphi \sqcup \psi \qquad \psi \vdash \varphi \sqcup \psi \qquad \frac{\varphi \vdash \gamma \quad \psi \vdash \gamma}{\varphi \sqcup \psi \vdash \gamma}$$

$\top$ is the unit of $\sqcap$ and $\bot$ the unit of $\sqcup$. ◀

In the following remark we will analyze how far we may extend the speculative interpretation of linear logic by chemical and physical phenomena.

308 Remark SPECULATIVE PHYSICAL AND CHEMICAL INTERPRETATION

The ***positive fragment*** allows a chemical interpretation: Deductions are balanced with respect to the number of occurrences of constants on the left and the right side of $\vdash$. Similarly, chemical reactions leave the number of participating atoms constant. There are certain chemical reactions which may take place in both directions. For example, a HCl molecule in a solution may dissociate into H^+ and Cl^- and recombine again. Other reactions have a preferred direction. Similarly, there are symmetric deductions ($\simeq$) and asymmetric ones ($\vdash$). This suggests some kind of "chemical potential" which is associated with the formulae of a deduction and which is responsible for its asymmetry.

The additives and the exponentials allow the generation and deletion of tokens, so they no longer correspond to chemical reactions but to phenomena of elementary particle physics

[Ste93]: The token $\alpha \sqcap \beta$ represents a particle which ***non-deterministically decays*** into either particle α or particle β. The generation and consumption of "meaningless" symbols like **1** may be compared with the generation and consumption of virtual particles. The token $\alpha \sqcup \beta$ may be used to model the ***uncertainty in the measurement process***. Before the measurement the token is $\alpha \sqcup \beta$. This tells us that the measurement process will reveal either property α or property β in the moment the measurement is actually carried out. However, neither the existance of the property nor the outcome of the measurement is determined in state $\alpha \sqcup \beta$. Only after the measurement we know its outcome, α, or β. The measurement process itself may be written as $(\alpha \sqcup \beta) \multimap (\alpha \sqcap \beta)$. ◀

All claims that certain deductions do not exist, require a lengthy decision procedure, searching over all subformulae of the deduction. This proof search was carried out with the help of the MALL prover implemented in PROLOG by SHANKAR [LS94].

Appendix C

Category Theory

Category theory has been called "abstract nonsense" by both its friends and its detractors. Perhaps what this phrase suggests to both is that category theory has relatively more form than content, compared to other areas of mathematics. Its friends claim this as a virtue, in contrast to the excessive concreteness and representation dependence of set theoretic foundations, and the relatively poor guidance for discovering elegant and coherent theories that they provide.

J. A. GOGUEN, *A Categorical Manifesto*
Mathematical Structures in Computer Science **1**(1991), pp. 49–67

This appendix presents a brief review of those concepts which are required for a categorical interpretation of process theory. It is not intended as an introduction to category theory and therefore neither motivations nor examples are provided. Its purpose, however, is to give a review of the employed concepts and notations and to provide a reference for those tedious technical details which often are left to the reader in the categorical literature. For motivations, examples, proofs, and other material on category theory the reader is referred to [BW90b], [Mac88], [Poi92], [RB88], [Pre75], [Par69].

The connections between category theory and process theory are outlined in Chapter 7.

In Section 1, a small *intuitive review* of category theory is provided for the non-mathematically inclined reader.

Section 2 provides some *basic definitions*.

In Section 3, the constructions of a *product* and a *coproduct* are presented and connected via the notion of *duality*.

In Section 4, *functors* and in Section 5, *natural transformations* are introduced. Section 6 provides a detailed analysis of the properties of the *product functor* using the language of functors and natural transformations.

Section 7 discusses the representation of natural transformations and presents the YONEDA *lemma*.

Section 8 introduces the concept of *cartesian closed categories*.

Finally, Section 9 discusses *symmetric monoidal closed categories*.

C.1 Category Theory for the Perplexed

Category theory is a discipline of pure mathematics which may serve as a framework for an *understanding of mathematical structure*. Similar to elementary set theory, it provides a *variety of constructions* which may be applied to many different objects. Therefore, category theory also provides a *uniform understanding* of many mathematical concepts which seem unrelated on a concrete level of understanding but share a lot of abstract properties.

Initially being an abstract and non-computational field of mathematics in the tradition of BOURBAKI, category theory lately has become an *important tool for theoretical and applied computer science* which is taught also within the framework of functional programming, compare [RB88]. There even is a *machine concept*, the so-called *categorical abstract machine*, which has been directly derived from category theory and serves as an important implementation technique for functional programming languages [Cou90b].

A category consists of *objects* which can be represented as vertices of a graph, and of directed *arrows* between these vertices. Whenever there exists an arrow from an object A to an object B, this indicates a structural relationship between these objects. Two objects may show no structural relationship in which case there is no arrow between them. On the other hand, two objects may be structurally related in many different ways which is represented by one arrow for every kind of relationship. The arrows often are special kinds of functions and are also called *morphisms*. This motivates a notation well known from functions: If f is a morphism from A to B, this can be written as $f : A \to B$. In this case, A is called the *domain* and B the *codomain* of f.

Similar to functions, morphisms may be composed with other morphisms: For example, if a morphism $f : A \to B$ describes a structural relationship of A to B, and if $g : B \to C$ describes one of B to C, then the *composition* of f and g can be formed: This is a morphism denoted by $g \circ f : A \to C$ which describes a structural relationship of A to C. Not all morphisms may be composed: In the composition $g \circ f$, the codomain of f must be equal to the domain of g. For this reason, the composition of morphisms cannot be defined as a binary operation on the set of all morphisms but it exhibits the peculiar structure of a family of composition operators: For every triple (A, B, C) of objects, there is a composition operator which composes a morphism $f : A \to B$ with a morphism $g : B \to C$ to a morphism $g \circ f : A \to C$.

Every object X is structurally related to itself by a special identity morphism id_X. Composing a morphism with such an identity morphism does not change the morphism.

Category theory is a technique of *describing structural relationships between objects by reasoning about morphisms*. Whenever *structural properties* are to be analyzed and implementational details are to be neglected, category theory is an appropriate language.

C.2 Basic Definitions

This section contains some basic definitions on category theory.

309 Definition CATEGORY
A *category* $\mathcal{C} = (\mathfrak{O}, \mathfrak{M}, \circ)$ consists of

(1) a class $\mathfrak{O}$ whose elements are called *objects*,

(2) for every pair X, Y of objects, a class $\mathfrak{M}(X, Y)$ whose elements are called *morphisms*, and

(3) for every morphism $f \in \mathfrak{M}(X, Y)$ and $g \in \mathfrak{M}(Y, Z)$ a *composed morphism* $g \circ f \in \mathfrak{M}(X, Z)$.

A category must satisfy the following axioms:

(tc) *Typing condition*:
The classes $\mathfrak{M}(X, Y)$ of morphisms are *pairwise disjoint*.

(id) *Existence of identity transitions*:
For every object X there exists an *identity morphism* $i \in \mathfrak{M}(X, X)$ which satisfies the following two properties: For arbitrary objects Y and morphisms $f \in \mathfrak{M}(Y, X)$, the equation $i \circ f = f$ holds and for arbitrary objects Y and morphisms $g \in \mathfrak{M}(X, Y)$, the equation $g \circ i = g$ holds.

(ass) *Associativity of composition*:
For $A, B, C, D \in \mathfrak{O}$, $f \in \mathfrak{M}(A, B)$, $g \in \mathfrak{M}(B, C)$, and $h \in \mathfrak{M}(C, D)$, the equation $(h \circ g) \circ f = h \circ (g \circ f)$ holds.

◀

Technically, $\mathfrak{M} = (\mathfrak{M}(X, Y))_{X,Y \in \mathfrak{O}}$ is a pairwise disjoint family of classes, indexed by pairs of states. $\circ$ is a family of composition operators, with the individual operator acting in the following way: $\circ_{(X,Y,Z)} : \mathfrak{M}(Y, Z) \times \mathfrak{M}(X, Y) \to \mathfrak{M}(X, Z)$.

310 Definition DOMAIN, CODOMAIN, AND TYPE
If $f \in \mathfrak{M}(X, Y)$ is a morphism, then the object X is called the *domain* $\text{dom}(f) := X$ and object Y is called the *codomain* $\text{codom}(f) := Y$ of f. f is called a morphism *from* the object X *to* the object Y which is denoted by $f : X \to Y$. The pair (X, Y) consisting of the domain and the codomain of a morphism is also called its *type*. ◀

For every morphism $f \in \bigcup_{X,Y \in \mathfrak{O}} \mathfrak{M}(X, Y)$, the domain X and the codomain Y are well-defined since the morphism classes $\mathfrak{M}(X, Y)$ are pairwise disjoint.

In the following, capital latin letters X, Y are used for objects, small latin letters f, g for morphisms and calligraphic letters $\mathcal{C}, \mathcal{D}$ for categories.

311 Proposition UNIQUENESS OF IDENTITY
In a category $\mathcal{C}$, for every object X there exists exactly one morphism i, which has the properties of the identity morphism, as stated in Definition 309. ◀

Due to Proposition 311, we may talk of *the* identity morphism for an object X and the notation id_X may be adopted for this morphism.

The category of sets provides a simple example for categories:

312 Example CATEGORY OF SETS
The category *Set* $= (\mathfrak{O}, \mathfrak{M}, \circ)$ of all sets contains as *objects* the class $\mathfrak{O}$ of all sets. For

two sets X and Y a *morphism* from X to Y is a function $f : X \to Y$ from X to Y. The class $\mathfrak{M}(X,Y)$ is the set of all functions $f : X \to Y$. The *composition* of two morphisms $f \in \mathfrak{M}(X,Y)$ and $g \in \mathfrak{M}(Y,Z)$ is the composition $g \circ f$ of the functions $f : X \to Y$ and $g : Y \to Z$; it is a function of the type $g \circ f : X \to Z$. ◀

313 Proposition CHARACTERIZATION OF SPECIAL FUNCTIONS
For a function $f : X \to Y$, the following conditions are equivalent:

(1) f is *injective.*

(2) f has the *left cancellation property*:
For every two functions $g, h : Z \to X$, if $f \circ g = f \circ h$ then $g = h$.

(3) f has a *left inverse*:
There is a function $g : Y \to X$ with $g \circ f = id_X$.

Furthermore, the following conditions are equivalent:

(1) f is *surjective.*

(2) f has the *right cancellation property*:
For every two functions $g, h : Y \to Z$, if $g \circ f = h \circ f$ then $g = h$.

(3) f has a *right inverse*:
There is a function $g : Y \to X$ with $f \circ g = id_Y$.

A function f is bijective iff it has a left and a right inverse. Then, the left and right inverse functions are equal and this function is called the *inverse* of f. ◀

Above, the conditions (1) are formulated in the language of set theory. All the other properties can be formulated in arbitrary categories and thus allow a generalization of the notions of specifal functions:

314 Definition SPECIAL MORPHISMS
Let $C = (\mathfrak{O}, \mathfrak{M}, \circ)$ be a category. A morphism $f : X \to Y$ is called a

monomorphism, iff it has the *left cancellation property*: For every two morphisms $g, h : Z \to X$, if $f \circ g = f \circ h$ then $g = h$.

coretraction, iff it has a *left inverse*: There is a morphism $g : Y \to X$ with $g \circ f = id_X$.

epimorphism, iff it has the *right cancellation property*: For every two morphisms $g, h : Y \to Z$, if $g \circ f = h \circ f$ then $g = h$.

retraction, iff it has a *right inverse*: There is a morphism $g : Y \to X$ with $f \circ g = id_Y$.

bimorphism, iff f is an *epimorphism* and a *monomorphism*.

isomorphism, iff it has a *left* and a *right inverse* and these are *equal*.

◀

In the light of this definition, Proposition 313 may be reformulated as follows:

315 Proposition CHARACTERIZATION OF SPECIAL FUNCTIONS
For a morphism $f : X \to Y$ in the category of sets, the following conditions are equivalent:

(1) f is *injective.*

(2) f is a *monomorphism.*

(3) f is a *coretraction.*

Furthermore, the following conditions are equivalent:

(1) f is *surjective.*

(2) f is an *epimorphism.*

(3) f is a *retraction*

And therefore:

(1) f is *bijective.*

(2) f is a *bimorphism.*

(3) f is an *isomorphism.*

◀

In an arbitrary category several notions for the important case of an isomorphism are equivalent:

316 Proposition CHARACTERIZATION OF ISOMORPHISMS
In a category C for a morphism $f : X \to Y$ the following conditions are equivalent:

(1) f is an *isomorphism.*

(2) f is a *retraction* and a *coretraction.*

(3) f is an *epimorphism* and a *coretraction.*

(4) f is a *monomorphism* and a *retraction.*

◀

317 Proposition ON SPECIAL MORPHISMS
In a category C for a morphism $f : X \to Y$ the following conditions hold:

(1) If f is an *isomorphism* then it is a *bimorphism.*

(2) If f is a *retraction* then it is an *epimorphism.*

(3) If f is a *coretraction* then it is a *monomorphism.*

◀

In many categories a morphism is an isomorphism iff it is a bimorphism. This is not true in an arbitrary category:

318 Definition BALANCED CATEGORIES
A category C is called *balanced,* iff every bimorphism[1] is an isomorphism[2]. ◀

[1] i.e. monomorphism and epimorphism.
[2] i.e. retraction and coretraction.

In the definition of a category the objects and morphisms are *classes* and not sets. The reason is that there are interesting constructions for which a definition using sets would lead to contradictions like those known from the "set of all sets". Similar problems arise in the representation theory of categories and with the "category of all categories". Special notions of categories have been defined in order to avoid these difficulties. The described problems are of a set theoretical nature and play an important role for applications of the YONEDA Lemma 363. The reader is referred to [Pre75] and [LS86] for the details.

319 Definition SMALL, LOCALLY SMALL, AND CONCRETE CATEGORIES
A category $\mathcal{C} = (\mathfrak{O}, \mathfrak{M}, \circ)$ is called

small, iff the class $\mathfrak{O}$ of objects and the classes $\mathfrak{M}(X, Y)$ of morphisms are sets.

locally small, iff the classes $\mathfrak{M}(X, Y)$ of morphisms are sets.

concrete, iff its objects are sets, its morphisms are functions between those sets and its compositions are compositions of functions.

◀

In the following proposition the role of injective, surjective, and bijective functions in concrete categories is further analyzed:

320 Proposition ON INJECTIVE, SURJECTIVE, AND BIJECTIVE MORPHISMS
Let f be a morphism in a concrete category. Then

f coretraction $\Rightarrow$ f injective

f injective $\Leftrightarrow$ f monomorphism

f retraction $\Rightarrow$ f surjective

f surjective $\Leftrightarrow$ f epimorphism

f isomorphism $\Rightarrow$ f bijective

f bijective $\Leftrightarrow$ f bimorphism

◀

321 Definition ISOMORPHIC OBJECTS
Two objects X and Y in a category $\mathcal{C}$ are called *isomorphic*, iff there exists an isomorphism from X to Y. ◀

322 Proposition ISOMORPHY IS EQUIVALENCE RELATION
The relation of *isomorphy* of two objects is an *equivalence* relation. ◀

As usual, the equivalence classes of a category are also called *isomorphism classes* of the category. The *quotient* of a category is the class of all its equivalence classes.

Intuitively, isomorphic objects are identical with respect to the contemplated structure. In classical algebraic structure theory, various special kinds of morphisms are studied. If $f : X \to Y$ is a morphism from object X to object Y, the following four situations commonly occur.

(1) X and Y have identical structure and differ only in unimportant details. In this case X and Y are called *isomorphic* and f is called an *isomorphism*.

(2) X is part of the structure of Y, but Y has some additional structure, which is not present in X. In this case f is called a *monomorphism*.

(3) By collapsing the more complicated structure of X one finally obtains the entire structure of Y. In this case f is called an *epimorphism*.

(4) By collapsing the structure of X one finally obtains a part of the structure of Y. This is the general situation, when we have an arbitrary morphism. It is also called a *homomorphism*, as a synonym.

323 Definition INITIAL AND TERMINAL OBJECTS
An object $X \in \mathfrak{O}$ in a category $(\mathfrak{O}, \mathfrak{M}, \circ)$ is called

an *initial object*, iff from the initial object there is exactly one arrow to every object of the category.

a *terminal object*, iff from every object of the category there is exactly one arrow to the terminal object.

◀

Initial and terminal objects are unique up to isomorphism:

324 Proposition UNIQUENESS OF INITIAL AND TERMINAL OBJECTS
If X and Y are two initial (terminal) objects of a category, then they are isomorphic. ◀

Initial and terminal objects are in a unique structural relation with all other objects of the category. An initial object can be "found as part of" every object, in this sense it is the simplest object of the category. A terminal object is "the smallest object" to which every other object can be mapped. All, even the most complex objects of the category, can be "collapsed" to the terminal object.

C.3 Products, Coproducts, and Duality

Throughout this section, let $\mathcal{C}$ be be a category and $\mathcal{X} = (X_i)_{i \in I}$ a family of objects of $\mathcal{C}$.

325 Definition PRODUCT
A *product* $(P, (\pi_i)_{i \in I})$ *of* $\mathcal{X}$ consists of an object P and a family $(\pi_i)_{i \in I}$ of morphisms $\pi_i \in \mathfrak{M}(P, X_i)$ with the following property: For every object Q and family $(q_i)_{i \in I}$ of morphisms $q_i \in \mathfrak{M}(Q, X_i)$ there exists a unique morphism $f \in \mathfrak{M}(Q, P)$ such that $\forall i \in I: \ \pi_i \circ f = q_i$.

This unique morphism f is denoted by $< q_i >_{i \in I}$, or by $< q_1, q_2, \ldots, q_n >$ if the set I is finite. ◀

This condition can be expressed by the commuting diagram[3] in Figure C.1.

A category is said to *contain products*, iff for every family $\mathcal{X}$ of objects there exists a product of $\mathcal{X}$.

[3]Morphisms whose unique existence is postulated in a definition will be represented by dotted arrows.

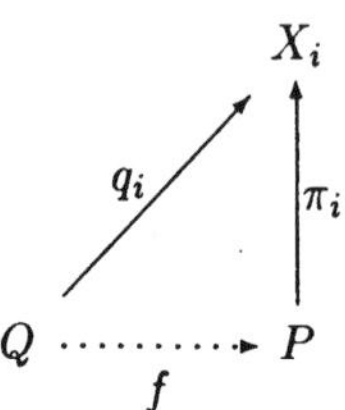

Figure C.1: Definition of the Product.

A category is said to *contain finite products*, iff for every family $\mathcal{X}$ of objects with finite index set I there exists a product of $\mathcal{X}$.

The following proposition states that products are unique up to isomorphism:

326 Proposition UNIQUENESS OF PRODUCTS
If $(P,(p_i)_{i\in I})$, $p_i : P \to X_i$ and $(Q,(q_i)_{i\in I})$, $q_i : Q \to X_i$ are two products of a family $\mathcal{X} = (X_i)_{i\in I}$, then there exists a unique isomorphism $f : P \to Q$ such that $\forall i \in I : \; p_i = q_i \circ f$. ◀

This uniqueness justifies the use of a notation $(\otimes_{i\in I} X_i, (\pi_i)_{i\in I})$ for *the* product. Finite products will be written as $X_1 \otimes X_2 \otimes \ldots \otimes X_n$.

327 Proposition BINARY, FINITE, AND EMPTY PRODUCTS
If a category contains a product of the family with *empty index set*, then the product object is a *terminal object* of the category.

If a category contains a *terminal object* and *all binary products*[4], then it contains *all finite* products. ◀

Every finite product can be defined by binary products or by the empty product. A product of four objects A, B, C and D, for example, can be defined by $A\otimes B\otimes C\otimes D = A\otimes(B\otimes(C\otimes D))$. In Proposition 357 it will be demonstrated that the product is associative, commutative, and contains the terminal object as neutral element. Hence, the order in which a finite product of objects is formed is irrelevant.

For the case of *binary products* the commuting diagram from Definition 325 is presented in Figure C.2.

328 Proposition PRODUCT IMPOSES STRUCTURE ON MORPHISM CLASSES
Let $(P,(\pi_i)_{i\in I})$ be a product of $\mathcal{X}$. Then for every object A there is a bijection between the morphism sets $\mathfrak{M}(A,P)$ and $\times_{i\in I}\mathfrak{M}(A,X_i)$. ◀

For finite products this means that there is a bijection between the morphism sets $\mathfrak{M}(A, X_1 \otimes \ldots \otimes X_n)$ and $\mathfrak{M}(A,X_1) \times \ldots \times \mathfrak{M}(A,X_n)$. This bijection is given by $f \mapsto (\pi_i \circ f)_{i\in I}$ or $f \mapsto (\pi_1 \circ f, \pi_2 \circ f, \ldots, \pi_n \circ f)$.

329 Definition COPRODUCT
A *coproduct* $(C,(\kappa_i)_{i\in I})$ of $\mathcal{X}$ consists of an object C and a family $(\kappa_i)_{i\in I}$ of morphisms $\kappa_i \in \mathfrak{M}(X_i, C)$ with the following property: For every object Q and family $(q_i)_{i\in I}$ of morphisms $q_i : \mathfrak{M}(X_i,Q)$ there exists a unique morphism $f \in \mathfrak{M}(C,Q)$ such that $\forall i \in I : \; f \circ \kappa_i = q_i$.

[4] i. e. all products of families with an index set of two elements.

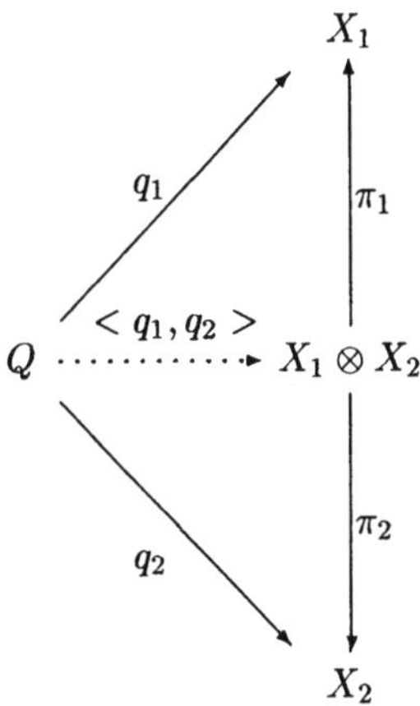

Figure C.2: Commuting Diagram for the Binary Product.

This unique morphism f is denoted by $[q_i]_{i \in I}$, or by $[q_1, q_2, \ldots, q_n]$ if the set I is finite. ◀
This condition can be expressed by the commuting diagram in Figure C.3.

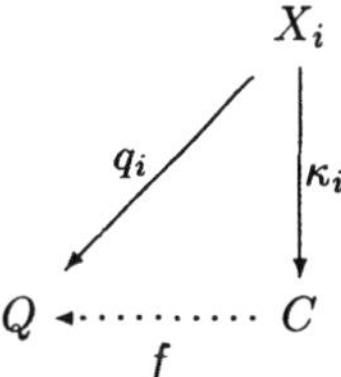

Figure C.3: Definition of the Coproduct.

A category is said to *contain coproducts*, iff for every family $\mathcal{X}$ of objects there exists a coproduct of $\mathcal{X}$.

A category is said to *contain finite coproducts*, iff for every family $\mathcal{X}$ of objects with finite index set I there exists a coproduct of $\mathcal{X}$.

The following proposition states that coproducts are unique up to isomorphism:

330 Proposition UNIQUENESS OF COPRODUCTS
If $(P, (p_i)_{i \in I})$, $p_i : X_i \to P$ and $(Q, (q_i)_{i \in I})$, $q_i : X_i \to Q$ are two coproducts of a family $\mathcal{X} = (X_i)_{i \in I}$, then there exists a unique isomorphism $f : Q \to P$ such that $\forall i \in I : p_i = f \circ q_i$.
◀

This uniqueness justifies the use of a notation $(\oplus_{i \in I} X_i, (\kappa_i)_{i \in I})$ for *the* coproduct. Finite coproducts will be written as $X_1 \oplus X_2 \oplus \ldots \oplus X_n$.

331 Proposition BINARY, FINITE, AND EMPTY COPRODUCTS

If a category contains a coproduct of the family with *empty index set*, then the coproduct object is an *initial object* of the category.

If a category contains an *initial object* and all *binary coproducts*[5], then it contains all finite coproducts. ◀

Every finite coproduct can be defined by binary coproducts or by the empty coproduct. A coproduct of four objects A, B, C and D, for example, can be defined by $A \oplus B \oplus C \oplus D = A \oplus (B \oplus (C \oplus D))$. The coproduct is associative, commutative, and has the initial object as neutral element. The order in which a finite coproduct of objects is formed is irrelevant.

For the case of binary coproducts the commuting diagram from Definition 329 is presented in Figure C.4.

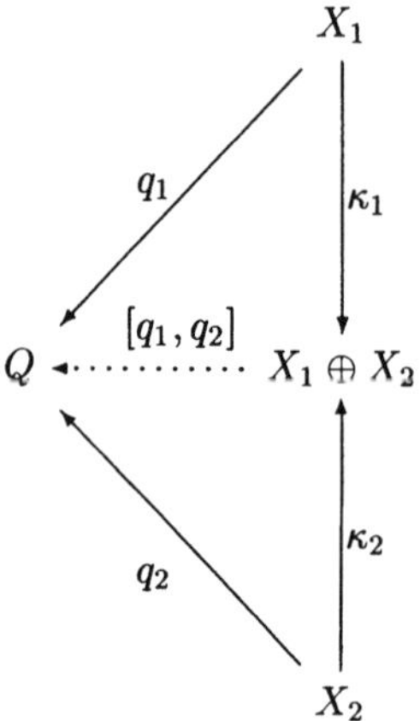

Figure C.4: Commuting Diagram for the Binary Coproduct.

The *dual category* $\mathcal{C}^*$ of the category $\mathcal{C}$ is the category $\mathcal{C}$ with "the directions of all arrows reversed":

332 Definition DUAL CATEGORY
The *dual category* $\mathcal{C}^*$ of a category $\mathcal{C}$ is the category $\mathcal{C}^* := (\mathfrak{O}_{\mathcal{C}^*}, \mathfrak{M}_{\mathcal{C}^*}, \circ_{\mathcal{C}^*})$ with

$$\mathfrak{O}_{\mathcal{C}^*} := \mathfrak{O}_{\mathcal{C}}$$

$$\mathfrak{M}_{\mathcal{C}^*}(A,B) := \mathfrak{M}_{\mathcal{C}}(B,A)$$

$$\circ_{\mathcal{C}^*} : \mathfrak{M}_{\mathcal{C}^*}(B,C) \times \mathfrak{M}_{\mathcal{C}^*}(A,B) \to \mathfrak{M}_{\mathcal{C}^*}(A,C)$$

◀

For $f \in \mathfrak{M}_{\mathcal{C}^*}(A,B)$ and $g \in \mathfrak{M}_{\mathcal{C}^*}(B,C)$ one obtains $f \in \mathfrak{M}_{\mathcal{C}}(B,A)$ and $g \in \mathfrak{M}_{\mathcal{C}}$ and hence $g \circ_{\mathcal{C}^*} f := f \circ_{\mathcal{C}} g$.

The concepts of initial and terminal objects are dual to each other:

[5] i. e. all coproducts of families with an index set of two elements.

333 Proposition DUALITY OF INITIAL AND TERMINAL OBJECTS
If an object is an initial object in a category, then it is a terminal object in the dual category. Similarly, if an object is a terminal object in a category, then it is an initial object in the dual category. ◀

The concepts of products and coproducts are dual to each other:

334 Proposition DUALITY OF PRODUCTS AND COPRODUCTS
If $(P, (\pi_i)_{i \in I})$ is a product of a family $\mathcal{X}$ in a category, then it is a coproduct of this family in the dual category. Similarly, coproducts in a category correspond to products in the dual category. ◀

This duality is reflected by the similarity of the above paragraphs on products and coproducts.

C.4 Functors

335 Definition COVARIANT FUNCTOR
Let $\mathcal{C}_1$ and $\mathcal{C}_2$ be two categories. A *covariant functor* $\mathcal{F} : \mathcal{C}_1 \to \mathcal{C}_2$ from $\mathcal{C}_1$ to $\mathcal{C}_2$ is a pair $(\mathcal{F}, \mathcal{F}_{A,B})$ consisting of

(1) a function $\mathcal{F} : \mathfrak{O}_1 \to \mathfrak{O}_2$ between the *objects* of the categories, and

(2) a family $\mathcal{F}_{A,B} : \mathfrak{M}_1(A,B) \to \mathfrak{M}_2(\mathcal{F}(A), \mathcal{F}(B))$ of functions between the *morphisms* of the categories,

such that the following identities hold:

$$\forall A \in \mathfrak{O}_1 : \ \mathcal{F}_{A,A}(id_A) = id_{\mathcal{F}(A)}$$

$$\forall A, B, C \in \mathfrak{O}_1, \ g : A \to B, \ f : B \to C : \ \mathcal{F}(f \circ_1 g) = \mathcal{F}(f) \circ_2 \mathcal{F}(g)$$

The class of all covariant functors from $\mathcal{C}_1$ to $\mathcal{C}_2$ is denoted by $\mathbf{Fun}(\mathcal{C}_1, \mathcal{C}_2)$. ◀

A *covariant functor* can be illustrated by the commuting diagrams in Figure C.5: Applying the functor $\mathcal{F}$ to the left commuting diagram yields the right commuting diagram.

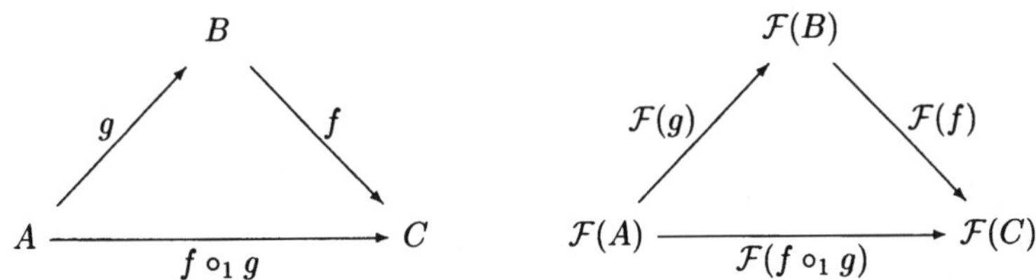

Figure C.5: Definition of a Covariant Functor.

A *contravariant functor* is the dual concept to the covariant functor:

336 Definition CONTRAVARIANT FUNCTOR
Let $\mathcal{C}_1$ and $\mathcal{C}_2$ be two categories. A *contravariant functor* $\mathcal{F} : \mathcal{C}_1 \to \mathcal{C}_2$ from $\mathcal{C}_1$ to $\mathcal{C}_2$ is a pair $(\mathcal{F}, \mathcal{F}_{A,B})$ consisting of

(1) a function $\mathcal{F} : \mathfrak{O}_1 \to \mathfrak{O}_2$ between the *objects* of the categories, and

(2) a family $\mathcal{F}_{A,B} : \mathfrak{M}_1(B, A) \to \mathfrak{M}_2(\mathcal{F}(A), \mathcal{F}(B))$ of functions between the *morphisms* of the categories,

such that the following identities hold:

$$\forall A \in \mathfrak{O}_1 : \ \mathcal{F}_{A,A}(id_A) = id_{\mathcal{F}(A)}$$

$$\forall A, B, C \in \mathfrak{O}_1, \ g : B \to A, \ f : C \to B : \ \mathcal{F}(g \circ_1 f) = \mathcal{F}(f) \circ_2 \mathcal{F}(g)$$

The class of all contravariant functors from $\mathcal{C}_1$ to $\mathcal{C}_2$ is denoted by $\mathbf{Fun}^*(\mathcal{C}_1, \mathcal{C}_2)$. ◀

A *contravariant functor* can be illustrated by the commuting diagrams in Figure C.6: Applying the functor $\mathcal{F}$ to the left commuting diagram yields the right commuting diagram.

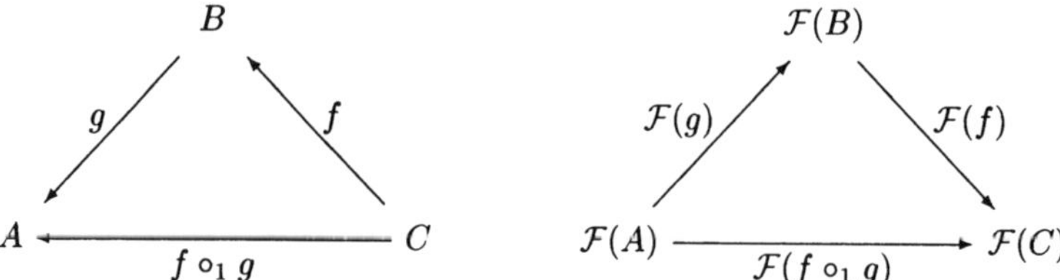

Figure C.6: Definition of a Contravariant Functor.

There exists an *identity functor*:

337 Definition IDENTITY FUNCTOR

The *identity functor* $\mathcal{I}_{\mathcal{C}} : \mathcal{C} \to \mathcal{C}$ of a category is defined on objects by

$$\begin{array}{llll} \mathcal{I}_{\mathcal{C}}: & \mathfrak{O} & \to & \mathfrak{O} \\ & X & \mapsto & X \end{array}$$

and on morphisms by

$$\begin{array}{llll} \mathcal{I}_{\mathcal{C}}: & \mathfrak{M}(X,Y) & \to & \mathfrak{M}(X,Y) \\ & f & \mapsto & f \end{array}$$

◀

It is possible to define a *composition* of functors:

338 Definition COMPOSITION OF FUNCTORS

The *composition* of a functor $\mathcal{F} : \mathcal{C} \to \mathcal{D}$ with a functor $\mathcal{G} : \mathcal{D} \to \mathcal{E}$ is the functor $\mathcal{G} \circ \mathcal{F} : \mathcal{C} \to \mathcal{E}$ defined by “pointwise” composition:

On the objects, the composed functor is defined by

$$\begin{array}{llll} \mathcal{G} \circ \mathcal{F}: & \mathfrak{O}_{\mathcal{C}} & \to & \mathfrak{O}_{\mathcal{E}} \\ & X & \mapsto & \mathcal{G}(\mathcal{F}(X)) \end{array}$$

and on the morphisms it is defined by

$$\begin{array}{llll} \mathcal{G}\circ\mathcal{F}: & \mathfrak{M}_{\mathcal{C}}(X,Y) & \to & \mathfrak{M}_{\mathcal{E}}(\mathcal{G}(\mathcal{F}(X)),\ \mathcal{G}(\mathcal{F}(Y))) \\ & f & \mapsto & \mathcal{G}(\mathcal{F}(f)) \end{array}$$

◀

A functor $\mathcal{F} : \mathcal{C} \to \mathcal{D}$ maps the categorical structure of a category $\mathcal{C}$ to a category $\mathcal{D}$, and may be interpreted as morphism in the category of small[6] categories.

339 Definition CATEGORY OF SMALL CATEGORIES
The category *Cat* of small categories has as objects all *small categories*. If $\mathcal{C}$ and $\mathcal{D}$ are two objects in this category, then the class $\mathfrak{M}_{\mathrm{Cat}}(\mathcal{C},\mathcal{D})$ of *morphisms* from $\mathcal{C}$ to $\mathcal{D}$ is the class $\mathbf{Fun}(\mathcal{C},\mathcal{D})$ of *all covariant functors* from $\mathcal{C}$ to $\mathcal{D}$. ◀

340 Definition SPECIAL FUNCTORS
A functor $\mathcal{F} : \mathcal{C}_1 \to \mathcal{C}_2$ is called

> *full*, iff it is surjective on the morphisms, i.e. all functions $\mathcal{F}_{A,B} : \mathfrak{M}_1(A,B) \to \mathfrak{M}_2(\mathcal{F}(A),\mathcal{F}(B))$ are surjective.
>
> *faithful*, iff it is injective on the morphisms, i.e. all functions $\mathcal{F}_{A,B} : \mathfrak{M}_1(A,B) \to \mathfrak{M}_2(\mathcal{F}(A),\mathcal{F}(B)$ are injective.
>
> a *full embedding*, iff it is full, faithful, and injective on objects, i.e. $\mathcal{F} : \mathfrak{O}_1 \to \mathfrak{O}_2$ is injective.

◀

341 Definition FUNTORIAL ISOMORPHISM
A functor $\mathcal{F} : \mathcal{C}_1 \to \mathcal{C}_2$ is called a *functorial isomorphism*, iff there exists a right inverse functor $\mathcal{G} : \mathcal{C}_2 \to \mathcal{C}_1$, such that $\mathcal{F}\circ\mathcal{G} = \mathcal{I}_{\mathcal{C}_2}$ and a left inverse functor $\mathcal{H} : \mathcal{C}_1 \to \mathcal{C}_2$ such that $\mathcal{H}\circ\mathcal{F} = \mathcal{I}_{\mathcal{C}_1}$. ◀

Hence, a functor between two small categories is a functorial isomorphism, iff it is an isomorphism in the category *Cat* of small categories.

342 Definition ISOMORPHIC CATEGORIES
Two categories are called *isomorphic*, iff there exists a functorial isomorphism between them. ◀

343 Remark DUALITY AND VARIANCE
A functor $\mathcal{F} : \mathcal{C} \to \mathcal{D}$ is a covariant functor, iff it can be interpreted as a contravariant functor on the dual of its source category $\mathcal{C}$; i.e. the functor $\mathcal{F}^+ : \mathcal{C}^* \to \mathcal{D}$ which is defined by

$$\begin{array}{lllll} \mathcal{F}^+: & \mathfrak{O}_{\mathcal{C}^*} & \to & \mathfrak{O}_{\mathcal{D}} & \mathcal{F}^+(A) := A \\ \mathcal{F}^+: & \mathfrak{M}_{\mathcal{C}^*}(A,B) & \to & \mathfrak{M}_{\mathcal{D}}(\mathcal{F}^+(A),\ \mathcal{F}^+(B)) & \mathcal{F}^+(f) := f \end{array}$$

is a contravariant functor. Also the dual version of this remark holds. ◀

Bifunctors and multifunctors are functors whose value depend on two or more objects and morphisms. Remark 343 is applied to define their variances:

344 Definition COVARIANT–COVARIANT BIFUNCTOR[7]
A *bifunctor* from $\mathcal{C}_1$ and $\mathcal{C}_2$ to $\mathcal{C}$ which is *covariant in both arguments* consists of a pair of functions operating on the objects and morphisms of the two categories $\mathcal{C}_1$ and $\mathcal{C}_2$ with results in a third category $\mathcal{C}$:

[6]The restriction to small categories is necessary for set-theoretical reasons, compare [Pre75].
[7]For short: A co-co bifunctor.

$$\mathcal{F} : \mathfrak{O}_1 \times \mathfrak{O}_2 \to \mathfrak{O}$$

$$\mathcal{F} : \mathfrak{M}_1(A_1, B_1) \times \mathfrak{M}_2(A_2, B_2) \to \mathfrak{M}_C(\mathcal{F}(A_1, A_2), \mathcal{F}(B_1, B_2))$$

with $A_1, B_1 \in \mathfrak{O}_1$ and $A_2, B_2 \in \mathfrak{O}_2$.

The following identities must hold:

$$\mathcal{F}(id_A, id_B) = id_{\mathcal{F}(A,B)} \qquad \text{for } A \in \mathfrak{O}_1 \text{ and } B \in \mathfrak{O}_2$$

$$\mathcal{F}(g_1 \circ f_1, g_2 \circ f_2) = \mathcal{F}(g_1, g_2) \circ \mathcal{F}(f_1, f_2)$$

with

$$f_1 : A_1 \to B_1 \qquad f_2 : A_2 \to B_2 \qquad \mathcal{F}(f_1, f_2) : \mathcal{F}(A_1, A_2) \to \mathcal{F}(B_1, B_2)$$

$$g_1 : B_1 \to C_1 \qquad g_2 : B_2 \to C_2 \qquad \mathcal{F}(g_1, g_2) : \mathcal{F}(B_1, B_2) \to \mathcal{F}(C_1, C_2)$$

and finally

$$g_1 \circ f_1 : A_1 \to C_1 \qquad g_2 \circ f_2 : A_2 \to C_2$$

$$\mathcal{F}(g_1 \circ f_1, g_2 \circ f_2) = \mathcal{F}(g_1, g_2) \circ \mathcal{F}(f_1, f_2) : \mathcal{F}(A_1, A_2) \to \mathcal{F}(C_1, C_2)$$

◀

With the help of Remark 343 the definitions of bifunctors with other variance properties may be deduced. This shall be illustrated by the definition of a bifunctor which is *contravariant in its first argument and covariant in its second argument*:

345 Definition CONTRAVARIANT–COVARIANT BIFUNCTOR[8]

A *bifunctor* from C_1 and C_2 to C, which is *contravariant in its first and covariant in its second argument*, consists of a pair of functions operating on the objects and morphisms of the two categories C_1 and C_2 with results in a third category C:

$$\mathcal{F} : \mathfrak{O}_1 \times \mathfrak{O}_2 \to \mathfrak{O}$$

$$\mathcal{F} : \mathfrak{M}_1(B_1, A_1) \times \mathfrak{M}_2(A_2, B_2) \to \mathfrak{M}_C(\mathcal{F}(A_1, A_2), \mathcal{F}(B_1, B_2))$$

with $A_1, B_1 \in \mathfrak{O}_1, A_2, B_2 \in \mathfrak{O}_2$.

The following identities must hold:

$$\mathcal{F}(id_A, id_B) = id_{\mathcal{F}(A,B)} \qquad \text{for } A \in \mathfrak{O}_1 \text{ and } B \in \mathfrak{O}_2$$

$$\mathcal{F}(f_1 \circ g_1, g_2 \circ f_2) = \mathcal{F}(g_1, g_2) \circ \mathcal{F}(f_1, f_2)$$

with

[8]For short: A contra-co bifunctor.

$$f_1 : B_1 \to A_1 \qquad f_2 : A_2 \to B_2 \qquad \mathcal{F}(f_1, f_2) : \mathcal{F}(A_1, A_2) \to \mathcal{F}(B_1, B_2)$$

$$g_1 : C_1 \to B_1 \qquad g_2 : B_2 \to C_2 \qquad \mathcal{F}(g_1, g_2) : \mathcal{F}(B_1, B_2) \to \mathcal{F}(C_1, C_2)$$

and finally

$$f_1 \circ g_1 : C_1 \to A_1 \qquad g_2 \circ f_2 : A_2 \to C_2$$

$$\mathcal{F}(f_1 \circ g_1, g_2 \circ f_2) = \mathcal{F}(g_1, g_2) \circ \mathcal{F}(f_1, f_2) : \mathcal{F}(A_1, A_2) \to \mathcal{F}(C_1, C_2)$$

◀

The concept of a bifunctor is easily generalized to the concept of a *multifunctor* by allowing more than two arguments.

From now on all functors implicitely are assumed to be covariant, unless the contrary is mentioned.

C.5 Natural Transformations

346 Definition NATURAL TRANSFORMATIONS OF CO FUNCTORS
Let $\mathcal{C}_1$ and $\mathcal{C}_2$ be two categories and let $\mathcal{F}_1$ and $\mathcal{F}_2$ be two covariant functors, both from $\mathcal{C}_1$ to $\mathcal{C}_2$.

A *natural transformation (functorial morphism)* $\mathbf{T} : \mathcal{F}_1 \to \mathcal{F}_2$ from functor $\mathcal{F}_1$ to functor $\mathcal{F}_2$ is a family $(\mathbf{T}_A)_{A \in \mathfrak{O}_1}$ of morphisms $\mathbf{T}_A \in \mathfrak{M}_2(\mathcal{F}_1(A), \mathcal{F}_2(A))$, such that

$$\forall A, B \in \mathfrak{O}_1 : \ \forall f \in \mathfrak{M}_1(A, B) : \ \ \mathbf{T}_B \circ \mathcal{F}_1(f) = \mathcal{F}_2(f) \circ \mathbf{T}_A$$

The class of all natural transformations from $\mathcal{F}_1$ to $\mathcal{F}_2$ is denoted by $\mathbf{Nat}(\mathcal{F}_1, \mathcal{F}_2)$. ◀

The above condition can be represented by the commuting diagram in Figure C.7.

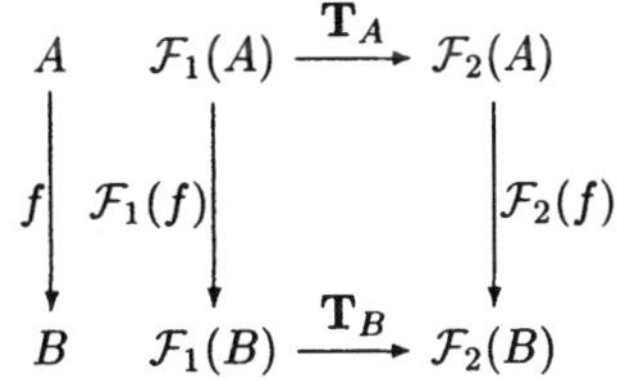

Figure C.7: Natural Transformation of Covariant Functors.

Similar as in Remark 343, the notion of a natural transformation of contravariant functors requires only dualizing the category which forms the domain of the functor $\mathcal{F}_1$:

347 Definition NATURAL TRANSFORMATIONS OF CONTRA FUNCTORS
Let $\mathcal{C}_1$ and $\mathcal{C}_2$ be two categories and let $\mathcal{F}_1$ and $\mathcal{F}_2$ be two contravariant functors, both from $\mathcal{C}_1$ to $\mathcal{C}_2$.

A *natural transformation (functorial morphism)* $\mathbf{T} : \mathcal{F}_1 \to \mathcal{F}_2$ from functor $\mathcal{F}_1$ to functor $\mathcal{F}_2$ is a family $(\mathbf{T}_A)_{A \in \mathfrak{O}_1}$ of morphisms $\mathbf{T}_A \in \mathfrak{M}_2(\mathcal{F}_1(A), \mathcal{F}_2(A))$, such that

$$\forall A, B \in \mathfrak{O}_1 : \forall f \in \mathfrak{M}_1(B, A) : \quad \mathbf{T}_B \circ \mathcal{F}_1(f) = \mathcal{F}_2(f) \circ \mathbf{T}_A$$

The class of all natural transformations from $\mathcal{F}_1$ to $\mathcal{F}_2$ is denoted by $\mathbf{Nat}(\mathcal{F}_1, \mathcal{F}_2)$. ◀

The above condition can be represented by the commuting diagram in Figure C.8.

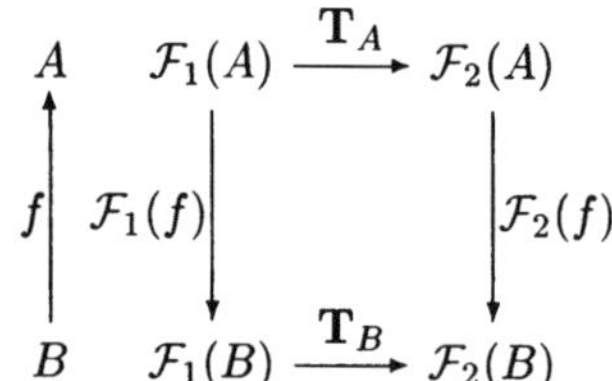

Figure C.8: Natural Transformation of Contravariant Functors.

There exists an *identity natural transformation*:

348 Definition IDENTITY NATURAL TRANSFORMATION
The *identity natural transformation* from the functor $\mathcal{F}$ to itself is given by the family $\mathbf{id}_{\mathcal{F}} = id_{\mathcal{F}(A)}$ of identity morphisms. ◀

It is possible to define a composition of natural transformations:

349 Definition COMPOSITION OF NATURAL TRANSFORMATIONS
Let $\mathcal{F}_1$, $\mathcal{F}_2$ and $\mathcal{F}_3$ be three functors, all from $\mathcal{C}$ to $\mathcal{D}$. The *composition* of a natural transformation $\mathbf{T} = (\mathbf{T}_A)_{A \in \mathfrak{O}_\mathcal{C}}$ from $\mathcal{F}_1$ to $\mathcal{F}_2$ with a natural transformation $\mathbf{S} = (\mathbf{S}_A)_{A \in \mathfrak{O}_\mathcal{C}}$ from $\mathcal{F}_2$ to $\mathcal{F}_3$ is a natural transformation $\mathbf{S} \circ \mathbf{T} = (\mathbf{S}_A \circ_\mathcal{D} \mathbf{T}_A)_{A \in \mathfrak{O}_\mathcal{C}}$ from $\mathcal{F}_1$ to $\mathcal{F}_3$. ◀

Hence, a category of covariant functors can be defined:

350 Definition COVARIANT FUNCTOR CATEGORY
Let $\mathcal{C}$ and $\mathcal{D}$ be two categories. The *covariant functor category* $\mathit{Funct}(\mathcal{C}, \mathcal{D})$ contains as objects all *covariant functors* from $\mathcal{C}$ to $\mathcal{D}$, i.e. $\mathfrak{O}_{\mathit{Funct}(\mathcal{C},\mathcal{D})} = \mathbf{Fun}(\mathcal{C}, \mathcal{D})$. If $\mathcal{F}$ and $\mathcal{G}$ are two objects of this category then the class $\mathfrak{M}_{\mathit{Funct}(\mathcal{C},\mathcal{D})}(\mathcal{F}, \mathcal{G})$ from $\mathcal{F}$ to $\mathcal{G}$ is the class $\mathbf{Nat}(\mathcal{F}, \mathcal{G})$ of all *natural transformations* from $\mathcal{F}$ to $\mathcal{G}$. Composition and identities are defined as above. ◀

Similarly, a category of contravariant functors can be defined by dualizing:

351 Definition CONTRAVARIANT FUNCTOR CATEGORY
Let $\mathcal{C}$ and $\mathcal{D}$ be two categories. The *contravariant functor category* $\mathit{Funct}^*(\mathcal{C}, \mathcal{D})$ contains as objects all *contravariant functors* from $\mathcal{C}$ to $\mathcal{D}$, i.e. $\mathfrak{O}_{\mathit{Funct}^*(\mathcal{C},\mathcal{D})} = \mathbf{Fun}^*(\mathcal{C}, \mathcal{D})$. If $\mathcal{F}$ and $\mathcal{G}$ are two objects of this category then the class $\mathfrak{M}_{\mathit{Funct}^*(\mathcal{C},\mathcal{D})}(\mathcal{F}, \mathcal{G})$ from $\mathcal{F}$ to $\mathcal{G}$ is the class $\mathbf{Nat}(\mathcal{F}, \mathcal{G})$ of all *natural transformations* from $\mathcal{F}$ to $\mathcal{G}$. Composition and identities are defined as above. ◀

352 Definition NATURAL ISOMORPHISM
A natural transformation $\mathbf{T} : \mathcal{F}_1 \to \mathcal{F}_2$ is called a natural isomorphism, iff there exists a left inverse natural transformation $\mathbf{S} : \mathcal{F}_2 \to \mathcal{F}_1$ such that $\mathbf{S} \circ \mathbf{T}$ is the identity natural transformation $\mathbf{id}_{\mathcal{F}_1}$ of $\mathcal{F}_1$, and a right identity natural transformation $\mathbf{U} : \mathcal{F}_1 \to \mathcal{F}_2$, such that $\mathbf{T} \circ \mathbf{U}$ is the identity natural transformation $\mathbf{id}_{\mathcal{F}_2}$ of $\mathcal{F}_2$. ◀

Hence, a natural transformation between two functors is a natural isomorphism, iff it is an isomorphism in the functor category.

353 Proposition POINTWISE NATURAL ISOMORPHISM
A natural transformation $\mathbf{T} = (\mathbf{T}_A)_{A\in\mathfrak{O}}$ is a *natural isomorphism*, iff it consists only of isomorphisms $\mathbf{T}_A \in \mathfrak{M}_2(\mathcal{F}_1(A), \mathcal{F}_2(A))$. ◀

Natural transformations can also be defined for bifunctors and multifunctors:

354 Definition NATURAL TRANSFORMATIONS OF CO-CO BIFUNCTORS
Let $\mathcal{C}_1$, $\mathcal{C}_2$ and $\mathcal{C}$ be three categories and $\mathcal{F}_1$ and $\mathcal{F}_2$ two co-co bifunctors of the type $\mathcal{C}_1 \times \mathcal{C}_2 \to \mathcal{C}$.

A *natural transformation* from $\mathcal{F}_1$ to $\mathcal{F}_2$ is a family $\mathbf{T} = (\mathbf{T}_{A_1,A_2})_{A_1\in\mathfrak{O}_1,A_2\in\mathfrak{O}_2}$ of morphisms $\mathbf{T}_{A_1,A_2} : \mathcal{F}_1(A_1, A_2) \to \mathcal{F}_2(A_1, A_2)$ such that

$$\forall A_1, B_1 \in \mathfrak{O}_1, A_2, B_2 \in \mathfrak{O}_2, f_1 : A_1 \to B_1, f_2 : A_2 \to B_2 :$$

$$\mathbf{T}_{B_1,B_2} \circ \mathcal{F}_1(f_1, f_2) = \mathcal{F}_2(f_1, f_2) \circ \mathbf{T}_{A_1,A_2}$$

This can be illustrated by the commuting diagram in Figure C.9. ◀

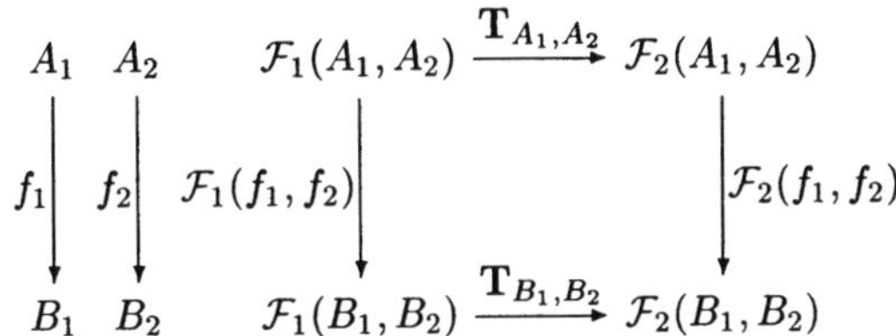

Figure C.9: Natural Transformation of Co-Co Bifunctors.

For contravariant-covariant bifunctors this is similar and an application of Remark 343 yields:

355 Definition NATURAL TRANSFORMATIONS OF CONTRA-CO BIF.
Let $\mathcal{C}_1$, $\mathcal{C}_2$ and $\mathcal{C}$ be three categories and $\mathcal{F}_1$ and $\mathcal{F}_2$ two contra-co bifunctors of the type $\mathcal{C}_1 \times \mathcal{C}_2 \to \mathcal{C}$.

A *natural transformation* from $\mathcal{F}_1$ to $\mathcal{F}_2$ is a family $\mathbf{T} = (\mathbf{T}_{A_1,A_2})_{A_1\in\mathfrak{O}_1,A_2\in\mathfrak{O}_2}$ of morphisms $\mathbf{T}_{A_1,A_2} : \mathcal{F}_1(A_1, A_2) \to \mathcal{F}_2(A_1, A_2)$ such that

$$\forall A_1, B_1 \in \mathfrak{O}_1, A_2, B_2 \in \mathfrak{O}_2, f_1 : B_1 \to A_1, f_2 : A_2 \to B_2 :$$

$$\mathbf{T}_{B_1,B_2} \circ \mathcal{F}_1(f_1, f_2) = \mathcal{F}_2(f_1, f_2) \circ \mathbf{T}_{A_1,A_2}$$

This can be illustrated by the commuting diagram in Figure C.10. ◀

This concept may be generalized to multifunctors.

C.6 Properties of the Product Functor

In this section the theory of the previous two sections is applied for analyzing the product construction.

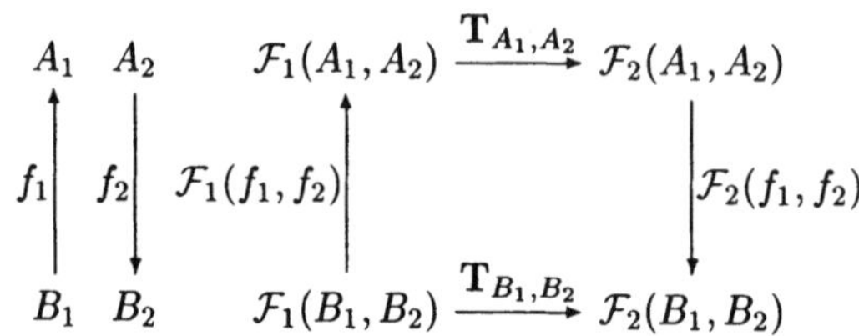

Figure C.10: Natural Transformation of Contra-Co Bifunctors.

356 Proposition PRODUCT CONSTRUCTION IS FUNCTOR
Let $\mathcal{C}$ be a category containing products. Then the construction of the *product* from n objects is a *covariant multifunctor* in n arguments. It is defined on objects by

$$\begin{array}{llll} \otimes: & \mathfrak{O} \times \ldots \times \mathfrak{O} & \rightarrow & \mathfrak{O} \\ & (X_1, \ldots, X_n) & \mapsto & X_1 \otimes \ldots \otimes X_n \end{array}$$

and on morphisms by

$$\begin{array}{llll} \otimes: & \mathfrak{M}(A_1, B_1) \times \ldots \times \mathfrak{M}(A_n, B_n) & \rightarrow & \mathfrak{M}(A_1 \otimes \ldots \otimes A_n, B_1 \otimes \ldots \otimes B_n) \\ & (f_1, \ldots, f_n) & \mapsto & f_1 \otimes \ldots \otimes f_n := \\ & & & (f_1 \circ \pi_1), \ldots, (f_n \circ \pi_n) \end{array}$$

PROOF:
In the above proposition, the following *morphisms* occur:

$$f_i : A_i \rightarrow B_i$$

$$\pi_i : A_1 \otimes A_2 \otimes \ldots \otimes A_n \rightarrow A_i$$

$$f_i \circ \pi_i : A_1 \otimes A_2 \otimes \ldots \otimes A_n \rightarrow B_i$$

$$(f_1 \pi_1) \otimes (f_2 \circ \pi_2) \otimes \ldots \otimes (f_n \circ \pi_n) : A_1 \otimes A_2 \otimes \ldots \otimes A_n \rightarrow B_1 \otimes B_2 \otimes \ldots \otimes B_n$$

The property that $\otimes$ is a *functor* requires that the equation

$$id_{A_1 \otimes \ldots \otimes A_n} = id_{A_1} \otimes \ldots \otimes id_{A_n}$$

holds and that for $f_i : A_i \rightarrow B_i$ and $g_i : B_i \rightarrow C_i$ also the following equation is satisfied:

$$(g_1 \circ f_1) \otimes \ldots \otimes (g_n \circ f_n) = (g_1 \otimes \ldots \otimes g_n) \circ (f_1 \otimes \ldots \otimes f_n)$$

In the following a the proof of the second equation for the case $n = 2$ is provided: From the four morphisms

$$f_1 : A_1 \rightarrow B_1 \qquad f_2 : A_2 \rightarrow B_2$$

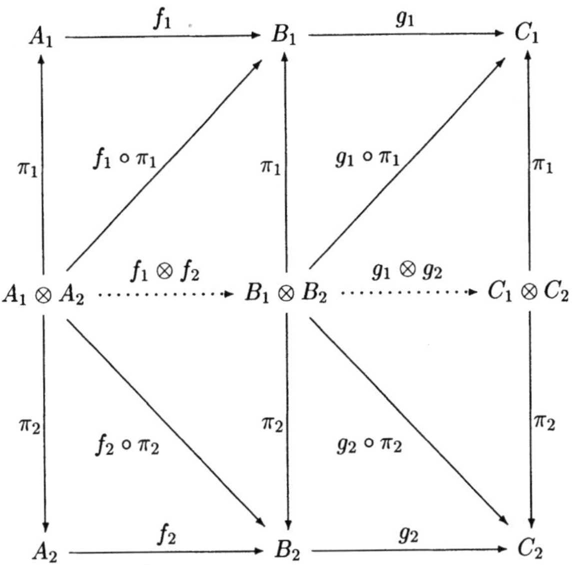

Figure C.11: Proof that Binary Product is Bifunctor, Part 1.

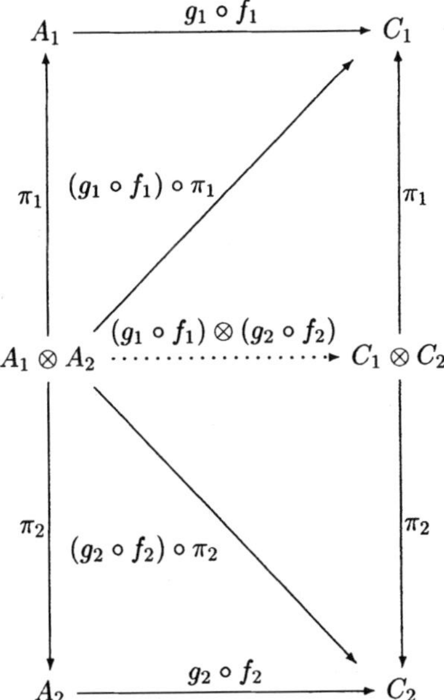

Figure C.12: Proof that Binary Product is Bifunctor, Part 2.

$$g_1 : B_1 \to C_1 \qquad g_2 : B_2 \to C_2$$

the two commuting diagrams in Figures C.11 and C.12 can be immediately constructed by using the definition of a product.

Hence, two morphisms from $A_1 \otimes A_2$ to $C_1 \otimes C_2$ are obtained. The uniqueness condition of the second diagram is indicated by the dotted arrow and can be obtained directly from the definition of the product. It allows to deduce the following equation:

$$(g_1 \circ f_1) \otimes (g_2 \circ f_2) = (g_1 \otimes g_2) \circ (f_1 \otimes f_2)$$

◀

357 Theorem PROPERTIES OF FINITE PRODUCTS
The product functor is *commutative*, *associative*, and has the terminal objects as *neutral elements*:

Commutativity:
There exists a natural isomorphism **S** from the bifunctor

$$\begin{array}{lll} \otimes : & \mathfrak{O} \times \mathfrak{O} & \to \mathfrak{O} \\ & (A, B) & \mapsto A \otimes B \\ \otimes : & \mathfrak{M}(A_1, B_1) \times \mathfrak{M}(A_2, B_2) & \to \mathfrak{M}(A_1 \otimes A_2, B_1 \otimes B_2) \end{array}$$

to the bifunctor

$$\begin{array}{lll} \otimes' : & \mathfrak{O} \times \mathfrak{O} & \to \mathfrak{O} \\ & (A, B) & \mapsto A \otimes' B := B \otimes A \\ \otimes' : & \mathfrak{M}(A_1, B_1) \times \mathfrak{M}(A_2, B_2) & \to \mathfrak{M}(A_1 \otimes' A_2, B_1 \otimes' B_2) = \\ & & \quad \mathfrak{M}(A_2 \otimes A_1, B_2 \otimes B_1) \end{array}$$

such that $S_{B,A} \circ S_{A,B} = id_{A \otimes B}$.

This situation is illustrated by the commuting diagram in Figure C.13.

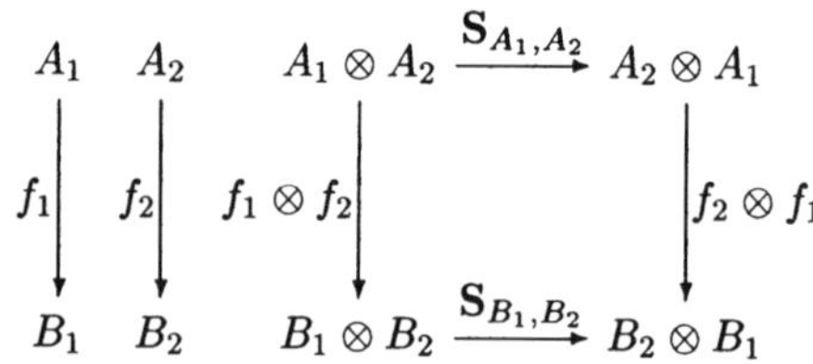

Figure C.13: Commutativity of the Product Functor.

Associativity:
There exists a natural isomorphism **A** from the trifunctor

$$\begin{array}{lll} \otimes_3^r : & \mathfrak{O} \times \mathfrak{O} \times \mathfrak{O} & \to \mathfrak{O} \\ & (A, B, C) & \mapsto A \otimes (B \otimes C) \\ \otimes_3^r : & \mathfrak{M}(A_1, B_1) \times \mathfrak{M}(A_2, B_2) \times \mathfrak{M}(A_3, B_3) & \to \mathfrak{M}(A_1 \otimes (A_2 \otimes A_3), B_1 \otimes (B_2 \otimes B_3)) \end{array}$$

to the trifunctor

$$\begin{array}{llll} \otimes_3^l : & \mathfrak{O} \times \mathfrak{O} \times \mathfrak{O} & \to & \mathfrak{O} \\ & (A, B, C) & \mapsto & (A \otimes B) \otimes C \\ \otimes_3^l : & \mathfrak{M}(A_1, B_1) \times \mathfrak{M}(A_2, B_2) \times \mathfrak{M}(A_3, B_3) & \to & \mathfrak{M}((A_1 \otimes A_2) \otimes A_3, (B_1 \otimes B_2) \otimes B_3) \end{array}$$

This situation is illustrated by the commuting diagrams in Figure C.14.

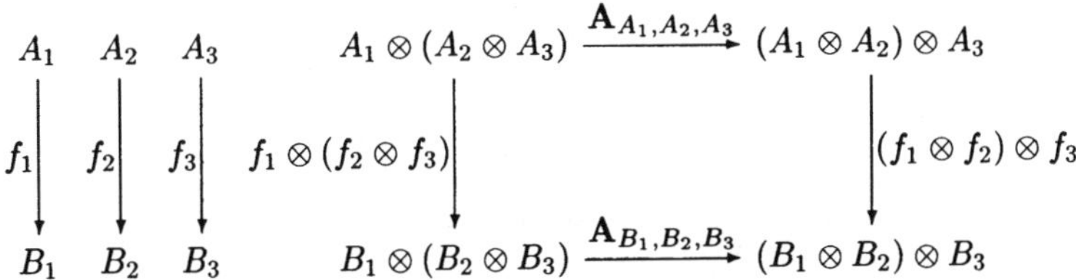

Figure C.14: Associativity of the Product Functor.

Neutral Element:
For every terminal object T there exists a natural isomorphism $\mathbf{L}^T$ from the functor defined by

$$\begin{array}{llll} T \otimes \_ : & \mathfrak{O} & \to & \mathfrak{O} \\ & A & \mapsto & T \otimes A \\ id_T \otimes \_ : & \mathfrak{M}(A, B) & \to & \mathfrak{M}(T \otimes A, T \otimes B) \\ & f & \mapsto & id_T \otimes f \end{array}$$

to the identity functor **id** of the category, and a natural isomorphism $\mathbf{R}^T$ from the functor

$$\begin{array}{llll} \_ \otimes T : & \mathfrak{O} & \to & \mathfrak{O} \\ & A & \mapsto & A \otimes T \\ \_ \otimes id_T : & \mathfrak{M}(A, B) & \to & \mathfrak{M}(A \otimes T, B \otimes T) \\ & f & \mapsto & f \otimes id_T \end{array}$$

to the identity functor **id** of the category.

This situation is illustrated by the commuting diagrams in Figure C.15.

These four natural transformations may be chosen in a way that the ***first*** and ***second trigonal*** diagram, the ***pentagonal***, and the ***hexagonal*** diagram, shown in Figures C.16, C.17, C.18, and C.19 commute.

The natural isomorphisms **S**, **A**, **L**, and **R** are called the *canonical natural isomorphisms* and are associated with every category containing finite products.

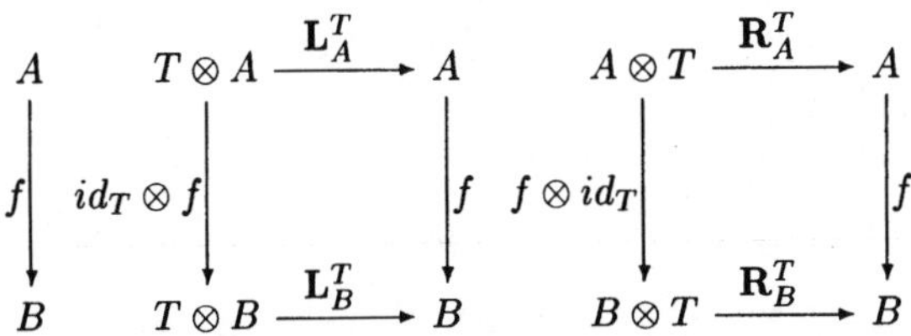

Figure C.15: Neutral Element of the Product Functor.

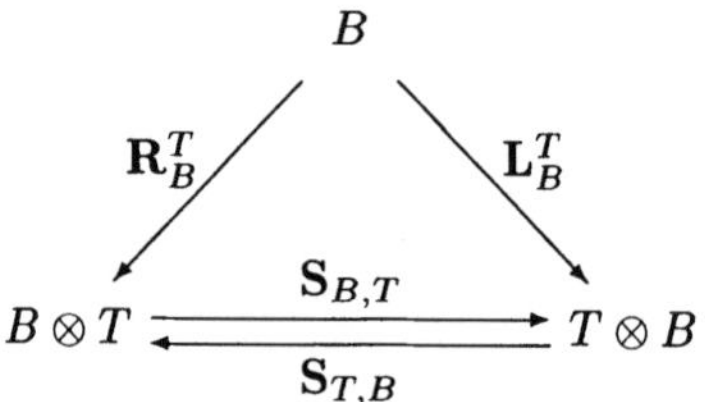

Figure C.16: First Trigonal Diagram.

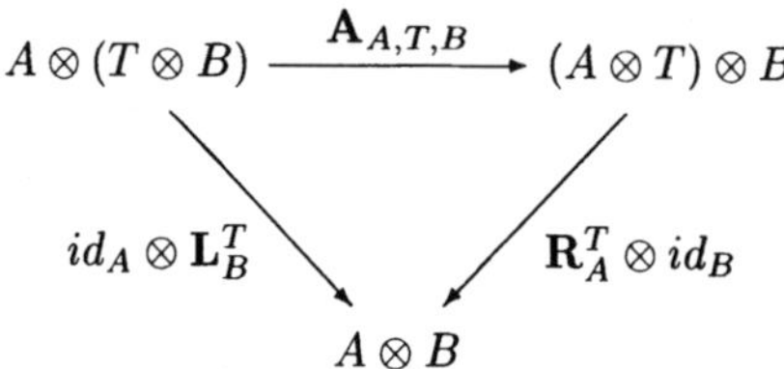

Figure C.17: Second Trigonal Diagram.

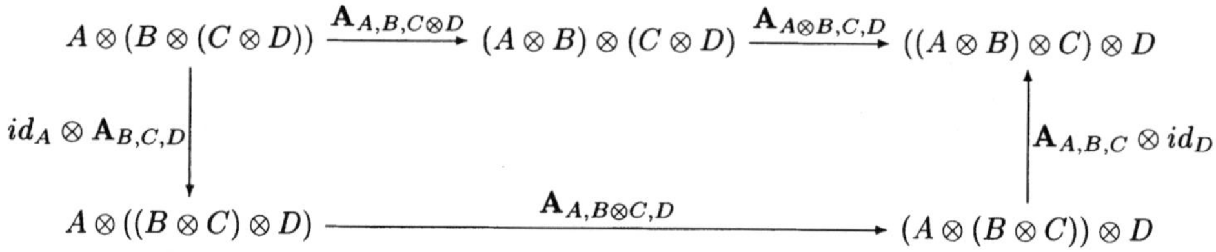

$$\begin{array}{ccccc}
A \otimes (B \otimes (C \otimes D)) & \xrightarrow{\mathbf{A}_{A,B,C\otimes D}} & (A \otimes B) \otimes (C \otimes D) & \xrightarrow{\mathbf{A}_{A\otimes B,C,D}} & ((A \otimes B) \otimes C) \otimes D \\
\downarrow {\scriptstyle id_A \otimes \mathbf{A}_{B,C,D}} & & & & \uparrow {\scriptstyle \mathbf{A}_{A,B,C} \otimes id_D} \\
A \otimes ((B \otimes C) \otimes D) & & \xrightarrow{\mathbf{A}_{A,B\otimes C,D}} & & (A \otimes (B \otimes C)) \otimes D
\end{array}$$

Figure C.18: Pentagonal Diagram.

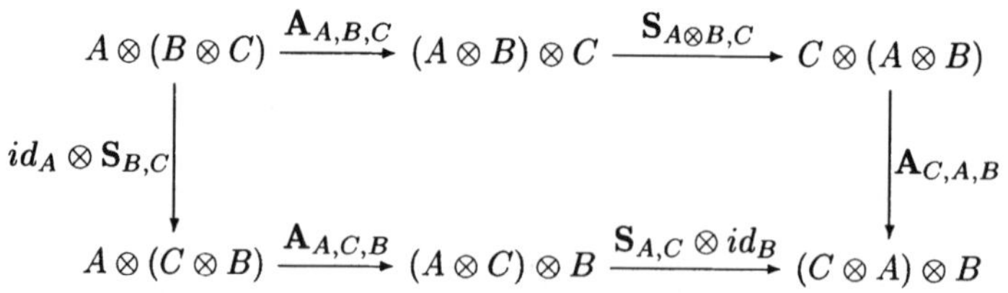

$$\begin{array}{ccccc}
A \otimes (B \otimes C) & \xrightarrow{\mathbf{A}_{A,B,C}} & (A \otimes B) \otimes C & \xrightarrow{\mathbf{S}_{A\otimes B,C}} & C \otimes (A \otimes B) \\
\downarrow {\scriptstyle id_A \otimes \mathbf{S}_{B,C}} & & & & \downarrow {\scriptstyle \mathbf{A}_{C,A,B}} \\
A \otimes (C \otimes B) & \xrightarrow{\mathbf{A}_{A,C,B}} & (A \otimes C) \otimes B & \xrightarrow{\mathbf{S}_{A,C} \otimes id_B} & (C \otimes A) \otimes B
\end{array}$$

Figure C.19: Hexagonal Diagram.

PROOF:
The proof of these properties is conceptually very easy, but requires lengthy diagram chasing as indicated above. ◀

The statements of this theorem can be interpreted intuitively. For the commutativity, this reads as follows:

The products $A_1 \otimes A_2$ and $A_2 \otimes A_1$ are ***isomorphic*** via the isomorphism $\mathbf{S}_{A_1,A_2}$ and the products $B_1 \otimes B_2$ and $B_2 \otimes B_1$ are isomorphic via an isomorphism $\mathbf{S}_{B_1,B_2}$. Now assume structural relationships which are expressed by the morphisms $f_1 : A_1 \to B_1$ and $f_2 : A_2 \to B_2$. The product bifunctor $\otimes$ produces a structural relationship for the products, expressed by the morphism $f_1 \otimes f_2 : A_1 \otimes A_2 \to B_1 \otimes B_2$. This relationship $f_1 \otimes f_2 : A_1 \otimes A_2 \to B_1 \otimes B_2$ can be ***translated*** into a relationship $f_2 \otimes f_1 : A_2 \otimes A_1 \to B_2 \otimes B_1$ with the help of above isomorphisms.

C.7 Representation of Categories

The category *Set* of all sets is well understood and thus one studies in which way arbitrary categories may be represented by set constructions. In this section, let C be a ***locally small*** category.

358 Definition COVARIANT HOM-FUNCTOR
For an object $X \in \mathfrak{O}$ the ***covariant hom-functor*** with respect to this object is a covariant functor $Hom : \mathfrak{O} \to Set$ from the category C into the category *Set* of all sets. It is is defined on the objects by

$$\begin{array}{rccl} Hom(X): & \mathfrak{O} & \to & \mathfrak{O}_{Set} \\ & A & \mapsto & Hom(X)(A) := \mathfrak{M}_C(X,A) \end{array}$$

and on the morphisms by

$$\begin{array}{rlcl} Hom(X): & \mathfrak{M}_C(A,B) & \to & \mathfrak{M}_{Set}(Hom(X)(A),\ Hom(X)(B)) = \\ & & & \mathfrak{M}_{Set}(\mathfrak{M}_C(X,A),\ \mathfrak{M}_C(X,B)) \\ & f & \mapsto & [g \mapsto f \circ g] \end{array}$$

$$Hom(X)(A) = \mathfrak{M}_C(X,A)$$

$$Hom(X)(f)(g) = f \circ g \in \mathfrak{M}_C(X,B) \qquad f : A \to B \qquad g : X \to A$$

$$Hom(X)(f) \in \mathfrak{M}_{Set}(\mathfrak{M}_C(X,A),\ \mathfrak{M}_C(X,B))$$

◀

The contravariant hom-functor is obtained by dualizing:

359 Definition CONTRAVARIANT HOM-FUNCTOR
For an object $X \in \mathfrak{O}$ the ***contravariant hom-functor*** with respect to this object is a contravariant functor $Hom^* : \mathfrak{O} \to Set$ from the category C into the category *Set* of all sets. It is is defined on the objects by

$$\begin{array}{rrcl} Hom^*(X): & \mathfrak{O} & \rightarrow & \mathfrak{O}_{Set} \\ & A & \mapsto & Hom^*(X)(A) := \mathfrak{M}_C(A,X) \end{array}$$

and on the morphisms by

$$\begin{array}{rlcl} Hom^*(X): & \mathfrak{M}_C(A,B) & \rightarrow & \mathfrak{M}_{Set}(Hom^*(X)(B),\ Hom^*(X)(A)) = \\ & & & \mathfrak{M}_{Set}(\mathfrak{M}_C(B,X),\ \mathfrak{M}_C(A,X)) \\ & f & \mapsto & [g \mapsto g \circ f] \end{array}$$

$$Hom^*(X)(A) = \mathfrak{M}_C(A,X)$$

$$Hom^*(X)(f)(g) = g \circ f \in \mathfrak{M}_C(B,X) \qquad f: A \rightarrow B \qquad g: B \rightarrow X$$

$$Hom^*(X)(f) \in \mathfrak{M}_{Set}(\mathfrak{M}_C(B,X),\ \mathfrak{M}_C(A,X))$$

◀

A functor is called ***representable***, iff it looks like[9] a hom-functor.

360 Definition REPRESENTABLE FUNCTOR
A covariant functor $\mathcal{F} : \mathcal{C} \rightarrow Set$ from a category $\mathcal{C}$ to the category of all sets is called ***representable***, iff it is naturally isomorphic to a suitable covariant hom-functor $Hom(X)$ with $X \in \mathfrak{O}$.

A contravariant functor $\mathcal{F} : \mathcal{C} \rightarrow Set$ from a category $\mathcal{C}$ to the category of all sets is called ***representable***, iff it is naturally isomorphic to a suitable contravariant hom-functor $Hom^*(X)$ with $X \in \mathfrak{O}$. ◀

This suggests to study how a category $\mathcal{C}$ may be mapped to the category of functors of the above type:

361 Definition CONTRAVARIANT YONEDA FUNCTOR
The contravariant YONEDA functor of $\mathcal{C}$ is a functor from the category $\mathcal{C}$ into the covariant functor category $Funct(\mathcal{C}, Set)$ from $\mathcal{C}$ to the category of sets:

$$Y : \mathcal{C} \rightarrow Funct(\mathcal{C}, Set)$$

On objects Y it is defined by

$$\begin{array}{rrcl} Y: & \mathfrak{O} & \rightarrow & \mathbf{Fun}(\mathcal{C}, Set) \\ & A & \mapsto & Hom(A) \end{array}$$

and on morphisms by

$$\begin{array}{rlcl} Y: & \mathfrak{M}(B,A) & \rightarrow & \mathbf{Nat}(Hom(A), Hom(B)) \\ & f & \mapsto & [h \mapsto h \circ f] \end{array}$$

◀

Hence, if $f : B \rightarrow A$ is a morphism in $\mathcal{C}$ then $Y(f)$ is a natural transformation from the covariant hom-functor $Hom(A)$ to the covariant hom-functor $Hom(B)$. Thus, $Y(f)$ is a family of morphisms indexed by objects of the domain category of the hom-functors. For an object $D \in \mathfrak{O}$ one obtains

[9] This means that it *is* a hom-functor up to an isomophism of functors, i.e. a natural isomorphism.

$$[Y(f)]_D : [Hom(A)](D) \to [Hom(B)](D)$$

$$[Y(f)]_D(h) = h \circ f \qquad \text{with } f : B \to A \qquad h : A \to D \qquad h \circ f : B \to D$$

The condition that $Y(f)$ is a natural transformation is represented by the commuting diagram in Figure C.20.

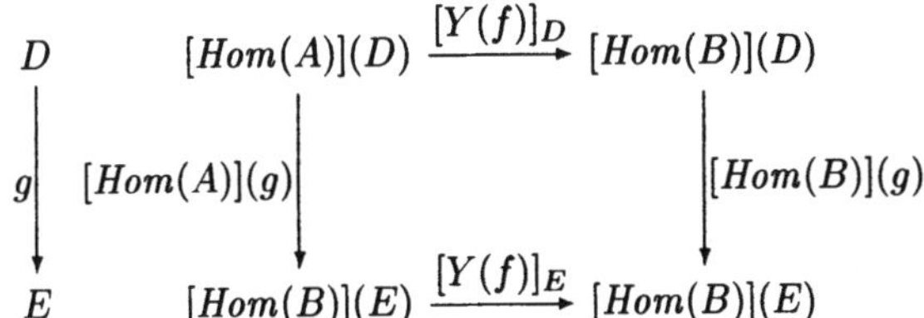

Figure C.20: Contravariant YONEDA Functor.

Here $g : D \to E$ is a morphism in $\mathcal{C}$ and one obtains for $h : A \to D$:

$$[Hom(B)](g)([Y(f)]_D(h)) = g \circ (h \circ f)$$

$$[Y(f)]_E(([Hom(A)](g)(h))) = (g \circ h) \circ f$$

and thus the above diagram indeed commutes due to the associativity of the composition.

The covariant YONEDA Functor is obtained by dualization:

362 Definition COVARIANT YONEDA FUNCTOR
The covariant YONEDA functor of $\mathcal{C}$ is a functor from the category $\mathcal{C}$ into the contravariant functor category $Funct^*(\mathcal{C}, Set)$ from $\mathcal{C}$ to the category of sets:

$$J : \mathcal{C} \to Funct^*(\mathcal{C}, Set)$$

On objects J it is defined by

$$J : \begin{array}{ccc} \mathfrak{O} & \to & \mathbf{Fun}^*(\mathcal{C}, Set) \\ A & \mapsto & Hom^*(A) \end{array}$$

and on morphisms by

$$J : \begin{array}{ccc} \mathfrak{M}(A,B) & \to & \mathbf{Nat}(Hom^*(A), Hom^*(B)) \\ f & \mapsto & [h \mapsto f \circ h] \end{array}$$

◀

Hence, if $f : A \to B$ is a morphism in $\mathcal{C}$ then $J(f)$ is a natural transformation from the contravariant hom-functor $Hom^*(A)$ to the contravariant hom-functor $Hom^*(B)$. Thus, $J(f)$ is a family of morphisms indexed by objects of the domain category of the hom-functors. For an object $D \in \mathfrak{O}$ one obtains

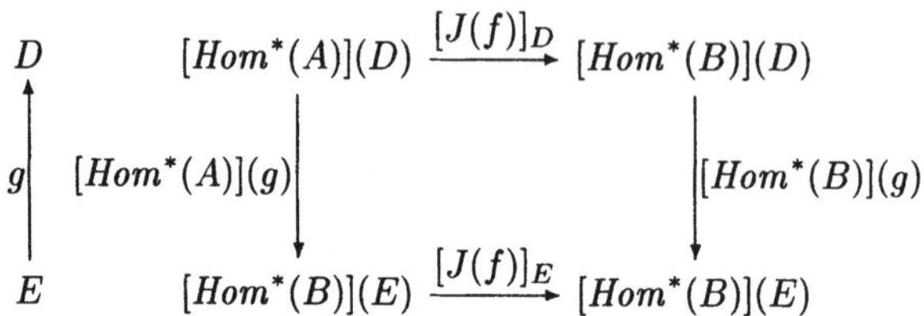

Figure C.21: Covariant YONEDA Functor.

$$[J(f)]_D : [Hom^*(A)](D) \to [Hom^*(B)](D)$$

$$[J(f)]_D(h) = f \circ h \qquad \text{with } f : A \to B \qquad h : D \to A \qquad h \circ f : D \to B$$

The condition that $Y(f)$ is a natural transformation is represented by the commuting diagram in Figure C.21.

Here $g : E \to D$ is a morphism in $\mathcal{C}$ and one obtains for $h : D \to A$:

$$[Hom^*(B)](g)([J(f)]_D(h)) = (f \circ h) \circ g$$

$$[J(f)]_E(([Hom^*(A)](g)(h))) = f \circ (h \circ g)$$

and thus the above diagram indeed commutes due to the associativity of the composition.

363 Theorem YONEDA LEMMA[10]
The contravariant and covariant YONEDA functors are *full* and *faithful*. ◀

The most important consequence of the YONEDA lemma is the following Theorem:

364 Theorem REPRESENTATION OF NATURAL TRANSFORMATIONS
Let A and B be objects and $\mathbf{T} : Hom(A) \to Hom(B)$ a natural transformation of the covariant hom-functors. Then there exists a unique morphism $f : B \to A$ such that for every object D the morphism $\mathbf{T}_D : [Hom(A)](D) \to [Hom(B)](D)$ can be represented by $\mathbf{T}_D(h) = h \circ f$. The natural transformation is a natural isomorphism, iff the morphism f is an isomorphism. ◀

C.8 Cartesian Closed Categories

By defining morphisms between certain objects, a structural relationship between objects is defined. In the previous sections this structural relationship was studied by *inventing higher levels of abstraction*, such as functors and natural transformations. Now this structural relationship will be studied without inventing new concepts but *within the framework of the analyzed objects*. This requires a transformation of morphisms which maps the relationship of objects back to the objects:

365 Definition EXPONENTIAL CONSTRUCTION
Let $\mathcal{C}$ be a category containing all finite products and let $B, C \in \mathfrak{O}$ be two objects. An

[10]Due to its important applications, the YONEDA lemma should rather be called a theorem.

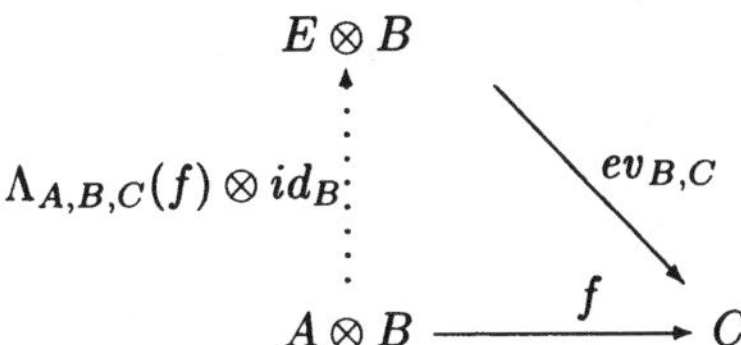

Figure C.22: Exponential Construction.

exponential construction of B *and* C consists of an object $E \in \mathfrak{O}$, called an *exponent* and a morphism $ev \in \mathfrak{M}(E \otimes B, C)$, called the ***evaluation morphism***, so that the following property is satisfied:

For every object A and morphism $f \in \mathfrak{M}(A \otimes B, C)$ there exists a unique morphism $\Lambda_{A,B,C}(f) \in \mathfrak{M}(A, E)$, called ***abstraction***, such that the following equation holds:

$$ev_{B,C} \circ (\Lambda_{A,B,C}(f) \otimes id_B) = f$$

◀

This condition can be expressed by the commuting diagram in Figure C.22.

A category is said to *contain exponentials*, iff for all objects B and C there exists an exponential construction of B to the base C.

The concept of an exponential construction generalizes the situation of the category of sets, which is described in the following example.

366 Example FUNCTIONS ON SETS AS EXPONENTIAL CONSTRUCTION
The category *Set* of all sets contains all finite products. If B and C are two sets then the class $\mathfrak{M}(B, C)$ of all morphisms from B to C equals the set of all functions from B to C. Hence, $\mathfrak{M}(B, C)$ is an object of the category *Set*. It is the exponential construction of B and C.

The ***evaluation morphism*** is provided by the function $ev_{B,C} : \mathfrak{M}(B, C) \times B \to C$ which applies an element f of $\mathfrak{M}(B, C)$ to an element b of B yielding an element $ev_{B,C}(f, b) := f(b)$ of C. Since the product of sets equals the product construction in the category *Set*, and since $ev_{B,C}$ is a function between sets, $ev_{B,C}$ is a morphism.

For every set A and function $f : A \times B \to C$ there exists a morphism $\Lambda_{A,B,C}(f) : A \to \mathfrak{M}(B, C)$ defined by $\Lambda_{A,B,C}(f)(a) := f(a, \bullet) = [b \mapsto f(a, b)]$. From the equation $ev_{B,C} \circ (\Lambda_{A,B,C}(f) \otimes id_B) = f$ the uniqueness of such a function $\Lambda_{A,B,C}(f)$ can immediately be deduced.

Thus, the operator $\Lambda_{A,B,C}$ transforms a function $f : A \times B \to C$ of two variables into a "higher order" function $\Lambda_{A,B,C}(f) : A \to \mathfrak{M}(B, C)$ of one variable. $[\Lambda_{A,B,C}(f)](a)$ is an element of $\mathfrak{M}(B, C)$ and thus a function from B to C. In functional programming this construction is called ***partial evaluation*** or CURRY***ing***.

The operator $\Lambda_{A,B,C}$ may be inverted: If $g : A \to \mathfrak{M}(B, C)$ is a function of one variable whose value is a function of another variable, then $\Lambda^{-1}_{A,B,C}(g)$ is a function of two variables which is defined by $[\Lambda^{-1}_{A,B,C}(g)](a, b) = ev_{B,C}(g(a), b) = [g(a)](b) \in C$. ◀

367 Proposition UNIQUENESS OF EXPONENTIALS
Exponential constructions are unique up to isomorphism. More precisely: If (E_1, ev_1) and (E_2, ev_2) are exponential constructions of A and B, then there exists a unique isomorphism $f : E_1 \to E_2$, such that $ev_2 \circ (f \otimes id_A) = ev_1$.

PROOF:
Since this presentation of the exponential construction and this proposition is not commonly found in the literature, a short proof will be presented:

Assume (E_1, ev_1) and (E_2, ev_2) are exponential constructions of A and B. Then for the morphism $ev_2 : E_2 \otimes A \to B$ there exists a unique $\Lambda_1(ev_2) : E_2 \to E_1$ such that $ev_1 \circ [\Lambda_1(ev_2) \otimes id_A] = ev_2$. Similarly, $ev_2 \circ [\Lambda_2(ev_1) \otimes id_A] = ev_1$. One observes

$$ev_1 \circ [id_{E_1} \otimes id_A] =$$

$$(ev_2 \circ [\Lambda_2(ev_1) \otimes id_A]) \circ [id_{E_1} \otimes id_A] =$$

$$ev_2 \circ ([\Lambda_2(ev_1) \circ id_{E_1}] \otimes [id_A \circ id_A]) =$$

$$ev_2 \circ (\Lambda_2(ev_1) \otimes id_A) =$$

$$ev_1 \circ [\Lambda_1(ev_2) \otimes id_A] \circ [\Lambda_2(ev_1) \otimes id_A] =$$

$$ev_1 \circ ([\Lambda_1(ev_2) \circ \Lambda_2(ev_1)] \otimes id_A)$$

Furthermore, $ev_1 \circ [id_{E_1} \otimes id_A] = ev_1$. Hence, $ev_1 \circ [id_{E_1} \otimes id_A] = ev_1 \circ ([\Lambda_1(ev_2) \circ \Lambda_2(ev_1)] \otimes id_A)$, from which it is possible to deduce by the uniqueness of the function g which satisfies $ev_1 \circ [g \otimes id_A] = g$, that $\Lambda_1(ev_2) \circ \Lambda_2(ev_1) = id_{E_1}$. Similarly, $\Lambda_2(ev_1) \circ \Lambda_1(ev_2) = id_{E_2}$, and thus $\Lambda_1(ev_2) : E_2 \to E_1$ and $\Lambda_2(ev_1) : E_1 \to E_2$ are isomorphisms. Furthermore, $ev_2 \circ [\Lambda_2(ev_1) \otimes id_A] = ev_1$. The uniqueness of this isomorphism again follows from the uniqueness condition in the definition of the exponential construction. ◀

This uniqueness justifies the use of a notation B^A or $A \Rightarrow B$ and $ev_{A,B}$ for *the* exponential construction of A and B.

The following proposition shows that the uniqueness condition in the definition of the exponential construction may be replaced by a second equation. Furthermore, the abstraction function is a bijection:

368 Proposition ALTERNATIVE EQUATIONAL CHARACTERIZATION
Let $\mathcal{C}$ be a category containing all finite products and let $B, C \in \mathfrak{O}$ be two objects. An object $E \in \mathfrak{O}$ and a morphism $ev_{B,C} \in \mathfrak{M}(E \otimes B, C)$ is an exponential construction of B and C, if and only if for every object A and morphism $f \in \mathfrak{M}(A \otimes B, C)$ there exists a morphism $\Lambda_{A,B,C}(f) \in \mathfrak{M}(A, E)$, such that the following two equations hold:

$$ev_{B,C} \circ (\Lambda_{A,B,C}(f) \otimes id_B = f$$

$$\Lambda_{A,B,C}(ev_{B,C} \circ (h \otimes id_B)) = h$$

Furthermore, Λ is a bijection. ◀

369 Definition SPECIAL CATEGORIES
A category is called

cartesian, iff it contains all finite products

cartesian closed, iff it contains all finite products and exponentials

bicartesian, iff it contains all finite products and finite coproducts

bicartesian closed, iff it contains all finite products, finite coproducts, and all exponentials

◀

In Proposition 356 it was shown that the binary product forms a co-co bifunctor. Similarly, the exponent forms a contra-co bifunctor:

370 Proposition EXPONENTIAL CONSTRUCTION AS BIFUNCTOR
Let $\mathcal{C}$ be a cartesian closed category. Then the construction which is defined on objects by

$$\begin{array}{llll} \Rightarrow: & \mathfrak{O} \times \mathfrak{O} & \rightarrow & \mathfrak{O} \\ & (A, B) & \mapsto & A \Rightarrow B \end{array}$$

and on morphisms by

$$\begin{array}{llll} \Rightarrow: & \mathfrak{M}(B, A) \times \mathfrak{M}(C, D) & \rightarrow & \mathfrak{M}(A \Rightarrow C, B \Rightarrow D) \\ & (g, h) & \mapsto & f \Rightarrow g := \Lambda_{A \Rightarrow C, B, D}(h \circ ev_{A,C} \circ (id_{A \Rightarrow C} \otimes g)) \end{array}$$

is a contra-co bifunctor. ◀

Let $\mathcal{C}$ be a category and $(B \Rightarrow C,\ ev_{B,C})$ the exponential construction for two objects B and C. Then for every object A there is a bijection between the morphism sets $\mathfrak{M}(A \times B, C)$ and $\mathfrak{M}(A, B \Rightarrow C)$. This bijection is provided by the abstraction $\Lambda_{A,B,C}$:

$$\Lambda_{A,B,C} : \mathfrak{M}(A \otimes B, C) \rightarrow \mathfrak{M}(A, B \Rightarrow C)$$

$$\Lambda^{-1}_{A,B,C} : \mathfrak{M}(A, B \Rightarrow C) \rightarrow \mathfrak{M}(A \otimes B, C)$$

The abstraction not only is a bijection but it is even a ***natural isomorphism***, hence it uniformly respects structural changes in the objects A, B, and C. Technically, this statement means the following:

371 Theorem ABSTRACTION Λ IS A NATURAL ISOMORPHISM[11] [12]
Let $\mathcal{L} : \mathcal{C} \times \mathcal{C} \times \mathcal{C} \rightarrow Set$ be the contra-contra-co trifunctor which is defined on objects by

$$\begin{array}{llll} \mathcal{L}: & \mathfrak{O} \times \mathfrak{O} \times \mathfrak{O} & \rightarrow & \mathfrak{O}_{Set} \\ & (A, B, C) & \mapsto & \mathfrak{M}_{\mathcal{C}}(A \otimes B, C) \end{array}$$

---

[11] Compare Remark 145 and Definition 147.

[12] This is *not* the version traditionally used in category theory. Instead, it is an application of the adjunction-with-parameters theorem from MCLANE [Mac88] to the usual form of adjunction statements on closed categories.

and on morphisms by

$$\begin{array}{llll}\mathcal{L}: & \mathfrak{M}_{\mathcal{C}}(A',A) \times \mathfrak{M}_{\mathcal{C}}(B',B) \times \mathfrak{M}_{\mathcal{C}}(C,C') & \rightarrow & \mathfrak{M}_{Set}(\mathfrak{M}_{\mathcal{C}}(A \otimes B, C),\ \mathfrak{M}_{\mathcal{C}}(A' \otimes B', C')) \\ & (f,g,h) & \mapsto & [u \mapsto h \circ u \circ (f \otimes g)]\end{array}$$

Let $\mathcal{R} : \mathcal{C} \times \mathcal{C} \times \mathcal{C} \rightarrow Set$ be the contra-contra-co trifunctor which is defined on objects by

$$\begin{array}{llll}\mathcal{R}: & \mathfrak{O} \times \mathfrak{O} \times \mathfrak{O} & \rightarrow & \mathfrak{O}_{Set} \\ & (A,B,C) & \mapsto & \mathfrak{M}_{\mathcal{C}}(A, B \Rightarrow C)\end{array}$$

and on morphisms by

$$\begin{array}{llll}\mathcal{R}: & \mathfrak{M}_{\mathcal{C}}(A',A) \times \mathfrak{M}_{\mathcal{C}}(B',B) \times \mathfrak{M}_{\mathcal{C}}(C,C') & \rightarrow & \mathfrak{M}_{Set}(\mathfrak{M}_{\mathcal{C}}(A, B \Rightarrow C),\ \mathfrak{M}_{\mathcal{C}}(A', B' \Rightarrow C')) \\ & (f,g,h) & \mapsto & [u \mapsto (g \Rightarrow h) \circ u \circ f]\end{array}$$

Then the family $\Lambda_{A,B,C} : \mathfrak{M}_{\mathcal{C}}(A \otimes B, C) \rightarrow \mathfrak{M}_{\mathcal{C}}(A, B \Rightarrow C)$ is a natural isomorphism from the functor $\mathcal{L}$ to the functor $\mathcal{R}$. ◀

The condition that Λ is a natural transformation corresponds precisely to the naturality condition which was required for higher order transition systems: Rewriting the naturality diagram from Remark 145 yields precisely the condition that Λ is a natural transformation. The respective diagram is shown in Figure C.23.

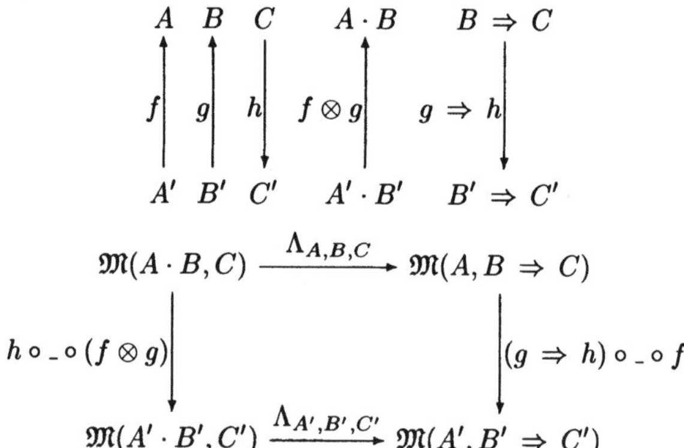

Figure C.23: Abstraction is Natural Isomorphism

C.9 Symmetric Monoidal Closed Categories

Cartesian closed categories provide an important example for higher order construction which is based on the product construction. The product construction is a very special bifunctor and there are many categories which do not contain products. Therefore, the question arises whether there exist similar higher order constructions which are based on weakened concepts of the product.

Property	Cartesian	Cartesian Closed	Monoidal	Symmetric Monoidal	Monoidal Closed	Symmetric Monoidal Closed
Decom	yes	yes	no	no	no	no
Unique	yes	yes	no	no	no	no
Term	yes	yes	no	no	no	no
Expo	no	yes	no	no	yes	yes
Comm	yes	yes	no	yes	no	yes
Assoc	yes	yes	yes	yes	yes	yes
Neu	yes	yes	yes	yes	yes	yes
1. Tri	yes	yes	no	yes	no	yes
2. Tri	yes	yes	yes	yes	yes	yes
Penta	yes	yes	yes	yes	yes	yes
Hexa	yes	yes	no	yes	no	yes

Table C.1: Generalized Product and Exponential Constructions.

372 Remark PROPERTIES OF BIFUNCTORS
Let $\mathcal{C}$ be a category and $\odot : \mathcal{C} \times \mathcal{C} \to \mathcal{C}$ a co-co bifunctor which in the following plays the role of a weakened product bifunctor $\otimes$. Category and bifunctor can satisfy a number of properties and there can be additional operators which satisfy one or more of the following properties:

Decom: For two objects A and B there exist morphisms $\pi_A : A \odot B \to A$ and $\pi_B : A \odot B \to B$.

Unique: For every object Q and morphisms $q_A : Q \to A$ and $q_B : Q \to B$ there exists a unique morphism $f : Q \to A \odot B$ such that $\pi_A \circ f = q_A$ and $\pi_B \circ f = q_B$. π_A and π_B are morphisms from property *Decom*.

Term: The category contains terminal objects.

Expo: The category contains all exponential constructions with respect to the bifunctor $\odot$. This is defined in more detail in Definition 365 and Proposition 368.

Comm: The bifunctor $\odot$ is commutative, as expressed by a natural isomorphism $\mathbf{S}$ in Theorem 357.

Assoc: The bifunctor $\odot$ is associative, as expressed by a natural isomorphism $\mathbf{A}$ in Theorem 357.

Neu: There is an object T and there are natural isomorphisms $\mathbf{L}^T$ and $\mathbf{R}^T$ as in Theorem 357. Refer to the note below concerning the correct quantifier for object T.

1. Tri: The first trigonal diagram from Theorem 357 commutes.

2. Tri: The second trigonal diagram from Theorem 357 commutes.

Penta: The pentagonal diagram from Theorem 357 commutes.

Hexa: The hexagonal diagram from Theorem 357 commutes.

For a correct interpretation of the definitions in Table C.1, note the following:

A *cartesian category* is a category which contains all finite products. It satisfies the properties Decom, Unique and Term by definition. As a consequence of the definition, products and terminal objects are unique up to isomorphism. Furthermore, there exist natural isomorphisms **S**, **A** so that the properties Comm and Assoc hold. For every terminal object T there are natural isomorphisms $\mathbf{R}^T$ and $\mathbf{S}^T$, such that the property Neu holds. All these natural isomorphisms make all listed diagrams commuting. Furthermore, a *cartesian closed category* has an exponential construction as characterized by Definition 365 or by the equivalent property from Proposition 368.

A *monoidal category* consists of a category $\mathcal{C}$, a natural isomorphism **A**, a special object T, and two natural isomorphisms $\mathbf{L}^T$ and $\mathbf{R}^T$, such that the properties Assoc, Neu, 2. Tri and Penta hold. In contrast to the cartesian category where only the existence of the respective objects is required, the natural isomorphisms and the neutral element are part of the definition of a monoidal category. A category $\mathcal{C}$ therefore may be furnished with the structure of a monoidal category in different ways, whereas a cartesian category carries a unique structure of a cartesian category, at least unique up to isomorphism.

A *symmetric monoidal category* additionally contains a natural isomorphism **S** and satisfies Comm, 1. Tri and Hexa.

A *monoidal closed category* additionally contains an exponential construction which is characterized by Definition 365 or by Proposition 368 which also holds in monoidal categories.

A *symmetric monoidal closed category* shares the properties of the latter two categories. ◀

The following important theorem is shown in [Mac88]:

373 Theorem COHERENCE FOR SYMMETRIC MONOIDAL CAT.
Every diagram in a symmetric monoidal category which is constructed from the components of the canonical isomorphisms and by using the product functor $\odot$ commutes. ◀

Unfortunately[13], an analogous theorem does not hold in symmetric monoidal closed categories:

374 Theorem COHERENCE FOR SYMMETRIC MONOIDAL CLOSED CAT.
Not every diagram in a symmetric monoidal closed category which is constructed from the components of the canonical isomorphisms, the evaluation morphism, the abstraction and by using the product functor $\odot$ and the exponential functor $\Rightarrow$ commutes. ◀

However, the coherence theorem *does hold* for symmetric monoidal closed categories if some small modifications are made in its statement. Compare [Jay90] for the algebraically involved details.

[13]For the importance of coherence theorems for process specification, compare Section 4.4.

Symbols

General

iff	As usual, synonymous for "if and only if"
$\blacktriangleleft$	End of a paragraph

Introduction

$*$	Operator of sequential composition
$+$	Operator of non-deterministic choice
$\|$	Operator of interleaving parallelism
δ	Deadlock in process algebras
$\longrightarrow$	Transition relation

Transition Systems

$(S, T, \longrightarrow)$	Ordinary transition system
$X \xrightarrow{t} Y$	Transition, state change
$\longrightarrow$	Transition relation
SEQ	Sequential constructor in OCCAM
PAR	Parallel constructor in OCCAM
$*$	Sequential composition in process algebra
$\|$	Parallel composition in process algebra
$+$	Choice operator in process algebra
$\succ$	Order relation on transitions
$(S, T, \longrightarrow, L, \mu)$	Labeled ordinary transition system
L	Set of labels
μ	Labeling function
$(S, T, \longrightarrow^*)$	Dual transition system
$\gg$	Relation of immediate causality
$\succeq$	Relation of causality

$(S_b, T_b, \longrightarrow_b, f_S, f_T)$	Behaviour
f_S, f_T	Observation functions
$\trianglerighteq$	Relation of extension of behaviours
$\mathbb{B}(\mathcal{P}$	Quotient of all behaviours of process $\mathcal{P}$
$(\mathbb{B}(\mathcal{P}), \trianglerighteq)$	Branching structure of process $\mathcal{P}$
$\circ$	Sequential composition
id_X	Identity transition of state X
$(S, T, \longrightarrow, \circ)$	Sequential transition system
$(S, \mathfrak{T}, \circ)$	Sequential transition system, alternate version
$\mathfrak{T}(X, Y)$	Set of transitions from state X to state Y
$\square$	Choice operator in guarded command language
$d_{\mathcal{B}}$	Fairness delay of behaviour $\mathcal{B}$

Distributed Transition Systems

$(S^\circ, \cdot, \varepsilon)$	Freely generated monoid of composite states
$(S, T, \longrightarrow)$	Distributed transition system
$(S, T, \longrightarrow, L, \mu)$	Labeled distributed transition system
$ind(X)$	Index of a state X
$rk(X)$	Rank of a state X
$def(X)$	Defect of a state X
$\trianglerighteq$	Relation of extension of behaviours
$\vec{\triangleright}$	Relation of immediate extension of behaviours
$\mathfrak{M}_{\mathcal{B}}$	Set of maximal cuts of behaviour $\mathcal{B}$
$\geqslant$	Cut order on the maximal cuts
T°	Bag of transitions
$\omega(t, X)$	Successor function of a process
$\mathcal{K}(t, X)$	Conflict set of transition t in state X
(t_1, t_2, X)	Confusion
$\circ$	Sequential composition
$\odot$	Parallel composition
$(S, \mathfrak{T}, \circ, \odot)$	First order transition system

Higher Order Transition Systems

$h(n, \bullet)$	Notation of partial evaluation
$ev_{B,C}$	Evaluation transition
$\Lambda_{A,B,C}$	Abstraction function
$\twoheadrightarrow$	Exponential operator on states
$\rightarrowtail$	Exponential operator on transitions
$(\hat{S}, \cdot, \twoheadrightarrow, \varepsilon)$	Algebra of higher order states
$(S, \mathfrak{T}, \circ, \odot, \twoheadrightarrow, \Lambda)$	Higher order transition system
$pl_{A,B}$	Placeholder transition

Process Specification Formalisms

$\models$	Model relation
$\vdash^1$	Labeled deduction
$\models^1$	Labeled model relation
$\mathcal{I}$	Interpretation function for states
$\mathcal{J}$	Interpretation function for transitions

Examples

module `<module name>` **is**	Begin of a module definition
endmodule	End of a module definition
sorts	Definition of sorts of an abstract data type
operators	Definition of the operator symbols
states	Definition state sorts
init	Specification of the initial state
variables	Definition of variable types
imports	Import of modules
daxioms	Data axioms
paxioms	Process axioms
if then	Conditional equation

Linear Logic

Most symbols which are used in linear logic are introduced in Tables B.1, B.2, B.3, and B.6.

$\vdash$	Deduction relation, various kinds, compare Table B.1
$\models$	Model relation
$\mathcal{L}$	Language
$\mathcal{M}$	Set of models
$\Vdash$	Consequence relation
$\mathcal{L}^*$	Set of sequences over the language $\mathcal{L}$
$\simeq$	Relation of equivalent linear logic formulae
$\#_c$	Number of occurrences of a symbol

Category Theory

$(\mathfrak{O}, \mathfrak{M}, \circ)$	Category
$\mathfrak{O}$	Class of objects of a category
$\mathfrak{M}$	Class(es) of morphisms of a category
$\mathfrak{M}(A, B)$	Class of morphisms of a category from A to object B

$f : A \to B$	Morphism from object A to object B
Set	Category of sets
$X \otimes Y$	Product construction
π_i	Projections of product construction
$< q_i >_{i \in I}$	Uniquely defined function in product construction
$X \oplus Y$	Coproduct construction
κ_i	Inclusions of coproduct construction
$[q_i]_{i \in I}$	Uniquely defined function in coproduct construction
$\mathcal{C}^*$	Dual category of the category $\mathcal{C}$
$\mathcal{F} : \mathcal{C} \to \mathcal{D}$	Functor from category $\mathcal{C}$ to category $\mathcal{D}$
$\mathbf{Fun}(\mathcal{C}_1, \mathcal{C}_2$	Class of all covariant functors from $\mathcal{C}_1$ to $\mathcal{C}_2$
$\mathbf{Fun}^*(\mathcal{C}_1, \mathcal{C}_2$	Class of all contravariant functors from $\mathcal{C}_1$ to $\mathcal{C}_2$
$\mathcal{I}_\mathcal{C}$	Identity functor of category $\mathcal{C}$
Cat	Category of small categories
$\circ$	Composition of functors
$\mathcal{F}^+$	Functor, dualized on its domain category
$\mathbf{T} : \mathcal{F}_1 \to \mathcal{F}_2$	Natural transformation from functor $\mathcal{F}_1$ to functor $\mathcal{F}_2$
$\mathbf{Nat}(\mathcal{F}_1, \mathcal{F}_2$	Class of all natural transformations from $\mathcal{F}_1$ to $\mathcal{F}_2$
$\mathbf{id}_\mathcal{F}$	Identity natural transformation of functor $\mathcal{F}$
$Funct(\mathcal{C}, \mathcal{D})$	Covariant functor category from $\mathcal{C}$ to $\mathcal{D}$
$Funct^*(\mathcal{C}, \mathcal{D})$	Contravariant functor category from $\mathcal{C}$ to $\mathcal{D}$
$\mathbf{S}$	Natural isomorphism for commutativity
$\mathbf{A}$	Natural isomorphism for associativity
$\mathbf{L}^T$, $\mathbf{R}^T$	Natural isomorphisms for neutrality
$Hom(X)$	Covariant hom-functor of object X
$Hom^*(X)$	Contravariant hom-functor of object X
Y	Contravariant YONEDA functor
J	Covariant YONEDA functor
$ev_{B,C}$	Evaluation morphism
$\Lambda_{A,B,C}$	Abstraction morphism
$B \Rightarrow C$	Exponential object
$\odot$	Bifunctor

Bibliography

[Abr90a] S. Abramsky. Computational Interpretations of Linear Logic. Research Report DOC 90/20, Imperial College, London, 1990.

[Abr90b] V. Abrusci. Sequent Calculus for Intuitionistic Linear Propositional Logic. In P. P. Petkov, editor, *Mathematical Logic*, pages 223–242, New York, London, 1990. Plenum Press.

[Abr91] S. Abramsky. Domain Theory in Logical Form. *Annals of Pure and Applied Logic*, 51:1–77, 1991.

[AFL90] P. C. Attie, J. R. Forman, and E. Levy. On Fairness as an Abstraction for the Design of Distributed Systems. In *10th Int. Conf. on Distributed Computing Systems*, pages 150–157. IEEE Computer Society Press, 1990.

[AG93] G. Attardi and M. Gaspari. Multilanguage Interoperability. Technical Report UBLCS-93-18, Laboratory for Computer Science, University of Bologna, 1993. Available by anonymous ftp from `ftp.cs.unibo.it`.

[AJ92] S. Abramsky and R. Jagadeesan. New Foundations for the Geometry of Interaction. Extended abstract; accepted for publication in LICS 92, 1992.

[AL91] A. Asperti and G. Longo. *Categories, Types, and Structures.* Foundations of Computing Series. MIT Press, 1991.

[Amb91] S. Ambler. *First Order Linear Logic in Symmetric Monoidal Closed Categories.* PhD thesis, University of Edinburgh, 1991.

[AO91] R. Apt and E. Olderog. *Verification of Sequential and Concurrent Programs.* Texts and Monographs on Computer Science. Springer, 1991.

[AP91] J.-M. Andreoli and R. Pareschi. Logic Programming with Sequent Systems – A Linear Logic Approach. In P. Schroeder-Heister, editor, *Extensions of Logic Programming*, number 475 in Lecture Notes in Artificial Intelligence, pages 1–30, 1991.

[Apt81] K. Apt. Ten Years of Hoare's Logic. *ACM Transactions on Programming Languages and Systems*, 3(4):431–483, 1981.

[AS87] B. Alpern and F. Schneider. Recognizing Safety and Liveness. *Distributed Computing*, 2:117–126, 1987.

[Asp90] A. Asperti. *Categorical Topics in Computer Science.* PhD thesis, University of Pisa, 1990.

[Avr88] A. Avron. The Semantics and Proof Theory of Linear Logic. *Theoretical Computer Science*, 57:161–184, 1988.

[Avr91] A. Avron. Simple Consequence Relations. *Information and Computation*, 92:105–139, 1991.

[Bar84] H. Barendregt. *The Lambda Calculus: Its Syntax and Semantics.* Studies in Logic and the Foundations of Mathematics. North-Holland, 1984.

[Bau90] B. Baumgarten. PETRI-*Netze.* BI Wissenschaftsverlag, 1990.

[BB90a] J. Baeten and J. Bergstra. Real Space Process Algebra. Technical Report P9005, Programming Research Group, University of Amsterdam, 1990.

[BB90b] J. Baeten and J. Bergstra. Real Time Process Algebra. Technical Report P8916b, Programming Research Group, University of Amsterdam, 1990.

[BB91] J. Baeten and J. Bergstra. A Survey of Axiom Systems for Process Algebras. Technical Report P9111, Programming Research Group, University of Amsterdam, 1991.

[BB92] G. Berry and G. Boudol. The Chemical Abstract Machine. *Theoretical Computer Science*, 96:217–248, 1992.

[BB93] J. Baeten and J. Bergstra. Non-Interleaving Process Algebra. Technical Report P9326, Programming Research Group, University of Amsterdam, 1993.

[BB94] J. Baeten and J. Bergstra. Graph Isomorphism Models for Non-Interleaving Process Algebra. Technical Report P9401, Programming Research Group, University of Amsterdam, 1994.

[BBK87] J. Baeten, J. Bergstra, and J. Klop. On the Consistency of Koomen's Fair Abstraction Rule. *TCS*, 51:129–176, 1987.

[BC88] G. Boudol and I. Castellani. Concurrency and Atomicity. *Theoretical Computer Science*, 59:22–84, 1988.

[BEM90] B. Bütler, R. Esser, and R. Mattmann. A Distributed Simulator for Higher Order PETRI Nets. In G. Rozenberg, editor, *Advances in* PETRI *Nets*, number 483 in Lecture Notes in Computer Science. Springer, 1990.

[Ber] J. Bergstra. Personal Communication.

[Ber89] J. Bergstra. Algebra of States and Transitions. Technical Report P8909, Programming Research Group, University of Amsterdam, 1989.

[BG82] R. Burstall and J. Goguen. Algebras, Theories and Freeness: An Introduction for Computer Scientists. In M. Broy and G. Schmidt, editors, *Theoretical Foundations of Programming Methodology*, pages 329–349. Reidel, 1982.

[BG93] M. Bezem and J. Groote. A Correctness Proof of a One-bit Sliding Window Protocol in μCRL. Logic Group Preprint Series 99, Department of Philosophy, University of Utrecht, 1993.

[BHG87] P. Bernstein, V. Hadzilacos, and N. Goodman. *Concurrency Control and Recovery in Database Systems.* Addison-Wesley, 1987.

[BHK89] J. Bergstra, J. Heering, and P. Klint. *Algebraic Specification.* Addison-Wesley, 1989.

[BHK90] J. Bergstra, J. Heering, and P. Klint. Module Algebra. *Journal of the ACM*, 37(2):335–372, 1990.

[BK86] J. Bergstra and J. Klop. Algebra of Communicating Processes. In J. deBakker, editor, *Mathematics and Computer Science*, volume 1 of *CWI Monographs*, pages 89–138. North Holland, 1986.

[BK88] J. Bergstra and J. Klop. Process Theory Based on Bisimulation Semantics. Technical Report P8824, Programming Research Group, University of Amsterdam, 1988.

[BK89] J. Bergstra and J. Klop. A Universal Axiom System for Process Specification. In M. Wirsing and J.A. Bergstra, editors, *Algebraic Methods: Theory, Tools and Applications*, volume 394 of *LCS*, pages 447–463. Springer, 1989.

[BK90] J. Bergstra and J. Klop. An Introduction to Process Algebra. In J. Baeten, editor, *Applications of Process Algebra*, number 17 in Cambridge Tracts in Theoretical Computer Science, pages 1–22. Cambridge University Press, 1990.

[BL70] G. Birkhoff and J. Lipson. Heterogeneous Algebras. *J. of Combinatorial Theory*, 8:115–133, 1970.

[BL94] M. Beaudouin-Lafon. *Object-Oriented Programming.* Chapman, 1994.

[BM77] J. Bell and M. Machover. *A Course in Mathematical Logic.* North-Holland, 1977.

[BMC64] C. Baier and M. Majster-Cederbaum. The Connection between an Event Structure Semantics and an Operational Semantics for TCSP. *Acta Informatica*, 31:81–104, 1964.

[Bor77] L. Borkowski. *Formale Logik.* Beck, 1977.

[BR83] S. Brookes and W. Rounds. Behavioural Equivalence Relations Induced by Programming Logics. In J. Diaz, editor, *Automata, Languages, and Programming 10th Colloquium Barcelona, Spain, July 1983*, number 154 in Lecture Notes in Computer Science, pages 97–108. Springer, 1983.

[Bro86] M. Broy. A Theory for Non-Determinism, Parallelism, Communication and Concurrency. *Theoretical Computer Science*, 45:1–61, 1986.

[BS92] G. Bellin and P. Scott. On the π Calculus and Linear Logic. Available by anonymous ftp from `theory.doc.ic.ac.uk`, 1992.

[BT90] J. Bergstra and J. Tucker. The Inescapable Stack: An Exercise in Algebraic Specification with Total Functions. Technical Report P8804b, Programming Research Group, University of Amsterdam, 1990.

[BW90a] J. Baeten and W. Weijland. *Process Algebra.* Number 18 in Cambridge Tracts in Theoretical Computer Science. Cambridge University Press, 1990.

[BW90b] M. Barr and C. Wells. *Category Theory for Computing Science.* Prentice-Hall International, 1990.

[Cap90] C. Cap. Distributed Systems with Data Replication. Technical Report IFI-TR-90.11, University of Zurich, Department of Computer Science, 1990.

[Cap93] C. Cap. *Theoretische Grundlagen der Informatik.* Springer, 1993.

[Cap94] C. Cap. Massive Parallelism with Workstation Clusters – Challenge or Nonsense? In *Proceedings of the High Performance Computing and Networking Conference HPCN94 Europe*, number 797 in Lecture Notes in Computer Science, pages 45–52. Springer, 1994.

[CBDM89] Y. Choquet-Bruhat and Y. DeWitt-Morette. *Analysis, Manifolds, and Physics.* North-Holland, 1989.

[Cha88] G. Chaitin. *Algorithmic Information Theory.* Number 1 in Cambridge Tracts in Theoretical Computer Science. Cambridge University Press, 1988.

[CL85] K. Chandy and L. Lamport. Distributed Snapshots: Determining Global States of Distributed Systems. *ACM Transactions Comput. Syst.*, 3(1):63–75, 1985.

[Coh87] D. Cohen. *Computability and Logic.* Ellis Howrood, 1987.

[Coh89] D. Cohen. *An Introduction to Hilbert Space and Quantum Logic.* Springer, 1989.

[Cou90a] B. Courcelle. Graph Rewriting: An Algebraic and Logic Approach. In J. van Leeuwen, editor, *Handbook of Theoretical Computer Science*, volume B, pages 193–242. Elsevier, 1990.

[Cou90b] G. Cousineau. The Categorical Abstract Machine. In G. Huet, editor, *Logical Foundations of Functional Programming*, University of Texas at Austin Year of Programming Series, pages 25–46. Addison-Wesley, 1990.

[CPS91] R. Cleaveland, J. Parrow, and B. Steffen. The Concurrency Workbench: A Semantics Based Tool for the Verification of Concurrent Systems. Aachener Informatikberichte 91–24, Department of Computer Science, University of Aachen, 1991.

[CS] C. Cap and V. Strumpen. Massively Parallel Computing in the Internet, Entry to the SuParCup'93. Unpublished Preprint, Department of Computer Science, University of Zurich, 1993. This entry won the first prize of the SuParCup'93.

[CS93] C. Cap and V. Strumpen. Efficient Parallel Computing in Distributed Workstation Environments. *Parallel Computing*, 19:1221–1234, 1993.

[CW85] L. Cardelli and P. Wegner. On Understanding Types, Data Abstraction and Polymorphism. *Computing Surveys*, 17(4):471–521, 1985.

[DG93] P. David and C. Guidal. Development of a Fault Tolerant Computer System for the Hermes Space Shuttle. In *Proceedings of the Twenty-third International Symposium on Fault-tolerant Computing FTCS23*, pages 641–646. IEEE Computer Society, 1993.

[Dij76] E. Dijkstra. *A Discipline of Programming*. Prentice-Hall, 1976.

[DP90] B. Davey and H. Priestley. *Introduction to Lattices and Order*. Cambridge Mathematical Textbooks. Cambridge University Press, 1990.

[DR89] V. Danos and L. Regnier. The Structure of Multiplicatives. *Archive for Mathematical Logic*, 28:181–203, 1989.

[Dro89] M. Droste. Event Structures and Domains. *Theoretical Computer Science*, 68:37–47, 1989.

[EFT86] H. Ebbinghaus, J. Flum, and W. Thomas. *Einführung in die mathematische Logik*. Wissenschaftliche Buchgesellschaft, Darmstadt, 1986.

[EGJMW91] M. Elvang-Gøransson, H. Jervell, E. Monteiro, and A. Waaler. Propositional Linear Logic: Wiring Semantics. Manuscript; submitted for publication to Information Processing Letters, 1991.

[EGLer] H. Ehrich, M. Gogolla, and U. Lipeck. *Algebraische Spezifikation abstrakter Datentypen*. Leitfäden und Monographien der Informatik. 1989, Teubner.

[EL92] E. Engeler and P. Läuchli. *Berechnungstheorie für Informatiker*. Teubner, 1992.

[EM90] H. Ehrig and B. Mahr. *Fundamentals of Algebraic Specification*. EATCS Monographs on Theoretical Computer Science. Springer, 1990.

[Eme90] E. Emerson. Temporal and Modal Logic. In J. van Leeuwen, editor, *Handbook of Theoretical Computer Science*, volume B, pages 995–1072. Elsevier, 1990.

[Eng86] E. Engeler. Modelling of Cooperative Processes. Technical Report 86-06, Eidgenössische Technische Hochschule, 1986.

[Eng93] E. Engeler. Sketch of a New Discipline of Modeling. In *Algorithmic Properties of Structures*, pages 231–241. World Scientific, 1993.

[EW90] U. Engberg and G. Winskel. PETRI Nets as Models of Linear Logic. In A. Arnold, editor, *CAAP '90*, number 431 in Lecture Notes in Computer Science, pages 147–161, 1990.

[EW94] U. Engberg and G. Winskel. Linear Logic on PETRI Nets. Technical Report RS-94-3, Department of Computer Science, University of Aarhus, 1994.

[Fek93] A. Fekete. Formal Models of Communication Services: A Case Study. *IEEE Computer*, pages 37–47, 1993.

[FH80] A. Field and P. Harrison. *Functional Programming.* Addison-Wesley, 1980.

[For] L. Fornera. Structural Analysis of SPECS PETRI Nets. Unpublished preprint, Landis & Gyr Corporate Research.

[Fra86] N. Francez. *Fairness.* Texts and Monographs in Computer Science. Springer, 1986.

[Fuc92] N. Fuchs. Hoare Logic, Executable Specifications and Logic Programs. Technical Report IFI-TR 92.02, Department of Computer Science, University of Zurich, 1992.

[Gal87] J. Gallier. *Logic for Computer Science.* Wiley, 1987.

[GCR94] S. Gerhart, D. Craigen, and T. Ralston. Experience with Formal Methods in Critical Systems. *IEEE Software*, pages 21–39, 1994.

[GdQ90] D. Gabbay and R. de Queiroz. Extending the Curry-Howard Interpretation to Linear, Relevant, and other Resource Logics. Manuscript, 1990.

[Gen91] H. Genrich. Predicate / Transition Nets. In K. Jensen and G. Rozenberg, editors, *High-level* PETRI *Nets*, pages 3–43. Springer, 1991.

[GG89] C. Gunter and V. Gehlot. Nets as Tensor Theories. In G. De Michelis, editor, *Proceedings of the 10-th International Conference on Application and Theory of* PETRI *Nets*, pages 174–191, 1989.

[Gir87a] J.-Y. Girard. Linear Logic. *Theoretical Computer Science*, 50:1–102, 1987.

[Gir87b] J.-Y. Girard. Linear Logic and Parallelism. In M. Venturini Zilli, editor, *Mathematical Models for the Semantics of Parallelism*, number 280 in Lecture Notes in Computer Science, pages 166–182. Springer, 1987.

[Gir89a] J.-Y. Girard. Geometry of Interaction 1: Interpretation of System F. In R. Ferro, C. Bonotto, S. Valentini, and A. Zanardo, editors, *Logic Colloquium '88*, pages 221–260, 1989.

[Gir89b] J.-Y. Girard. Towards a Geometry of Interaction. In J. Gray and A. Scedrov, editors, *Categories in Computer Science and Logic*, pages 69–108. American Mathematical Society, 1989.

[Gir90] J.-Y. Girard. Geometry of Interaction 2: Deadlock-free Algorithms. In P. Martin-Löf and G. Mints, editors, *COLOG-88*, number 417 in Lecture Notes in Computer Science, pages 76–93, 1990.

[GL91] U. Goltz and R. Loogen. Modelling Nondeterministic Concurrent Processes with Event Structures. *Fundamenta Informatica*, 14:39–73, 1991.

[GLT88] J.-Y. Girard, Y. Lafont, and P. Taylor. *Proofs and Types.* Number 7 in Cambridge Tracts in Theoretical Computer Science. Cambridge University Press, 1988.

[GM89] J. Goguen and J. Meseguer. Order-Sorted Algebra. Technical Report SRI-CSL-89-10, SRI International, 1989.

[Gog91] J. Goguen. A Categorical Manifesto. *Mathematical Structures in Computer Science*, 1:49–67, 1991.

[Gol84] R. Goldblatt. *Topoi – The Categorical Analysis of Logic.* Number 98 in Studies in Logic and the Foundations of Mathematics. North-Holland, 1984.

[Gom92] C. Gomard. A Self-Applicable Partial Evaluator for the Lambda Calculus: Correctness and Pragmatics. *ACM Transactions on Programming Languages and Systems*, 14(2):147–172, 1992.

[Goo92] G. Goos. Programmiertechnik zwischen Wissenschaft und industrieller Praxis. Überarbeitete Fassung eines Vortrages, gehalten am 17. Dezember 1992 bei IBM Böblingen zum 60. Geburtstag von Prof. Endres, 1992.

[Gor88] M. Gordon. *Programming Language Theory and Its Implementation: Applicative and Imperative Paradigms.* Prentice-Hall International, 1988.

[GP90] J. Groote and A. Ponse. The Syntax and Semantics of μCRL. Technical Report CS-R9076, CWI Amsterdam, 1990.

[GP91] J. Groote and A. Ponse. Proof Theory for μCRL. Technical Report CS-R9138, CWI Amsterdam, 1991.

[Gri89] D. Gries. *The Science of Programming.* Texts and Monographs in Computer Science. Springer, 1989.

[GTW78] J. Goguen, J. Thatcher, and E. Wagner. An Initial Algebra Approach to the Specification, Correctness, and Implementation of Abstract Data Types. In R. Yeh, editor, *Current Trends in Programming Methodology*, pages 80–149. Prentice-Hall International, 1978.

[GWMF93] J. Goguen, T. Winkler, J. Meseguer, and K. Futatsugi. Introducing OBJ. Unpublished preprint, Oxford University Computing Laboratory, 1993.

[Ham78] A. Hamilton. *Logic for Mathematicians.* Cambridge University Press, 1978.

[Hei92] M. Heiner. PETRI Net Based Software Validation, Prospects and Limitations. Technical Report TR-92-022, International Computer Science Institute, Berkeley, California, 1992.

[Hen88] M. Hennessy. *Algebraic Theory of Processes.* MIT Press, 1988.

[Hen91] R. Hennicker. Observational Implementation of Algebraic Specifications. *Acta Informatica*, 28:187–230, 1991.

[Hen92] M. Hennessy. Timed Process Algebras: A Tutorial. Lecture notes, International Summer School on Process Design Calculi, Marktoberdorf, 1992.

[Hes90] W. Hesselink. Axioms and Models of Linear Logic. *Formal Aspects of Computing*, 2:139–166, 1990.

[HG93] C. Hennebert and G. Guiho. SACEM: A Fault Tolerant System for Train Speed Control. In *Proceedings of the Twenty-third International Symposium on Fault-tolerant Computing FTCS23*, pages 624–628. IEEE Computer Society, 1993.

[HI93] M. Hennessy and A. Ingolfsdottir. Communicating Processes with Value-Passing and Assignments. *Formal Aspects of Computing*, 3:1–36, 1993.

[HL] M. Hennessy and H. Lin. Proof Systems for Message-Passing Process Algebras. Unpublished preprint, University of Sussex.

[HL93] I. Horebeek and J. Lewi. *Algebraic Specifications in Software Engineering.* Springer, 1993.

[HO80] G. Huet and D. Oppen. *Equations and Rewrite Rules, a Survey.* Academic Press, 1980.

[Hoa85] C. Hoare. *Communicating Sequential Processes.* Prentice-Hall International, 1985.

[Hog89] D. Hogrefe. *Estelle, LOTOS und SDL.* Springer, 1989.

[HU79] J. Hopcroft and J. Ullman. *Introduction to Automata Theory, Languages and Computation.* Addison-Wesley, 1979.

[Hue90] G. Huet. Cartesian Closed Categories and Lambda Calculus. In G. Huet, editor, *Logical Foundations of Functional Programming*, University of Texas at Austin Year of Programming Series, pages 7–24. Addison-Wesley, 1990.

[Hüt94] H. Hüttel. Undecidable Equivalences for Basic Parallel Processes. In M. Hagiya and J. Michell, editors, *Theoretical Aspects of Computer Science*, number 789 in Lecture Notes in Computer Science, pages 454–464. Springer, 1994.

[Inm84] Inmos. *Occam Programming Manual.* Prentice-Hall International, 1984.

[Inm88] Inmos. *Transputer Reference Manual.* Prentice-Hall, 1988.

[Jan88] M. Jantzen. *Confluent String Rewriting.* Number 14 in EATCS Monographs on Theoretical Computer Science. Springer, 1988.

[Jay90] C. Jay. The Structure of Free Closed Categories. *Journal of Pure and Applied Algebra*, 66:271–285, 1990.

[Jen91] K. Jensen. Coloured PETRI Nets: A High Level Language for System Design. In K. Jensen and G. Rozenberg, editors, *High-level* PETRI *Nets*, pages 44–122. Springer, 1991.

[Kan91a] M. Kanovich. The Horn Fragment of Linear Logic is NP-complete. ITLI Prepublication Series X-91-14, University of Amsterdam, 1991.

[Kan91b] M. Kanovich. The Multiplicative Fragment of Linear Logic is NP-complete. ITLI Prepublication Series X-91-13, University of Amsterdam, 1991.

[Kan94] M. Kanovich. PETRI Nets, Horn Programs, Linear Logic and Vector Games. In M. Hagiya and J. Michell, editors, *Theoretical Aspects of Computer Science*, number 789 in Lecture Notes in Computer Science, pages 642–666. Springer, 1994.

[Kel64] M. Kelly. On MacLane's Conditions for Coherence of Natural Associativities, Commutativities, etc. *Journal of Algebra*, 1:397–402, 1964.

[KK94] M. Künzli and R. Künzli. Konzeption und Implementierung der Sprache Petril. Semesterarbeit, Department of Computer Science, University of Zurich, 1994.

[KL94] J. Knight and B. Littlewood. Critical Task of Writing Dependable Software. *IEEE Software*, pages 16–20, 1994.

[KM84] M. Kelly and S. MacLane. Coherence in Closed Categories. *Journal of Pure and Applied Algebra*, 1:97–140, 1984.

[Kni89] K. Knight. Unification: A Multidisciplinary Survey. *ACM Computing Surveys*, 21(1):93–124, 1989.

[Krö87] F. Kröger. *Temporal Logic of Programs*. EATCS Monographs on Theoretical Computer Science. Springer, 1987.

[Laf88] Y. Lafont. The Linear Abstract Machine. *Theoretical Computer Science*, 59:157–180, 1988. Some corrections in volume 62(1988), pp. 327-328.

[Laf90] Y. Lafont. Interaction Nets. In *Seventeenth Annual ACM Symposium on Principles of Programming Languages*, pages 95–108. Association for Computing Machinery, 1990.

[Lam78] L. Lamport. Time, Clocks and the Ordering of Events in a Distributed System. *Communications of the ACM*, 21(7):558–565, 1978.

[LDP90] R. Lai, T. Dillon, and K. Parker. Verification Results for ISO FTAM Basic Protocol. In E. Brinksma, G. Scollo, and C. Vissers, editors, *Protocol Specification, Testing and Verification*, volume IX of *Proceedings on the Ninth Symposium on Protocol Specification, Testing and Verification, Enschede, 1989*, pages 223–234. North Holland, 1990.

[Led91] E. Lederer. *Two-Stage Programming of Parallel Systems: Model, Notation and Examples*. PhD thesis, University of Zurich, 1991.

[Lin92] P. Lincoln. *Computational Aspects of Linear Logic*. PhD thesis, Stanford University, 1992.

[Llo87] J. Lloyd. *Foundations of Logic Programming*. Springer, 1987.

[LMSS90] P. Lincoln, J. Mitchell, A. Scedrov, and N. Shankar. Decision Problems for Propositional Linear Logic. Technical Report SRI-CSL-90-08, SRI International, Computer Science Laboratory, 1990.

[LMSS92] P. Lincoln, J. Mitchell, A. Scedrov, and N. Shankar. Decision Problems for Propositional Linear Logic. *Annals Pure Appl. Logic*, 56:239–311, 1992. Special Volume dedicated to the memory of John Myhill.

[LS86] J. Lambek and P. Scott. *Introduction to Higher Order Categorical Logic*. Number 7 in Cambridge Studies in Advanced Mathematics. Cambridge University Press, 1986.

[LS91] Y. Lafont and T. Streicher. Games Semantics for Linear Logic. In *Proceedings of the Sixth Annual Symposium on Logic in Computer Science*, pages 43–50. IEEE Computer Society Press, 1991.

[LS92] P. Lincoln and A. Scedrov. First Order Linear Logic Without Modalities is NEXPTIME-Hard. Manuscript, 1992. Available by anonymous ftp from `ftp.cis.upenn.edu`, file pub/papers/scedrov/mall1.dvi.

[LS94] P. Lincoln and N. Shankar. Proof Search in First-order Linear Logic and Other Cut-free Sequent Calculi. Technical report, SRI International Computer Science Laboratory, Menlo Park, 1994. Available by anonymous ftp from `ftp.csl.sri.com`.

[LW92] P. Lincoln and T. Winkler. Constant-Only Multiplicative Linear Logic is NP-Complete. Manuscript, 1992. Available by anonymous ftp from `ftp.csl.sri.com`, file pub/lincoln/comult-npc.dvi.

[MA86] E. Manes and M. Arbib. *Algebraic Approaches to Program Semantics*. Texts and Monographs in Computer Science. Springer, 1986.

[Mac88] S. MacLane. *Categories for the Working Mathematician*. Number 5 in Springer Graduate Texts in Mathematics. Springer, 1988.

[Mäh93] M. Mähler. *Fairness in verteilten Systemen*. Nummer 246 in Fortschritt-Berichte VDI, Reihe 10. VDI Verlag, 1993.

[Man77] Y. Manin. *A Course in Mathematical Logic*. Number 53 in Springer Graduate Texts in Mathematics. Springer, 1977.

[Mau92] S. Mauw. *PSF, A Process Specification Formalism*. PhD thesis, University of Amsterdam, 1992.

[May81] E. Mayr. An Algorithm for the General PETRI Net Reachability Problem. In *Proceedings of the 13-th ACM Symp. on Theory of Computing, Milwaukee*, pages 238–246, 1981.

[McD82] D. McDermott. A Temporal Logic for Reasoning about Processes and Plans. *Cognitive Science*, 6:101–155, 1982.

[Mer92] S. Merz. *Temporal Logic as a Programming Language*. PhD thesis, Universität München, 1992.

[MG85] J. Meseguer and J. Goguen. Initiality, Induction and Computability. In *Algebraic MEthods in Semantics*. Cambridge University Press, 1985.

[Mil89] R. Milner. *Communication and Concurrency*. Prentice-Hall International, 1989.

[MM90] J. Meseguer and U. Montanari. PETRI Nets are Monoids. *Information and Computation*, 88:105–155, 1990.

[MOM89] N. Marti-Oliet and J. Meseguer. From PETRI Nets to Linear Logic. In D. Pitt, D. Rydeheard, P. Dybjer, A. Pitts, and A. Poigné, editors, *Category Theory and Computer Science*, number 389 in Lecture Notes in Computer Science, pages 313–340. Springer, 1989.

[MOM91] N. Marti-Oliet and J. Meseguer. From PETRI Nets to Linear Logic. *Mathematical Structures in Computer Science*, 1:66–101, 1991.

[MOO87] M. Maekawa, A. Oldehoeft, and R. Oldehoeft. *Operating Systems*. Benjamin-Cummings, 1987.

[MPW89a] R. Milner, J. Parrow, and D. Walker. A Calculus of Mobile Processes, Part 1. Technical Report ECS-LFCS-89-85, Department of Computer Science, University of Edinburgh, 1989.

[MPW89b] R. Milner, J. Parrow, and D. Walker. A Calculus of Mobile Processes, Part 2. Technical Report ECS-LFCS-89-86, Department of Computer Science, University of Edinburgh, 1989.

[MPW92] R. Milner, J. Parrow, and D. Walker. A Calculus of Mobile Processes, Part 1. *Information and Computation*, 100(1):1–40, 1992.

[MT92] K. Meinke and J. Tucker. Universal Algebra. In S. Abramsky, D. Gabbay, and T. Maibaum, editors, *Handbook of Logic in Computer Science*, volume 1 of *Oxford Science Publications*, pages 189–411. Clarendon Press, Oxford, 1992.

[MTW73a] C. Misner, K. Thorne, and J. Wheeler. *Gravitation*. Freeman, 1973.

[MTW73b] C. Misner, K. Thorne, and J. Wheeler. *Gravitation*. Freeman, 1973.

[Mul93] S. Mullender. *Distributed Systems*. ACM Press Frontier Series. ACM Press, 2 edition, 1993.

[MV93] S. Mauw and G. Veltink. *Algebraic Specification of Communication Protocols*. Number 36 in Cambridge Tracts in Theoretical Computer Science. Cambridge University Press, 1993.

[NPW81] M. Nielsen, G. Plotkin, and G. Winskel. PETRI-Nets, Event Structures, and Domains. *Theoretical Computer Science*, 13:85–108, 1981.

[NRT92] M. Nielsen, G. Rozenberg, and P. Thiagarajan. Elementary Transition Systems. *Theoretical Computer Science*, 96:2–33, 1992.

[Odi92] P. Odifreddi. *Classical Recursion Theory*. Number 125 in Studies in Logic and the Foundations of Mathematics. North-Holland, 1992.

[OK85] H. Ono and Y. Komori. Logics without the Contraction Rule. *Journal of Symbolic Logic*, 50(1):169–201, 1985.

[Ono90] H. Ono. Structural Rules and a Logical Hierarchy. In P. Petkov, editor, *Mathematical Logic*, pages 95–104. Plenum Press, New York, 1990.

[Par69] B. Pareigis. *Kategorien und Funktoren*. Mathematische Leitfäden. Teubner Stuttgart, 1969.

[Plo76] G. Plotkin. A Powerdomain Construction. *SIAM J. Computing*, 5(3):452–487, 1976.

[Plo83] G. Plotkin. An Operational Semantics for CSP. In D. Bjorner, editor, *Formal Description of Programming Concepts*, pages 199–225. North-Holland, 1983.

[Poi92] A. Poigne. Basic Category Theory. In S. Abramsky, D. Gabbay, and T. Maibaum, editors, *Handbook of Logic in Computer Science*, volume 1 of *Oxford Science Publications*, pages 413–640. Clarendon Press, Oxford, 1992.

[Pop84] K. Popper. *Nummer 4 in Logik der Forschung*. Die Einheit der Gesellschaftswissenschaften. Paul Siebeck, 1984.

[Pra94a] V. Pratt. Event Spaces and Their Linear Logic. Unpublished preprint, Computer Science Department, Stanford, 1994.

[Pra94b] V. Pratt. Linear Logic for Generalized Quantum Mechanics. Unpublished preprint, Computer Science Department, Stanford, available by anonymous ftp from `ftp.cs.stanford.edu`, 1994.

[Pre75] G. Preuß. *Grundbegriffe der Kategorientheorie*. Nummer 739 in BI Hochschultaschenbücher. Bibliographisches Institut BI, 1975.

[RB88] D. Rydeheard and R. Burstall. *Computational Category Theory*. Prentice-Hall International, 1988.

[Rei87] W. Reisig. *Das Verhalten verteilter Systeme*. Nummer 170 in GMD Berichte. Oldenburg, 1987.

[Rei91] W. Reisig. PETRI Nets and Algebraic Specifications. In K. Jensen and G. Rozenberg, editors, *High-level* PETRI *Nets*, pages 137–170. Springer, 1991.

[Rei92] W. Reif. Verification of Large Software Systems. In *Foundations of Software Technology and Theoretical Computer Science*, number 652 in Lecture Notes in Computer Science. Springer, 1992.

[Rei93] B. Reinwald. *Workflow-Management in verteilten Systemen: Entwurf und Betrieb geregelter arbeitsteiliger Anwendungssysteme*. Teubner, 1993.

[Ric85] L. Richter. *Betriebssysteme*. Teubner, 1985.

[Roo91] D. Roorda. *Resource Logics: Proof-theoretical Investigations*. PhD thesis, University of Amsterdam, 1991.

[SA91] A. Sperschneider and G. Antoniou. *Logic – A Foundation for Computer Science*. International Computer Science Series. Addison-Wesley, 1991.

[Sce90] A. Scedrov. A Brief Guide to Linear Logic. *Bulletin of the European Assoc. for Theoretical Computer Science*, 41:154–165, 1990.

[Sch91] H. Schellinx. Some Syntactical Observations on Linear Logic. *Journal of Logic and Computation*, 1(4):537–559, 1991.

[Seb93] R. Sebesta. *Concepts of Programming Languages*. Benjamin-Cummings, 1993.

[Sin89] M. Singhal. Deadlock Detection in Distributed Systems. *IEEE Computer*, 22(11):37–48, 1989.

[SL87] A. Shankar and S. Lam. Time-dependent Distributed Systems: Proving Safety, Liveness, and Real-time Properties. *Distributed Computing*, 2:61–79, 1987.

[Smi88] F. Smith. *Propositional Dynamic Logic of Regular Programs with Tests, Converse and Infinite Computations*. Number 10 in Carleton-Ottawa Mathematical Lecture Note Series. Department of Mathematics and Statistics, Carleton University, Ottawa, 1988.

[Smy78] M. Smyth. Power Domains. *Journal of Computer and System Sciences*, 16:23–36, 1978.

[SS78] R. Sexl and H. Schmidt. *Raum – Zeit – Relativität*. Vieweg, 1978.

[Ste93] G. Sterman. *An Introduction to Quantum Field Theory*. Cambridge University Press, 1993.

[Sti92] C. Stirling. Modal and Temporal Logic. In S. Abramsky, D. Gabbay, and T. Maibaum, editors, *Handbook of Logic in Computer Science*, volume 2 of *Oxford Science Publications*, pages 478–551. Clarendon Press, Oxford, 1992.

[Sun83] G. Sundholm. Systems of Deduction. In D. Gabbay and F. Guenther, editors, *Handbook of Philosophical Logic*, volume 1, pages 133–188. Reidel, 1983.

[TK87] S. Toueg and R. Koo. Checkpointing and Rollback-Recovery for Distributed Systems. *IEEE Transactions on Software Engineering*, 13(1):23–31, 1987.

[Tro] A. Troelstra. Personal Communication.

[Tro92] A. Troelstra. *Lectures on Linear Logic*. CSLI Lecture Notes. Center for the study of language and information, 1992.

[Tro94] A. Troelstra. Errata to the Lectures on Linear Logic. Available from the author by electronic mail at anne@fwi.uva.nl, 1994.

[Vaa89] F. Vaandrager. *Algebraic Techniques for Concurrency and their Application*. PhD thesis, University of Amsterdam, 1989.

[Vaa90] F. Vaandrager. Two Simple Protocols. In J. Baeten, editor, *Applications of Process Algebra*, number 17 in Cambridge Tracts in Theoretical Computer Science, pages 23–44. Cambridge University Press, 1990.

[vBB93] J. van Benthem and J. Bergstra. Logic of Transition Systems. Technical Report P9308, Programming Research Group, University of Amsterdam, 1993.

[vBvES93] J. van Benthem, D. van Eijck, and V. Stebletsova. Modal Logic, Transition Systems, and Processes. Technical Report CS-R9321, Amsterdam Stichting Mathematisch Centrum, 1993.

[Wad90] P. Wadler. Linear Types Can Change the World. In M. Broy and C. Jones, editors, *Programming Concepts and Methods*. North-Holland, 1990.

[Wad94] P. Wadler. A Syntax for Linear Logic. In S. Brookes et al., editors, *Mathematical Foundations of Programming Semantics*, number 802 in Lecture Notes in Computer Science, pages 513–529. Springer, 1994.

[Wat92] O. Watanabe. *Kolmogorov Complexity and Computational Complexity*. EATCS Monographs on Theoretical Computer Science. Springer, 1992.

[Wec91] W. Wechler. *Universal Algebra for Computer Scientists*. Number 25 in EATCS Monographs on Theoretical Computer Science. Springer, 1991.

[Wei89] D. Weiss. *Occam 2, Das Referenzhandbuch*. Hanser, 1989.

[Win80] G. Winskel. *Events in Computation*. PhD thesis, University of Edinburgh, 1980. Technical Report CST-10-80.

[Win87] G. Winskel. Event Structures, PETRI-Nets. In *Applications and Relationships to Other Models of Concurrency*, number 255 in Lecture Notes in Computer Science, pages 325–392. Springer, 1987.

[Win88] Glynn Winskel. An introduction to event structures. In *Linear Time, Branching Time and Partial Order in Logics and Models for Concurrency*, number 354 in Lecture Notes in Computer Science, page 364. Springer, 1988.

[Win90] J. Wing. A Specifier's Introduction to Formal Methods. *IEEE Computer*, pages 8–22, 1990.

[Wir90] M. Wirsing. Algebraic Specification. In J. van Leeuwen, editor, *Handbook of Theoretical Computer Science*, volume B, pages 675–788. Elsevier, 1990.

[Yet90] D. Yetter. Quantales and (Noncommutative) Linear Logic. *Journal of Symbolic Logic*, 55(1):41–64, 1990.

Index

Appelrath/Boles/
Claus/Wegener
Starthilfe
Informatik

Von Prof. Dr.
Hans-Jürgen Appelrath
Dipl.-Inform. **Dietrich Boles**
Universität Oldenburg
Prof. Dr. **Volker Claus**
Universität Stuttgart
und Prof. Dr. **Ingo Wegener**
Universität Dortmund

1998. 158 Seiten mit 34 Bildern.
16,2 x 22,9 cm.
Kart. DM 24,80
ÖS 181,– / SFr 22,–
ISBN 3-519-00241-8

Die Teubner-Starthilfen erleichtern den Übergang vom Abitur ins Studium. Für die Informatik stellt sich dabei die Frage: Wie hilft man beim Einstieg in dieses Fach, über das – etwa im Gegensatz zu Mathematik und Physik – vor allem bei Schülern und Schülerinnen oftmals falsche Vorstellungen herrschen? Die Autoren der »Starthilfe Informatik« wählen den Weg, dem Leser zunächst die zentralen Begriffe »Algorithmus« und »Datenstrukturen« bzgl. Darstellungsformen, Effizienz und Programmiermethodik näherzubringen. Eine Einführung in die objektorientierte Softwareentwicklung und ein Überblick über Kerngebiete der Praktischen Informatik runden den Band ab. Mit diesem Wissen sollte der inhaltliche Einstieg ins Informatikstudium problemlos gelingen.

Preisänderungen vorbehalten.

B. G. Teubner Stuttgart · Leipzig

Zeitfracht Medien GmbH
Ferdinand-Jühlke-Straße 7
99095 Erfurt, Deutschland
produktsicherheit@kolibri360.de